Visual Valuation

Implementing Valuation Modeling and Geographic Information Solutions

Edited by Mark R. Linné, MAI, SRA
with Michelle M. Thompson, PhD

Readers of this text may also be interested in the following publications from the Appraisal Instittute:

- *The Appraisal of Real Estate*, 13th edition
- *GIS in Real Estate: Integrating, Analyzing, and Presenting Locational Information*
- *A Guide to Appraisal Valuation Modeling*
- *Mathematics for Real Estate Appraisers*
- *Practical Applications in Appraisal Valuation Modeling*

Appraisal Institute®
Professionals Providing Real Estate Solutions

Visual Valuation

Implementing Valuation Modeling and Geographic Information Solutions

Edited by Mark R. Linné, MAI, SRA
with Michelle M. Thompson, PhD

Appraisal Institute • 550 West Van Buren • Chicago, IL 60607 • www.appraisalinstitute.org

The Appraisal Institute advances global standards, methodologies, and practices through the professional development of property economics worldwide.

Reviewers: Frank J. Lucco, SRA
R. Wayne Pugh, MAI
John A. Schwartz, MAI
Jeffrey Swango, SRA

Chief Executive Officer: Frederick H. Grubbe
Director, Marketing & Member Resources: Hope Atuel
Senior Manager, Publications: Stephanie Shea-Joyce
Senior Technical Writer: Michael McKinley
Technical Book Editor: Emily Ruzich
Manager, Book Design/Production: Michael Landis

Printed in the United States of America

Library of Congress Cataloging-in-Publication Data

Visual valuation : implementing valuation modeling and geographic information solutons/ edited by Mark R. Linné, with Michelle M. Thompson

p. cm

ISBN 978-1-935328-10-0

1. Real property–Valuation. 2. Geograhic information systems. I. Linné, Mark R. II. Thompson, Michelle M., 1960-

HD1387.V57 2010

333.33'20285–dc22

2010023131

Table of Contents

Foreword

The 20 contributors to *Visual Valuation: Implementing Valuation Modeling and Geographic Information Solutions* approach the issues confronting the valuation profession from diverse viewpoints. However, all the contributors to this book would no doubt agree that information technology–the latest generation of geographic information systems (GIS), valuation modeling applications, and predictive analytics software–is an essential set of tools underused by appraisers.

The valuation scenarios collected in this new book range from the approachable (incorporating GIS technology into an individual's land appraisal practice) to the intimidating (developing a multistate valuation model that integrates a variety of valuation tools and data sources within a major lender's business processes). In each case, the authors find a way to leverage existing technologies to address the issues under investigation and to map out a path forward.

The three sections of the book target readers at different levels of sophistication. Early chapters instruct the technological neophyte trying to get up to speed on the issues. Later chapters address the power user who is comfortable with practical applications of the technology. The final chapters explore the academic arena, where experts apply the most sophisticated modeling techniques in original research work.

Each chapter can be read on its own or as part of a larger argument about the evolution of appraisal practice in the twenty-first century.

As appraisers step out of depressed real estate markets, with collateral valuation under scrutiny from all sides, *Visual Valuation* provides a timely treatment of contemporary concerns. The Appraisal Institute is proud to present this new book and hopes that the discussion it provokes will be a fruitful examination of how valuation professionals can use new technologies to adapt to a challenging and shifting business environment.

Leslie P. Sellers, MAI, SRA
2010 President
Appraisal Institute

Preface

Mark R. Linné, MAI, SRA, Editor

There can be no doubt that the valuation profession in all of its various incarnations and manifestations throughout the United States is in flux.

It is difficult to truly institute meaningful change. Forces both endemic and ingrained sustain the status quo, no matter how damaged, ill-advised, and ultimately unfulfilling, to continue in its present form.

Future generations may very well question what valuation professionals did to make the profession better and what actions that we, as a community of professionals, took to change the manifest problems that continue to plague our industry. The actions we take today can enhance the ability of valuation professionals to retain their place as trusted advisors to their clients.

One key step in addressing the present situation is to speak loudly and with clarity on the issues of the day and propose a better way of doing things. This book, with discussions of theory, practical case studies, and academic contributions, runs the gamut of differing perspectives and different levels of complexity. Beginning with an overview that provides context to the reader, the text moves to concrete examples of innovation and ends with the cutting edge of valuation theory. No matter the experience level of the reader, this text has something for everyone and something new to aspire to.

The development of this text was a progressive process in which think pieces, practitioner-oriented case studies, and academic chapters were all joined with a common theme: the application and integration of two powerful analytic tools–valuation modeling and geographic information systems (GIS)–and how their use and integration could enhance the appraisal process in an environment of readily available data. The topics of data, data standards, education, GIS, and valuation modeling are all explored from the perspective of best practices in the valuation process and with the goal of empowering valuation professionals to leverage their skill set and meet the needs of a changing market.

The editors believe that this text is unique in its integration of both practitioner and academic content, and the end result provides an opportunity for a broad base of readers to gain a strong understanding of the potentials that exists in a future characterized by ever-greater access to and use of data and analysis.

Acknowledgments

The road from conception to publication is a long and arduous one. Without the input, support, editing, and general feedback from a broad range of friends, colleagues, and stakeholders, the road would have been far less navigable.

This book is the product of multiple authors who gave of their time and energy to provide something of value to the profession. For their efforts and patience, I thank them. For their contributions to the body of knowledge of the appraisal profession, I am hopeful that their efforts are recognized for providing the impetus to move the profession in a direction that I believe is important and relevant.

Many of my colleagues were very supportive of the time and energy required to finish this text over a three-year period. Of particular note, I would like to thank Jeff Bradford, CEO of Bradford Technologies, for providing the support and opportunity to work on meaningful projects and to take an idea and move it to an incarnation that helps to change an industry in meaningful ways. The support of colleagues James Regnere and Kim Hartmark was also critical in this effort.

The help and support of a talented professional staff from the Appraisal Institute should be given its due. The efforts of Stephanie (Tep) Shea-Joyce who provided the proper mixture of cajoling, stern words (when necessary), and kind support have been a joy for more than a decade, and helped to move the author group to completion. Michael McKinley provided remarkable editing acumen–focusing on the nuances of expression and providing insight and thoughtful feedback to make the text what it is in the finished product. Michael Landis provided great graphic and design elements to the book–and a great cover.

Finally, and most critically, I would like to thank my family. The time spent on writing and editing that allowed the progression from manuscript to finished book is one that directly subtracts from the time spent on family. My wife, Marjorie, and my children, Heather, Paul, Jennifer, and Katherine, have my eternal thanks for allowing their husband and father the time required to finish his responsibilities on this text. Their inspiration and support helped to make this manageable.

Our goal–in the beginning and to the end–was to provide a text that is meaningful and relevant to the profession. Our hope is that readers find it to be both.

Mark R. Linné, MAI, SRA, Editor

About the Authors

Mark R. Linné, MAI, SRA, CAE, FRICS, has been a leader in the design, deployment, and adoption of interactive valuation technologies for more than a decade. Mr. Linné is an author, editor, speaker, columnist, inventor, AVM expert, data standards proponent, software developer, and veteran appraiser who has focused on technology, data, and valuation modeling and their roles in appraisal. Mr. Linné is the coauthor of two previous books for the Appraisal Institute, *A Guide to Appraisal Valuation Modeling* and *Practical Applications in Appraisal Valuation Modeling*. Mr. Linné serves as executive vice president for education and analytics for AppraisalWorld, Inc.

Michelle Thompson, PhD, is an assistant professor in the University of New Orleans department of planning and urban studies. She has developed curricula in the areas of market valuation, geographic information systems, and urban studies and has collaborated with public and private organizations to develop training and research opportunities for community mapping projects. Dr. Thompson was a licensed residential real estate appraiser but now focuses on teaching and applied research using public participation geographic information services.

Richard Bernknopf, PhD, is an economist with the USGS Western Geographic Science Center in Menlo Park, California. His research focuses on demonstrating the relevance (value to society) of natural science information and the transformation of that information into a form compatible with decision-making processes. Dr. Bernknopf has published papers and books on economics, geology, geophysics, hydrology, and planning. Dr. Bernknopf has been a consulting professor and codirector of the Center for Earth Science Information Research at Stanford University and on the editorial board of *The Professional Geographer*. Currently, he is the codirector of the Spatial Integration Laboratory for Urban Systems (SILUS) at the University of Pennsylvania.

John Cirincione, SRA, CRE, FRICS, serves as the director of business development for JVI Solutions located in Lake Mary, Florida. He holds a number of volunteer leadership positions related to open data standards in appraisal and is an individual member of the Open Geospatial Consortium (OGC) with interests in geospatial information and geoprocessing service standards involved with real estate, appraisal, and mortgage. Mr. Cirincione graduated from The Ohio State University in Columbus, Ohio.

John Connelley is the appraisal systems manager for Farm Credit Services of America in Omaha, Nebraska. His primary responsibilities include helping to create and maintain the various software systems used by the appraisal team. Mr. Connelley graduated from the University of Nebraska at Lincoln with a degree in agricultural economics.

Ann-Margaret Esnard, PhD, is a professor of urban and regional planning and the director of the Visual Planning Technology Lab at Florida Atlantic University. Her expertise encompasses GIS/spatial analysis, coastal vulnerability assessment, land use planning, and disaster planning. She has written on topics that include geospatial technologies, GIS education, public participation GIS, vulnerability assessments of coastal and flood hazards, population displacement from catastrophic disasters, and disaster recovery. Dr. Esnard has designed GIS course curricula and has taught GIS to students and professionals with diverse academic backgrounds since the early 1990s.

Kevin C. Gillen, PhD, is a research fellow with the Institute for Urban Research at the University of Pennsylvania and a vice president of Econsult Corporation. With a background in urban economics and real estate finance, he is pursuing research in the applied work of the analysis of real estate developments and operation of real estate markets, including their fiscal, economic, and financial implications. Prior to attending Wharton, he worked for the Federal Reserve Board of Governors and HUD's Federal Housing Finance Agency (FHFA).

Elizabeth Green is a principal consultant with rel-e-vant, a consulting firm specializing in real estate finance technology and data-centric solutions. Ms. Green has over 25 years of experience in the residential mortgage industry in data analysis, system integration, and software product development for property valuation, loan origination, loan servicing, and secondary marketing solutions. As a collaborative business solutions architect in both the mortgage and valuation industries, she has been a driving force as an industry subject matter expert and designer behind many innovations. Ms. Green also serves as the chair of the MISMO Property and Valuation Services Workgroup.

James Gresham is currently starting up a new business venture in Littleton, Colorado. He holds a master's degree in mineral economics from Pennsylvania State University, with additional training in econometrics and mathematical modeling from Colorado School of Mines. Mr. Gresham has been the lead developer and architect for innovative valuation software for commercial and residential use. In addition, he has over 20 years of experience in developing real property databases and related models.

Michael W. Ireland, CAE, is the assessor in Bloomington, Illinois, a position he has held since 1980. He has developed and implemented his own computer-assisted mass appraisal system (CAMA) and integrated CAMA with a geographic information system (GIS). Mr. Ireland was a member of a task force of the Appraisal Standards Board that updated Standard 6 of USPAP and is a senior instructor for the International Association of Assessing Officers (IAAO).

Kirk Manker is the vice president/chief appraiser of Farm Credit Services of America, where he leads the real estate and chattel appraisal activities for the Farm Credit System Association. Mr. Manker graduated from South Dakota State University with degrees in agricultural business and commercial economics.

Christopher A. Miner, MAI, SRA, MRICS, is the chief operating officer of Greenfield Advisors. Mr. Miner has written and taught for the University of Connecticut, the Appraisal Institute, and other real estate associations. He has developed GIS applications for site evaluation and selection, valuation, market analysis, and land appraisal. Mr. Miner has developed sales forecasting and market planning models for retail and health care companies and has led teams performing complex valuation and market analyses for projects in many states.

Patrick M. O'Connor is an international leader in the field of computer-assisted appraisal. Mr. O'Connor was the chief assessor of New York City and chief commercial/industrial appraiser for USAID's Russian Federation Property Tax Pilot Project. He has provided consulting services, performed real estate analysis, and developed computer-assisted valuations using econometric models in Canada, England, Northern Ireland, the Russia Federation, and the United States. Mr. O'Connor is president of O'Connor Consulting Inc., which provides consultant services in ratio studies, market analysis, and computer-assisted mass appraisal (CAMA) valuations for assessment authorities and AVMs for private-sector clients.

Milton Ospina is the NAVTEQ business development manager responsible for supporting GIS enterprise implementations of NAVTEQ map data products, services, and solutions. Previously, Mr. Ospina was the ESRI urban and regional planning and economic development solutions manager. Mr. Ospina travels frequently in the U.S and abroad to speak about the powerful role and benefits of GIS in government, the power of GIS coupled with up-to-date business and demographic data for economic and business development, and the role public participation GIS (PPGIS) and GIS-based planning support systems (PSS) play in improving the quality of life of our communities and in enhancing the planning process.

Ana Puszkin-Chevlin, PhD, is a real estate and environmental planning consultant for nonprofit organizations and private companies. Her expertise lies in coastal hazards resiliency, vulnerability assessment, land use planning and spatial analysis, and real estate market research. Since 2004, her academic research has focused on coastal land use planning policy in Florida and neighboring Atlantic and Gulf of Mexico states. Previously, Dr. Puszkin-Chevlin had been a special projects research officer for Columbia University's Earth Institute and the assistant director of the graduate program in urban planning at Columbia University.

Sumeeta Srinivasan, PhD, teaches geographic information systems and spatial analysis at Harvard University where she is affiliated with the Center for Geographic Analysis. Her research interests include the study of urbanization and its effects on infrastructure and the use of spatial methods in human geography. She is particularly interested in studying these topics in the context of South and East Asia.

James Russell Thimgan is a partner in Thimgan and Associates. He is a mass appraisal and valuation modeling expert who works with municipalities and assessment jurisdictions to value commercial and residential properties for assessment purposes. Mr. Thimgan is also an author who has contributed to the International Association of Assessing Officer's 2010 text, *Property Assessment Valuation.*

Francoise Vermeylen is the director of the Cornell Statistical Consulting Unit (CSCU). She coordinates the statistical consulting services provided to researchers at Cornell University and has more than 20 years of experience providing statistical consulting to researchers in various fields ranging from epidemiology to agricultural sciences. She consults on a broad range of statistical methods including, but not limited to, linear and generalized linear models, multilevel models, categorical data analysis, survival analysis, and multivariate statistics.

Susan Wachter, PhD, is a professor of real estate and finance at Wharton, where she holds the Richard B. Worley Professor of Financial Management Chair. She founded and currently serves as director of Wharton's Geographical Information Systems Lab. She is the cofounder and codirector of the Institute for Urban Research at Penn and the Centre for Asset Securitization in Asia at Singapore Management University. Dr. Wachter served as Assistant Secretary for Policy Development and Research (PD&R) at the U.S. Department of Housing and Urban Development (HUD), a presidentially appointed and Senate confirmed position, from 1998 to 2001.

Anne M. Wein, PhD, is an operations research analyst employed by the USGS Western Geographic Science Center in Menlo Park, California. She holds a docorate in decision sciences from the Graduate School of Business of Stanford University (1988). She has analyzed agricultural, manufacturing, computer, environmental, geologic, and economic systems from positions in industry, academia, and government. Her current activities entail coordination of economic consequences for natural hazard scenarios and design of a methodology for assessing national carbon sequestration potential in ecosystems.

Introduction

Mark R. Linné, MAI, SRA, Editor

The following chapter overviews are provided to give readers a sense of the content of each chapter and the editor's thoughts on how each chapter provides some insight into the manner in which valuation modeling, data and data standards, and geographic information systems can provide practical applications to professionals and others interested in the future of the valuation profession.

The 15 chapter have been broken into three sections. The first section provides some timely think pieces on the nature of the profession and the direction that it is going. The second section includes a number of practitioner-focused case studies and some more sophisticated tools that appraisers can use in their search for technological solutions. The final section consists of several chapters that are more academic in tone and content and that explore cutting-edge applications and tools that meld geographic information, real estate data, and advanced modeling to provide a better, more supportable outcome for valuation.

Section I
Reimagining Valuation: Thoughts on the Future of Appraisal

Section I provides an introduction to the theme of change and a forward-looking perspective on the challenges and opportunities that face the appraisal profession. The chapters provide insight into how practitioners can adapt to the needs of clients and the opportunities to take advantage of a new landscape of valuation that promises to transform the existing paradigm.

Chapter 1
Valuation 2.0: The Changing Landscape of Appraisal

The first chapter sets the stage for the future of valuation, considering the concept of an evolutionary change from the status quo to what we call *Valuation 2.0*. This chapter suggests that the world is changing rapidly and that specific change is rushing towards the appraisal profession and the clients that valuation professionals serve. In the same manner that the Web changed from its earliest incarnation in the 1990s to the present, changes in valuation space will transform the manner in which valuation is perceived, analyzed, and delivered to users of valuation services. Driving much of this change is the issue of data standards, and this chapter sets the stage for several later chapters that address the availability of more data than ever before and the need for new tools and that explore what is possible in a data-rich valuation environment.

Historic context is next offered in the chapter, as automated valuation models (AVMs) and appraiser-assisted valuation products are discussed along with an introduction into some of the underlying crisis points in collateral valuation (e.g., as broker price opinions, or BPOs, begin to emerge as an alternative to traditional valuation). The concept of AVM-enabled appraisal is introduced and evaluated from the perspective of how this concept can serve to meet the financial sector as it seeks a flight to greater quality and transparency in the valuation of real estate.

The chapter concludes with a positive prognosis for the profession, suggesting that the best days of the appraisal profession are not behind, but rather, are clearly awaiting appraisers in the near future. It will be the responsibility of appraisers to be vigilant and aware of the changes, ultimately taking advantage of the brave new world that awaits.

Chapter 2
Leveraging Valuation 2.0: Examining the Changing Appraisal Landscape

Chapter 2 moves the thesis of Valuation 2.0 forward, suggesting that new tools, analytics, and products are called for in a world data is more prevalent and more accessible, where data not only is more prevalent and more accessible, but also requires a commensurate level of interpretation to serve as a more meaningful representation of the marketplace. Appraisers who take advantage of and recognize the opportunities that can result from a more detailed and complete understanding of the market will have a competitive advantage. This chapter introduces the concept of *incrementalism*, which suggests that moderate enhancements in existing reporting to clients through the introduction of more sophisticated analysis might be the best strategy for appraisers to take in serving client needs.

Diving deeper, Chapter 2 examines the use of advanced analytics from the perspective of its use in interactive valuation models (IVMs), appraisal review and validation, legal and expert testimony, and mark-to-market portfolio valuations. The pros and cons of working with AVMs are then examined, and the use of traditional "black box" AVMs is ultimately rejected after an examination of the potential downside of using products over which the appraiser has little meaningful control.

Next, the chapter moves to a discussion of what tools might be acceptable to appraisers and then proposes the concept of a *Valuescape* that would harness the collaborative aspects of appraisers working for the common good. The combination of appraisal collaboration, open-source applications, and advanced tools becomes the rallying cry for a movement that seeks to help the profession meet the needs of a changing valuation and financial services environment, which ultimately appraisers can leverage to their collective advantage.

Chapter 3
Open Data Standards for Real Estate, Appraisal, and Mortgage Banking: Implications for the Industry

The first two chapters of the book introduced the concept of advanced data availability. Chapter 3 specifically focuses on presenting the concept of how data standards that are open to all stakeholders can enhance the productivity of all sectors of real estate and the financial services sector and can enable all of the players in collateral valuation to mitigate risk and have a greater understanding of the market.

The importance of a common language for data is defined and compared to the other advances occurring in automated credit decisions, tools and electronic mortgages.

The authors suggest that a new set of metrics may be necessary to explain and expand on historical valuation techniques in the face of greater data availability. This chapter provides background and context for the data standards movement and then examines how both valuation practitioners and the academic community can work together to bring advanced technology and analytics to mainstream appraisers. Vendors, practitioners, and academia all have a vested interest in meeting the challenges presented by data and technology, and the benefits of working together for the common good of the valuation profession are examined.

Chapter 4
The Valuation Professional of the Future: Training for Innovation

Chapter 4 discusses the foundational statistical elements required to provide appraisers with an understanding of the education needs of the profession to ensure competence. The requirements of the Uniform Standards of Professional Appraisal Practice (USPAP) and the ongoing efforts of government-sponsored enterprises such as Fannie Mae and Freddie Mac to implement an infusion of standardized data provide a landscape that will require additional competence from the valuation community.

The bulk of the chapter examines the evolving requirements for a greater scientific basis for appraisal analysis, including the requirements for the use of regression analysis. The remainder of the chapter considers what is required to move statistical analysis and regression into the mainstream. The type of education that will be required is described, with an ideal curriculum suggested and presented. The chapter concludes with the suggestion that a collaborative effort of appraisers and academics is best suited to determine how advanced techniques can be implemented.

Chapter 5
Introduction to Geographic Information Systems

Chapter 5 begins the book's examination of geographic information systems (GIS) with background for readers new to the topic, providing information sufficient to more easily understand the role of GIS in the valuation process, which will be expanded upon in the chapters that follow. This chapter examines the basic tools of GIS, how data is created, the hardware and database requirements of a GIS environment, the changes that have occurred with contemporary GIS, and the manner in which what was once an academic and research specialty has become mainstreamed in the last decade.

Chapter 6
Integrating Geographic Information and Valuation Modeling for Real Estate

Building on the introduction to GIS in the previous chapter, Chapter 6 examines how GIS can be incorporated in the valuation process to empower and enable appraisers with a spatial perspective of data. Assessment offices across the country have used GIS in tandem with computer-assisted mass appraisal (CAMA) analysis to deliver ever-more accurate values for tax purposes. The natural integration of modeling techniques in real estate and GIS is demonstrated by the core data elements that each requires. The concept that GIS is 90% data is explored and used as a starting point for understanding how spatially enabled analysis provides analysts with a greater understanding of data and, more importantly, the markets that data represents from an entirely new perspective.

Chapter 7
GIS for Business and Economic Development: From Data to Analysis

Chapter 7 focuses on the ability of community-based organizations (CBOs) to use geographic information systems to support a community's quality of life and expand the data sources that can be used for economic development. This technology has been improved by significant advances in GIS knowledge, accessibility, and data availability among CBOs. The influence on economic development that community or parcel level data has, when tied to public data, is ever increasing. The new role that CBOs play as data providers has allowed public officials to enhance general data, making it more specific and accurate.

Both the assessment and appraisal professions will find relevant information in this chapter about how community-led data collection is used for valuation by "on-the-ground" property condition analysis. Valuation professionals should consider how valuation methodology in weak or emerging markets can be modified where "market value" is difficult to ascertain using traditional methods.

Section II
Implementing Innovation: Practical Applications of Advanced Technology

Section II provides insight into how valuation practitioners have incorporated valuation modeling and geographic information to enhance their practice, or to develop enterprise-level valuation systems that enhance the valuation process. This section focuses on explaining how tools can be used in a practical manner to benefit both practitioners and those who rely on professional valuation services.

Chapter 8
GIS and GPS Usage in Land Appraisal

Chapter 8 provides a hands-on case study of the manner in which GIS can be used in the valuation of vacant land. This chapter explores the uses in varying land types such as

- Acreage in transition areas
- Agricultural properties
- Forestry properties

Beginning with an overview of available GIS software, the chapter then covers the subject of data and data availability. An integrated example using the software and the data is then examined, and an evaluation of development limitations is then presented for the reader. A final section examines how Global Positioning System (GPS) technology can be used to enhance the valuation process.

Chapter 9
Developing Enterprise-Level Valuation Applications: Incorporating Appraisers in Innovation

Providing a conceptual integration and implementation of both GIS and valuation modeling, Chapter 9 is a comprehensive treatment of how appraisers can move far beyond traditional appraisal and consulting services, and actually help a client assess the collateral risk environment, examine alternatives, and then implement cutting-edge technologies to empower an entire enterprise.

The chapter specifically examines a large lender client and its need for a GIS-enabled valuation modeling technology. A team of appraisal and subject matter experts were able to bring advanced analytic solutions to the agricultural valuation field, enabling this $1 billion lender to implement some of the most advanced valuation technology in the world. The narrative examines how the enterprise's needs were assessed and how a solution was implemented that resulted in cost savings and a more efficient organization.

Chapter 10
Innovating the Approaches to Value: Incorporating Valuation Modeling Techniques into Traditional Appraisal Practice

Building on the experiences discussed in the previous chapter, Chapter 10 looks at the proposition that the current three approaches to value are still relevant but require an innovative perspective that incorporates advanced modeling techniques to leverage the data available in the market.

The chapter examines the challenges of building an enterprise-level application in tandem with the staff of the large agricultural lender introduced in the previous chapter. The chapter also looks at how the collaboration helped to create functionality that more completely met the needs of the organization, through the collaborative efforts of staff appraisers, management, valuation modelers, data experts, and GIS practitioners. Ultimately the use of the application permitted a broad-based application modeling of cost, depreciation, sales, and income data, effectively enhancing the use and applicability of the three traditional approaches to value.

The chapter concludes with an examination of the benefits to the entire enterprise, which included the realization on the part of management of the importance of the valuation process and the appraisal experts who safeguard the lending practice.

Chapter 11
Practical Integrations of Valuation Modeling and Geographic Information Systems: First Steps toward a More Robust Market Analysis and Valuation Methodology

Chapter 11 provides an overview of the modeling and GIS tools and techniques available to appraisers from the perspective of valuation professionals who have developed and continue to explore the manner in which these tools can be used to facilitate real estate market analysis. Beginning with a discussion of Excel, the chapter moves to other types of statistical tools and how they can be used gain insight into market analysis and real estate valuation. GIS software is examined next along with the enhancements to the valuation process that result from the incorporation of spatial analysis.

The remainder of the chapter examines the practical aspects of building valuation models and includes the basics of specification and calibration of the model, two fundamental tenets of model construction. The incorporation of regression analysis is explored, and the chapter provides an overview of two applications that have been specifically designed with an understanding of the needs of appraisers.

The chapter concludes with a discussion of appraisers building valuation models using the available tools.

Chapter 12
Innovation in Residential Valuation Tools: Integrating Data and Analytics

The question of how innovation enters the marketplace is the subject of Chapter 12, which details the development of a residential valuation tool that empowers appraisers amidst the large amounts of data that exists in the real estate market today. The chapter recounts the genesis of the project at a conference of industry thought leaders, as participants examined the best methods of providing a meaningful solution to leverage data and analytic tools that would empower appraisers. The type of functionality that such an application must possess is detailed, as a regression-based valuation tool is developed and deployed into the market. The chapter concludes with an overview of the functional elements of the application and a detailed analysis of its accuracy and credibility.

Section III
Pushing the Envelope: Advanced Case Studies from Academia

Section III provides insight into how the academic community uses both valuation modeling and geographic information to enhance the predictive capabilities of technology and real estate data. A number of case studies and projects are detailed that consider cutting-edge research, demonstrating the practical incorporation of both regression modeling and geographic information into the valuation process.

Chapter 13
Advanced Statistical Analysis of Real Estate Data Using Geoda

Chapter 13 provides an overview of the benefits of combining simple regression tools such as Excel with GIS tools such as Geoda to provide a greater understanding of market dynamics to the valuation practitioner. Recent research indicates that empirical models of hedonic pricing have begun to take into account the potential bias and loss of efficiency that can result when spatial effects such as location and proximity to geographic variables are ignored in the estimation process.

The residuals of an ordinary least squares (OLS) model for housing prices can be expected to have a relationship to their geographic location because several variables related to crime, public school quality, and proximity to amenities and disamenities are impossible to obtain at the disaggregated individual house level.

The availability of GIS software in estimating such models has improved the ability of hedonic models to include local information about the amenities that are available. In this chapter, GIS is used to estimate several environmental disamenities and amenities including proximity to highways, open spaces, and air pollution.

Chapter 14
Using Econometrics and Geographic Information Systems for Property Valuation: A Spatial Hedonic Pricing Model

Chapter 14 explores how to create a *decision support tool* (DST) to evaluate the hedonic pricing functionality of a statistical regression model based on the behavior of buyers and sellers who participate in actual real estate transactions. Regression analysis is a property valuation approach that relates the market price of a property to the values of its individual attributes, such as lot size, number of bedrooms, zoning classification, and environmental amenities. Differences in property prices are used to deduce what individual attributes are worth to participants (either negatively or positively) in a real estate market based on the market data analyzed by the model.

This chapter examines a conventional set of site characteristics along with additional independent variables derived from a GIS analysis to discuss the influence of location characteristics and measured distances of a parcel to specific amenities. While different approaches to real estate appraisal exist, such as sales comparison, income capitalization, and cost, this chapter presents a statistically rigorous method as an alternative. The chapter includes a description of the hedonic pricing approach, a review of past applications, and then the application for land valuation in Miami-Dade County, Florida.

The strengths and weaknesses of using the hedonic pricing approach for predicting real estate prices are described in the chapter summary. The addition of the explanatory variables derived from GIS analysis almost doubles the explanatory power of the econometric model for estimating land prices across the sample data set.

Chapter 15
Coastal Property Analysis through a Hazard Exposure Lens

The goal of Chapter 15 is to provide valuation professionals with an overview of coastal development and redevelopment issues and challenges from these multiple vantage points. In addition, the chapter demonstrates the manner in which GIS technology is being used to analyze coastal property through a hazard and disaster vulnerability lens. To fulfill this goal, concepts, analyses, and applications are presented at four scales:

- Parcel
- Portfolio
- Municipal
- Regulatory

Examples in the chapter are drawn from professional and academic sources, documented case studies, and the academic research of the authors.

This chapter will appeal to a broad range of professionals, such as land planners, environmental advocates, regulators, insurers, attorneys, appraisers, lenders, developers, and other real estate professionals, all of whom play key roles in the evaluation, planning, and development of coastal land. Although the chapter is written from an academic perspective, it is useful to valuation professionals who may not necessarily understand how their role in valuing pre- and post-disaster properties influences (albeit indirectly) the vulnerability of property and people to hurricanes and other coastal hazards. The chapter also shows why it is important that valuation professionals understand the variety of vantage points and types of spatial data and GIS analyses used by the broad range of professionals who shape and reshape coastal communities.

Section I

Reimagining Valuation: Thoughts on the Future of Appraisal

1 Valuation 2.0

The Changing Landscape of Appraisal

Mark R. Linné, MAI, SRA

Introduction

In 2010 the valuation profession faces changes as profound as any in its history. In the wake of the collapse of the subprime lending market, it is clear that collateral valuation matters to a healthy market economy. It is just as clear that today market data, standardization of data, tools for data analysis, and the real estate industry's search for quality are at a tipping point that can lead the valuation community to new opportunities.

In recent years, the only constant for virtually all business and organizational structures has been change, and the appraisal profession, which has historically resisted the ongoing changes within the financial services and information services sectors of the economy, is poised to catch up.

What are the lessons to be learned from the events of 2007-2009, and how will the valuation profession move forward to regain the public's trust and provide its clients with new products and services that enhance the accuracy of collateral valuation? In recent years, lenders have been relying on esoteric collateral valuation products, principally broker price opinions (BPOs), which financial services clients have increasingly chosen over traditional appraisals. The search for ways to satisfy current client needs will bring all of the disparate data, technology, and methods together in a paradigm shift, leading finally to a redefinition and evolution of the profession. I call this change *Valuation 2.0.*

Why *Valuation 2.0*? The term *Web 2.0* often refers to the perceived second generation of Internet applications, describing the manner in which information and users are connected. It represents a maturation of the Web, a greater focus on connectivity, and a change in the way software developers and end users take advantage of the networking capability of social media to connect themselves. It is about connecting data, the data standards that facilitate that connection, and the manner in which the Internet becomes the platform for the exchange of information. In the same way that the Internet has undergone a transformation from Web 1.0 to Web 2.0, current changes in the valuation profession will transform the manner in which valuation services are perceived, analyzed, and delivered.

Over the last decade, data analysis, including statistical modeling and techniques such as regression analysis, has increasingly found its way into unexpected segments of society. Indeed, the increasing and successful use of statistical modeling in a broad range of situations to improve performance and predictive capabilities has become mainstream. Several books, including *Super Crunchers: Why Thinking-by-numbers is the New Way to be Smart* by Ian Ayres and *Moneyball* by Michael Lewis, detail how every segment of society can benefit from a focus on the statistical analysis of data.

Moneyball looks at the growing statistical science of sports, now termed *cybersports*, which identifies how a scientific and statistical approach to baseball and other sports can improve management's ability to craft a winning formula. *Super Crunchers* demonstrates that data-driven decision-making is not just revolutionizing business or baseball. Rather, it is changing the manner in which health care reimbursements, education policy, and even tax regulations are crafted. Understanding data and the predictive power that it provides connects statistical analysis to better decision-making.

The power of statistical analysis in making better-informed business decisions can be seen in the success of Nate Silver of FiveThirtyEight.com and other sites that forecast baseball and election results and in the data mining Google and Amazon do to anticipate the book or compact disc a consumer ought to buy next based on past purchases or searches. These are tangible examples of how this type of analysis can be used to predict future activities based on a consideration of past performance.

What should appraisers be aware of as the future rushes towards them at breathtaking speed? How can appraisers position themselves? And what can practitioners and valuation futurists teach appraisers about how to position themselves in the marketplace to not only survive but to grow and prosper? These are the questions that this chapter, and ultimately this book, will seek to answer.

The Impact of Standards: It's All About Data

Real estate is the last digital frontier–one of the few major components of the economy yet to take advantage of the standardization of data and the technological innovation that have transformed other sectors of the economy. In his 2005 book *The World Is Flat: A Brief History of the Twenty-First Century*, *New York Times* columnist Thomas Friedman discusses how the world is being flattened by various *drivers* towards technology and standards. The book analyzes globalization, primarily in the early twenty-first century. The title is a metaphor for viewing the world as a level playing field in terms of commerce, where all competitors have an equal opportunity. The title also alludes to the perceptual shift required for countries, companies, and individuals to remain competitive in a global market where historical and geographical divisions are becoming increasingly irrelevant.

These same two drivers, data standardization and technological innovation, are flattening the real estate world and, by extension, the valuation terrain. Friedman looks at the influence of data and standards in the real estate market and suggests that because of the size of the market, real estate data and standards could have one of the biggest effects of any trend in any other market. Friedman even says that the effect of data and standards in the real estate market could be greater economically than any other event of the last century, "including the Internet."

In 2005, Gabe Minton, the former executive vice president of the Mortgage Bankers Association (MBA) of America, estimated that the overall drive towards data and process standardization in the next few years will be tectonic in scope. The drive towards eliminating inefficiencies in the real estate market could have an impact as great as $1 trillion and will simultaneously cause both chaos and opportunities for those who can provide meaningful solutions to the industry.

Perhaps the biggest trend that will affect the valuation profession is the realistic possibility that in the near future appraisers and assessors will have access to more data than they ever believed possible. For the first 75 years of the profession's history, appraisers and assessors existed in an environment characterized by a dearth of data. In 2010, there is a glut of data from many sources. Proprietary databases from commercial sources–e.g., FirstAmerican/Core Logic, Data-

Quick, and LPS on the residential side, and CoStar and LoopNet on the commercial side–represent the monetization of the assessment profession's public record database that now provides information on more than 100 million properties.

The potential shift in the valuation profession's relationship to market data leads to several questions:

- How will the profession cope when it has access to most, if not all, of the data within the market?
- Do valuation professionals have the tools, the education, and the methodologies to evaluate this cascade of information?
- What general trends are on the horizon, and what are the trends that are likely to have the greatest impact on the appraisal profession?
- How can appraisers prepare themselves to take advantage of the opportunities that will undoubtedly result from all of these forces?

At the 2007 MBA Multi-Family Asset Administration and Technology Conference, the topic of data standards was considered in a panel presentation aptly titled "The Property Data Revolution." Perhaps the most critical points made by members of the panel include the observations that data standards create order from chaos, that they give understanding to data, and finally that they put information into context. In other words, data standards will eliminate the worry about the "how" of moving data around between different participants in the mortgage transaction. In the near future, appraisers and assessors will be more concerned with the "why" of understanding data.

Standards will ensure that real estate and valuation data flows to all parties within a transaction, eliminating transactional friction and leading to a more efficient and transparent marketplace. It is clear in talking to lenders, vendors, and industry insiders that data standards will not only change how appraisers present data to their clients, but it will in all likelihood change the way appraisers value properties and perform analysis.

A Brief History of the Flattening of the Valuation World: Appraisers and AVMs

The use of *automated valuation models* (AVMs) has grown dramatically from their inception in the mid-1990s to 2010. Fannie Mae and Freddie Mac run an AVM automatically on every appraisal they receive as part of their appraisal review function, currently accounting for about half of all appraisals performed. Add to that total all the appraisals ordered by major lenders like Wells Fargo, Bank of America, and Chase, which all run an AVM on just about every loan they make. Combining the secondary mortgage market leaders and the major lenders means that a vast majority of the 20 million valuations performed in 2009 used an AVM in some fashion, as an initial indication of value, a review tool, or even a stand-alone valuation for smaller loans or in tandem with an appraisal or BPO. And AVMs are fast becoming the accepted tool to provide an additional indication of value for collateral valuation decisioning. After an initially slow acceptance by the market, AVMs are increasingly used to provide valuations across a broad spectrum of applications in the mortgage lending and related fields. While the economic downturn that began in 2008 has caused many users of AVMs to apply greater caution, the future will likely bring an increase in the use of and reliance on AVMs as accuracy increases and geographic coverage expands.

What are now known as AVMs were originally developed for the assessment sector and were known as *computer-assisted mass appraisal* (CAMA) techniques. These technologies and techniques enabled assessors to value thousands and even tens of thousands of properties using models developed primarily for the residential sector. Assessors were able to use these types of tools because they had access to the entirety of the public record database, containing detailed information on all properties. This advantage over appraisers and developers of AVMs enabled the assessment community to effectively use and develop significant advancements in understanding the nature of data and to develop their own internal valuation models, which in many cases rivaled or even exceeded the accuracy of less data-intensive AVMs developed for private commercial use. The newfound availability of data and the need for efficient and effective processing of realty data have helped CAMA technologies cross over from assessment practice into broader use by appraisers as well.

Lenders and other users of appraisal services have long sought to reduce the cost and turnaround time of appraisals. In the 1990s, appraisal valuation models began to emerge as a cost-effective alternative to the traditional residential appraisal. Early AVMs were able to target only limited markets where data was plentiful. Over the last decade, market acceptance of AVMs has increased dramatically, as "hit rates" and geographic coverage have improved. Even the Uniform Standards of Professional Appraisal Practice (USPAP) recognized the increasingly important role that AVMs play in the market when Advisory Opinion 19 (AO-19) was published in 2001. Historically, AVMs have been used in second mortgage and equity lending transactions, with only limited use in the first mortgage sector. The reluctance of lenders to use AVMs for first mortgages is due to uncertainty about the reliability of the product in high loan-to-value situations and in declining markets. However, as competitive pressures force lenders to process transactions more rapidly and at lower cost, some of the early reservations about AVM reliance have been mitigated.

Though AVMs are widely used these days, very little is generally known about the internal operations of the models by institutions that lend billions based on the value estimates generated by AVMs. Like lenders, appraisers have historically been highly suspicious of AVMs and hesitant to engage with the technology. Unfortunately, as AVMs have garnered an ever-increasing share of the marketplace, appraisers have been unable to offer a meaningful alternative, and the entire industry has essentially been bifurcated, with AVMs on one side and appraisers on the other.

AVMs were constructed to offer stand-alone collateral valuation solutions. The role of appraisers was never considered or even contemplated in the development of these tools in the 1990s and early 2000s. AVM vendors, cognizant of push-back from various stakeholders, attempted to meld the best features of an AVM with an appraiser, with the end-goal of providing a product that the market would consider more reliable. For many years, this attempt was not successful because it transferred an unacceptable risk to the appraiser, especially when the model specifications were still poorly understood.

It should be recognized that statistical methodology and mass appraisal technology have always been available to appraisers. However, there were no turnkey solutions. The construction of even a local AVM is remarkably difficult, requiring significant investment in specialized skills and data gathering that are beyond the reach of most valuation professionals. Later chapters of this book will provide a more detailed understanding of the cost and effort required to bring such a solution to market. Suffice it to say that an AVM is rarely, if ever, a solution that is developed by a single practitioner, given the effort, time, and money required.

Appraisers have long assumed that AVMs have stolen much of the business appraisers used to rely on. While it is true that AVMs have emerged over the last decade as a viable collateral valuation alternative in certain circumstances, other products, namely broker price opinions (BPOs), are responsible for the lion's share of the revenue loss, replacing form appraisals (URARs and 2055s) with cheaper, faster alternatives. For too long the financial markets have focused on BPOs to gauge the appropriateness of a loan and the underlying collateral. Most lenders will admit they use BPOs for general market information. However, in so doing, lenders have tended over time to use the ultimate value conclusions as well. The truth of the matter is that BPOs are being used in quantities that would shock and dismay appraisers. In 2009, more than 10 million BPOs were performed in the United States, and there are indications that such use will continue to increase.

However, as financial sector woes continue, there is a clear indication of a "flight to quality" in collateral valuation. The use of BPOs is on the way out or on the way in–depending on how the product is being used in the market. It is less clear what will replace them. Unless appraisers can provide a product that meets the lending industry's needs, they will once again be shut out.

What has recently emerged in the market is the concept of *alternative valuation products* (AVPs) that bring back the concept of blending data and new tools to provide more meaningful analysis. Some of these new tools seek to redress the mistakes of the past when vendors attempted to "force" the appraiser to use the AVM output and agree with its conclusions. The newer AVP products are more interactive in functionality and thus have been labeled *interactive valuation models* (IVMs). These products give the appraiser/user the ability to interact with the analysis more directly, which in turn gives the appraiser more control over the analysis and the valuation outcome.

Alternative valuation products have the potential to provide appraisers with the opportunity to meet a market need that, until recently, clients have sought to obtain from nonappraisal sources. While there is opportunity for appraisers–the accepted experts in collateral valuation–to join the conversation, they must be prepared to offer not just the traditional array of products, but also to demonstrate that they can meet their clients' needs with a broad range of efficiently produced and meaningful information. Quality will prevail if it can be delivered in a cost-effective manner, but it must come in the form of products that meet the needs of the market for speed and cost. By providing a solution for this client need, valuation professionals will emerge with the opportunity to dominate the market and will no longer have to share the field with purveyors of inferior and ultimately unsatisfying substitutes.

Appraisers performing URAR/Form 1004 residential appraisals for $400, and requiring five days to deliver, must consider how they can compete with a BPO provided in two days for $75-$100. This is the reality of the valuation services marketplace. Alternative valuation products that can effectively meet the needs of the "gap" market (i.e., price points greater than $100 but less than $400) will find demand from lenders and other users. Many lenders have indicated that they would be willing to use products that are of better quality than BPOs and involve an appraiser. Appraisers must use technology to maximize their efficiencies and deliver more client-driven products that will compete in the alternative valuation product marketplace.

My discussions with industry experts suggest that the availability of a credible, interactive, appraiser-influenced product would open a floodgate of acceptance by appraisers and also satisfy lender demands for a collateral valuation product that would use advanced tools and be interpreted by an appraiser. By embracing AVM and modeling technology, valuation professionals will be able to take advantage of their local market knowledge and reap substantial financial rewards. They will also be able to leverage this technology towards other valuation services, providing market opportunities throughout the range of scope of work of current appraisal practice. With markets in decline for the first time in more than 15 years, there is recognition that a solid, supportable value has merit and must be the ultimate goal.

The solution is to get valuation professionals more engaged in the system, the process, and the solution. Appraisers have an opportunity to reshape the discussion and to collaborate to develop a better product–a product that the market wants and needs.

Conclusions

As has been the case in all facets of the knowledge-based economy, data standards will be the catalyst for redefining real estate analysis as Valuation 2.0. As standards become mainstream and their use and adoption increases, members of the appraisal profession have three potential courses of action:

1. Appraisers can ignore the message, assuming that standards have a long way to go before their impact is truly felt.
2. They can participate in the process and help, as stakeholders, to shape and focus the standards that are developed.
3. Or, at the very least, they can be aware of and stay abreast of the progress of standards so that they can adapt to meet the challenges that standards adoption will bring.

The real solution to the challenges facing the appraisal profession lies in changing the current appraisal paradigm. Appraisers are capable of delivering more than a 1004 report solution to their clients. Appraisers must understand that they have the ability to deliver a broad spectrum of alternative valuation products that can meet a wide variety of needs. Clients of appraisers need data sifted, analyzed, and pushed to them. Alternative information should be suggested and presented. Adjustments should be extracted from the market and offered and interpreted by the appraiser. All of the essential data necessary for whole market analysis should be provided to ensure a more complete

within the appraisal process. One of the benefits of statistical analysis and advanced data analysis is that they allow appraisers to render market value opinions in a replicable fashion in a manner that more completely supports the appraisal analysis and conclusions. The tools also provide the intended user of an appraisal with the ability to understand the analysis, the quantification of the various valuation factors, and ultimately, understand the process and support for a valuation conclusion. The end result of this process will create a tremendous increase in the credibility of individual appraisers, appraisal assignments, and the profession as a whole.

When approaching the question of when and how to introduce technology into appraisal practice, many appraisers begin their journey by taking a small first step: reporting additional market data used in valuation conclusions. By performing expanded market analysis, supporting adjustments with statistical analysis, or adding graphical content to provide better visualization, appraisers are providing incremental improvements to what have largely become commoditized products.

The obvious use of advanced analytics is within the mortgage banking environment, but products with more advanced analytics could also be applied in the following areas:

- Interactive valuation models
- Appraisal review or validation
- Legal and expert testimony
- Replacing and providing alternatives to BPOs
- Mark-to-market portfolio valuations

The development of advanced analytics allows the appraisal profession, and those who rely on appraisal services, to enjoy a much higher degree of confidence. Transparent methods applied across the profession would increase both actual and perceived credibility of appraisers and their conclusions. The Uniform Standards of Professional Appraisal Practice (USPAP) establishes and maintains public trust in the appraisal process. Through the adoption of advanced statistical techniques that help to support and provide credibility to valuation analysis, critical credit decisions can be made on the basis of more supportable conclusions. USPAP is the basis of public trust in the profession and statistical analysis strengthens the public trust. Along with mandated systems of analysis and reporting, advanced analytics have the potential to strengthen the public's trust in appraisers like no other innovation has in the past 60 years.

Working with AVMs

Without any established process, appraisers may not have a clear path of how they can (or how they should) work with AVMs. In other words, how do appraisers incorporate AVMs into their practices? There is no single answer to this question. Some strengths of AVMs identified by appraisers in their practices include using AVMs for comparative analysis or as a checkpoint in validating a valuation. AVMs can also be useful in other ways, particularly in identifying potential comparables or framing a range of value to either confirm or disprove a valuation.

Vendors who provide valuation product solutions that incorporate traditional AVMs are hard-pressed to accommodate the appraiser mind-set. Rather than creating products that emphasize the role of the appraiser, the vendors create endless variations on the theme of forcing the square peg of traditional AVMs into the round hole of appraiser skill sets. An example can be seen in the previously detailed appraiser-assisted AVM, in which an appraiser was asked to examine an AVM and conclude if the conclusions of the report were considered credible. Often the solution is little more than trying to engage, coerce, or convince the appraiser to use traditional products to drive additional revenue to the vendor. Getting past this dynamic is a challenge for vendors and appraisers alike.

The greatest impetus for changing the current paradigm is that AVMs have a number of shortcomings that are difficult to effectively overcome. Appraisers responding to articles published in *Valuation* magazine have identified the weaknesses in traditional AVMs, which include

- A lack of transparency in the method that the AVM relies on to derive values
- Inability to determine condition problems with the subject
- Functional issues with subject
- External factors not recognized in the valuation
- Questions in the comparability of "comparable" sales
- Inability to recognize concessions or financing that negatively impact value
- No verifications of sales
- No true market analysis other than that evidenced through sales activity
- Lagging market indicators/currency of market information[2]

2. "Letters to the Editor," *Valuation* (Second Quarter 2008): 10-13.

All of these points are accurate and appropriate within the existing paradigm. In order to overcome these objections, appraisers must be able to use new tools that are not bound by the rules of traditional convention. New products require new rules for the game.

> **Appraisers Building AVMs**
> Many appraisers see the value in providing some type of valuation system that incorporates AVMs or valuation modeling technology. My involvement with two previous books published by the Appraisal Institute has put me in contact with a number of appraisers who, despite the barriers to entry, have actually built their own AVMs:
>
> - In one case, an appraiser took the text *A Guide to Appraisal Valuation Modeling* and, after contracting with a number of Russian statisticians, had the book translated into Russian and proceeded to build a cutting-edge AVM with some unique property listing features that could read the MLS comments section to identify potential upgrades or problems with a given property.
> - In another instance, Ernie Durbin, a leading residential appraiser in Cincinnati, adapted the information in both the previously referenced text and the follow-up text, *Practical Applications in Appraisal Valuation Modeling*, to design a robust valuation product that incorporated local real estate listing data with public record data to provide robust market analytics.
>
> References: Mark R. Linné, M. Steven Kane, and George Dell, *A Guide to Appraisal Valuation Modeling* (Chicago: Appraisal Institute, 2000) and Mark R. Linné, M. Steven Kane, and Jeffrey A. Johnson, *Practical Applications in Appraisal Valuation Modeling* (Chicago: Appraisal Institute, 2004).

Existing AVM products are, in and of themselves, problematic. Many appraisers recognize the difficulty of working under rules set by vendors who have little appreciation for an appraiser's regulatory environment. New tools are called for that aggregate data and analytics and that are more appropriate for appraisers to interact with. These tools are being conceptualized and built by those within and outside of the profession. Once an environment exists that overcomes that shortcoming, appraisers can offer the market a product that combines the best of local knowledge and analytics. Appraisers can clearly be most effective when interacting with valuation analytics developed within a robust data environment. The downside that many in the industry recognize with existing AVM products can be overcome with the involvement of a valuation professional.

Collaboration and the Development of the Valuescape

What happens when it becomes possible to quantify the meaning of location for the first time? What is the role of geographic information in ensuring that the information is referenced spatially in a manner that benefits its aggregation, analysis, and understanding? What is the role of valuers, valuation geographers, and spatial analysts in this segment of the economy? Opportunities for individuals and communities to take advantage of spatially referenced data, and the opportunities for greater understanding of the role valuation plays in all of our lives, should be at the forefront of how appraisers interact with spatial information. Spatially referencing fixed property data makes it easier to understand, relate, and use valuation information.

The shifting of the real estate terrain to include more data and spatially aware information creates an opportunity to aggregate and standardize the data layer of value, a data layer that I have termed the *Valuescape*. The Valuescape is the realization of the concept of data tied to a specific point on the earth. Data and information that is spatially tied to a specific point on the earth provides the ability to make spatially aware data available for analysis and utilization by appraisers and their clients. One question posed in the realm of the Valuescape is to examine what opportunities are available when it is possible to instantaneously know the value of the 90 million residential properties that make up the U. S. housing stock. What is the value of that information? How can communities benefit from knowing and using this information to promote social and community goals? The ability to pull any individual unique value or any grouping of properties provides opportunities for both public and private organizations to use this powerful information in meaningful and tangible ways. What type of analytics could fuel and, in turn, refine this remarkable aggregation of data?

The development of the valuation data layer and potential collaboration opportunities to further enhance the Valuescape will result in powerful new knowledge bases and the opportunity to benefit by tapping into the powerful valuation analytics that could be developed.

Imagining Collaborative Valuation

Today there is very little connectivity in data across a geographic landscape. Data is only available for a specific property within its specific environment. This paradigm is ready to shift. Imagine a valuation layer of data that covers the entire country: 90 million homes worth an estimated $15 trillion in market value, aggregated for the first time, where multiple valuation information intermediaries channel the data to consumers. This is the vision for the Valuescape: the aggregation of the data layer in a manner that permits its mining, refining, and delivery in a variety of formats to differentiated user groups.

Imagine linking those early adopters who already have the knowledge and critical understanding and the desire to create a massive and comprehensive national valuation infrastructure. These persons fuse the expertise of geographic information, spatial analysis, mass appraisal techniques, valuation modeling, and local expertise. The ability to link all of these skill sets in a collaborative environment that relies on an open-source strategy will reward contributors with opportunities to develop and drive further innovation.

This innovation will give the disenfranchised appraisal profession new purpose and opportunity in a collaborative role in the development of the Valuescape. For the first time, technology will be linked with valuation expertise to deliver rapid access to enhanced and accurate valuations of every property in the country. While there is certainly the possibility that individual privacy issues could be a concern, the benefit to all stakeholders of having all of the referenced data be public in nature is likely to outweigh individual objections.

This data availability will radically transform the nature of how value is established and delivered to users in the United States. By using "massively distributed collaboration," valuers and valuation geographers can achieve remarkable accuracy quickly and efficiently at a surprisingly low cost. Mitch Kapor, the inventor of Lotus 1-2-3, recognized the remarkable opportunity of this type of collaboration in a presentation at UC-Berkeley on November 9, 2005:

> The sudden and unexpected importance of the Wikipedia, a free online encyclopedia created by tens of thousands of volunteers and coordinated in a deeply decentralized fashion, represents a radical new modality of content creation by *massively distributed collaboration.* This talk will discuss the unique principles and values which have enabled the Wikipedia community to succeed and will examine the intriguing prospects for application of these methods to a broad spectrum of intellectual endeavors.[3]

An emerging grouping of activity centered around wikis, e-conferences, blogs, and other content-creating communities online provides venues to access, enhance, and refine data and knowledge. This trend can directly and profoundly impact the development of the Valuescape.

The concept of massively distributed collaboration could be undertaken on an ongoing basis. The benefit is that meaningful data can be archived, searched, and retrieved. Throughout this process, the underlying data can be enhanced in its accuracy and brought back into the database in an enhanced manner that provides additional value.

In an influential 2003 article in *Wired* magazine, executive editor Thomas Goetz wrote about the emerging open-source development model for business in the twenty-first century:

> Open source harnesses the distributive powers of the Internet, parcels the work out to thousands, and uses their piecework to build a better whole–putting informal networks of volunteer coders in direct competition with big corporations. It works like an ant colony, where the collective intelligence of the network supersedes any single contributor.[4]

The concept is straightforward and could be applied to real property valuation with far-reaching effects. Regional modeling experts could create all of the base models for 90 distinct metropolitan areas, representing most of the United States. Leveraging national and interna-

3. http://www.ischool.berkeley.edu/newsandevents/events/dls11092005/.
4. Thomas Goetz, "Open Source Everywhere," *Wired* (November 2003): 158-167, 208-209.

tional expertise, this process will permit local appraisers, Realtors, and other valuation professionals to examine, validate, and refine the valuations that aggregately make up the Valuescape. Furthermore, the data from the Valuescape could be made available to anyone who wishes to examine and interact in the development process.

It is critical that the profession be proactive in encouraging interaction with the Valuescape development. Potential collaborators must be identified and coupled with communication, education, and training. Courses covering automated valuation models (AVMs), introductory and advanced valuation modeling, and geographic information and spatial analysis topics are all critical. All of the education will enable students to achieve advanced technological competence that can be coupled with their local valuation expertise. (More information on this topic will be examined within this text as it relates to a potential statistics curriculum.)

The open-source approach to product development described by Goetz is about empowerment. Users must have a hand not just in shaping a potential product and the data, but in designing the product. This interaction will draw a wide variety of potential collaborators:

> Some will be genuinely motivated by the greater good, or find it satisfying to apply their professional knowledge to a broader audience, pro bono–style. And some will get to prove how smart they are.[5]

In a real estate valuation context, openness will make the data more useful. The Valuescape will be the ultimate free market–the free market of ideas. Commercial use should not be forbidden or discouraged, but rather encouraged to provide tangible benefit to these types of collaboration efforts.

Creating a collaborative environment permits a broader opportunity for innovation for potential uses. By opening up the availability of data, it will be possible to tap into and enhance a fast-growing valuation ecosystem. Following the lead of other mash-up oriented open-source environments such as Google, YouTube, and Amazon, the development of the Valuescape provides a vision of a collaborative community that would include software companies, data providers, and local community groups to further develop ancillary applications for end users. Additionally, other users could build free programs that allow Webmasters to easily pour public content and product information into their sites. Clearly there are benefits to both public and private stakeholders in this process.

The open-source approach can be a boon for all stakeholders. Goetz points out that licensing from a trusted collaborative project

---

5. Ibid.

saves money and leaves the technology open to further development.[6] By reducing the obstacles to innovate, collaborate, and prosper in the industry, dramatic innovation from the valuation mainstream will result. Collaboration can provide a potential source of knowledge materials from which the valuation culture and economy of the twenty-first century can be built. The opportunity for the Valuescape lies in making something new and finding a better way. "It isn't just about better software," Goetz concludes. "It's about better everything."[7]

Drawing Conclusions

Appraisers certainly face challenges as a profession when clients do not value their experience and objectivity. With the advent of greater data availability, advanced analytics, and a lending industry in need of new solutions, appraisers can find their true place and provide their clients with greater value than ever before.

Appraisers of the near future will be able to access property-level data that includes an aggregation of public and proprietary data. Layers of other relevant data will be provided to enable an evaluation of all physical and economic considerations that may influence the value of a given property. Appraisers will be able to define neighborhood boundaries in real time, import all of the data relevant to the assignment, and tap any historical data of their own or of others whose data has been made accessible. Appraisers will be able to run a series of analytic tools, including regression analysis, that will provide insight into the market, support the adjustments made, and finally provide a value that is confirmed and validated through a series of statistical measures. In real time, appraisers will be able to deliver a credible report, custom-designed to meet the specific needs of their clients.

Can appraisers by themselves change the course of an industry? While it is true that appraisers play a vital role in the typical real estate transaction, they are also the weakest player in the field relative to the other stakeholders such as the lender and broker. Change cannot come to the valuation profession if appraisers attempt to change in isolation.

Ultimately, it will be the job of appraisers, vendors, professional organizations, and clients to work together to bring innovations to market and accept those innovations as a meaningful improvement in how valuation meets the needs of the market. This is the true fulfillment of the promise of Valuation 2.0.

---

6. Ibid.
7. Ibid.

3 Open Data Standards for Real Estate, Appraisal, and Mortgage Banking

Implications for the Industry

Mark R. Linné, MAI, SRA, Elizabeth Green, and John Cirincione, SRA

Commerce depends on communication between diverse parties and between diverse information systems. Communication requires common systems of representing information, and standardization means agreeing on common systems. A significant challenge that has plagued the real estate industry is the lack of a common language for data. Terminology and core business elements have a bewildering array of nomenclature, with different descriptive field names and definitions applied to the same processes and data.

Many stakeholders would benefit from data standards efforts. Municipal planners, assessors, real estate developers, appraisers, architects, builders, utility companies, permitting agencies, building managers, maintenance contractors, owners, bankers, brokers, insurance companies, attorneys, CPAs, and first responders all need information about buildings and building site infrastructure. Most of these communities create new information on top of information provided by others.

Unfortunately, much of the information collected by others is not easily available so it needs to be collected again, an inherent redundancy. Even data in incompatible formats has value. Data could often be reused. All involved professionals develop data, deliver it to those who paid for it, file their copy away, and eventually discard it. The "common" data is not conveniently registered in an online catalog and stored online, behind an access control system, where it could be efficiently accessed by anyone who has a need and a right to access it. The reason for this lack of data sharing is that, until now, implementing shared library systems for building information has simply not been a practical goal except in limited situations. Over time, legacy systems, practices, and protocols have developed, further complicating the ability to integrate information and increasing the disparity of systems.

Integrating data from different types and brands of software has always been difficult, requiring batch conversion of files with significant manual intervention. In addition, different data producers have used different and incompatible data structures that define, for example, the encoding of "window" or "water main." Appropriate data sharing and archiving policies have not been put in place because data coordination has simply not been a high priority for the organizations that need to be involved. Long-term data management or governance policies need to be agreed upon by the high-level policy makers of different companies and agencies, who may not understand such issues or be accustomed to working together on them. In general, institutional barriers and inertia yield very slowly to data sharing and system interoperability initiatives, even when it is obvious that data sharing and interoperability will save time and money.

The activities of real estate-related professions involve *workflows*. That is, the flow of activities and interactions between multiple participants. This is true in a real estate transaction, which involves a seller, a buyer, agents of each, inspectors, appraisers, lenders, investors, and regulatory officials. Today, much of this "flow" involves digital information, and corporations and agencies of all kinds seek to engineer more efficient workflows that take full advantage of information technology. Where workflows involve the cooperation of multiple participants in independent organizations working on multiple tasks, the only reasonable approach to the workflow is to agree on an open architecture and an interoperable data standards environment. The technology and framework for a comprehensive set of open, consensus-derived information technology standards are readily available and can facilitate the movement of one step smoothly into the next. Virtually all components and documents can be integrated to enable workflows in an on-demand basis when there is systemic cooperation.

In recent years, groundbreaking developments in the rapid adoption of the Internet have occurred primarily because of the cooperation and adoption of open standards both in the programming languages and in the communications protocols and document/data delivery languages such as XML and HTML. Such standards are referred to as *horizontal standards* as they are not specific to any one industry but are fundamental to the tenets of the underlying technologies and the Internet. This is a key example is the importance of standards cooperation. The Worldwide Web Consortium, or W3C, was created to ensure compatibility and agreement among industry members in the adoption of new standards for Internet technologies. Prior to the consortium's creation, incompatible versions of HTML were offered by different vendors, increasing the potential for inconsistency between web pages. The consortium was created to get all those vendors to agree on a set of core principles and components that would be supported by everyone.

Overcoming Obstacles to Sharing Real Estate Information

Massive costs and time loss result from the lack of a standard "dictionary" or common semantics in real estate information. This specifically leads to redundant manual entry and intensive data translation on a system-by-system basis as an appraisal makes its way from the practitioner into the mortgage stream. The net result of these inefficiencies are higher costs, lower data integrity, and most importantly greater risk of errors, damaged data caused by translation inaccuracies, and the risk that the meaning and intent of the data will be lost or misconstrued. Furthermore, the lack of data standards serves to allow oligopolies to develop that constrain both appraisal practitioners and their clients from changing vendors and suppliers, and this inhibits innovation.

Open data standards are fundamental to the efficient, accurate, and seamless exchange of data. Many competing standards within the mortgage and real estate services industries exist for serializing, parsing, and describing data within a given industry. The purpose of an open data standard is to support common agreements that enable communications open to all. The term *open standard* implies that there are no fees associated with the use of the standard and that the standard is governed by a formalized, not-for-profit organization. It also means that the contributors and users of the standard cannot claim a copyright or charge a fee for the explicit use of the standard. A common misconception of a data standard is to think of it only as a file format specification. It is important to note that many specifications that are sometimes referred to as "standards" are proprietary and only available under restrictive contract terms (if they can be obtained at all) from the organization that owns the copyright on the specification.[1]

1. Elizabeth Green, Predictive Methods Conference, PMC2010.

Moving the appraisal industry beyond the struggles of data translation will create a new level of technological advancement that will improve the professional capabilities and relevance of appraisers, not the least of which is the ability of appraisers to electronically deliver vital appraisal expertise more rapidly and directly to their clients. This will mean instead of manually collecting and translating data and keying it in to form templates, appraisers can focus on analyzing and interpreting data to deliver improved valuations and a create a higher level of professional services. Reducing data translation and cycles removes friction from the overall process. And increased accuracy and precision in understanding appraisal results benefit the client and the appraiser as well as all downstream systems and processes.[2]

Open data standards encourage technology innovation. Advances in technology and data access can be rapidly made available to appraisers and appraisal clients. An instructive success story relates to the deployment of MISMO's open standard for requesting and receiving credit scores. (MISMO is discussed more extensively below.) Once the standard was established, the credit repositories were able to deploy it one time to the entire group of more than 30 vendor organizations that were involved in providing credit reports to the financial industry. Because of the cooperation on a standardized format for sending and receiving this information, lenders were able to integrate with the entire credit industry in one implementation, saving millions of dollars with efficiencies and ultimately bringing significant improvements to the underwriting of credit in the mortgage industry.

Intra-Industry Data Standardization

Industry-specific data standards have historically been developed by discrete trading partners to improve the flow of information. With the tremendous impact of modern technologies, access to data, and the Internet revolution as a whole, constituencies with common interests–such as lenders, title companies, insurers, and appraisers–need a more universal means of communicating efficiently, effectively, and accurately as all are stakeholders in the greater real estate lending marketplace.

When different entities use similar naming conventions for data fields to handle the same type of information, they are freed from the requirement to develop costly, dedicated data conversion systems in order to interface with others in the same industry. Such interfacing has become a priority in the real estate industry as the ongoing inte-

2. Elizabeth Green, "Empowering Appraisers with Data Standards," *LiveValuation Magazine* (April 2010).

gration with the public capital markets has increased the pressure for rapid, consistent, and accurate transfers of data.

The ability to exchange or integrate previously incompatible data is an additional benefit of standardization. In the mortgage industry, for example, the ability to have a common reference for data interpretation from the real estate listing through the appraisal, underwriting, loan delivery, and ultimately investor due diligence in the secondary market for mortgage-backed securities is a clear example of how cooperation and a consistent understanding of the data is vital to the flow of commerce. But an even more important benefit, beyond the improved efficiency achievable from replicable data and process methodology, is that data integrity and trust is dramatically increased.

As society becomes increasing interconnected, the cooperative use of a data standard outside of its primary community is an important consideration because interoperability extends the benefits of efficiency. To that end, different standards bodies can achieve interoperability in many ways, ranging from formal arrangements such as partnerships or alliances to simple agreements to adopt portions of one standard in another. Different policies and procedures do not necessarily prevent cooperation among standards bodies.

Standards organizations whose purviews overlap in real estate have much to gain (and their stakeholders have much to gain as well) from cooperating to provide seamless crossover and portability between real estate–related business sectors. Four significant standards organizations within the real estate industry are

1. Mortgage Industry Standards Maintenance Organization (MISMO)
2. Open Standards Consortium for Real Estate (OSCRE)
3. Real Estate Standards Organization (RESO)
4. Open Geospatial Consortium (OGC)

Mortgage Industry Standards Maintenance Organization (MISMO)

The mortgage industry, an important driver in the real estate industry, has been working towards data standardization through MISMO (www.mismo.org). Established in 1999, MISMO is a not-for-profit, wholly owned subsidiary of the Mortgage Bankers Association of America. MISMO's primary goal has been to finalize the residential and commercial standards that will facilitate an entirely electronic mortgage. The clients of appraisers have been the catalyst for standards to improve loan quality and production efficiencies as a direct result of the proliferation of the secondary mortgage market. Lenders and the investors who buy those loans have increasingly demanded more consistency, efficiency, and transparency within the appraisal process as the economy emerges from the subprime lending crisis of 2008.

Residential Lending

Much of MISMO's effort for the last several years has focused on the residential side of real estate through the Real Estate Property and Valuation Services Work Group (PaVS) due to the manufacturing-like process of residential lending. The common reference model and logical data dictionary of MISMO encompass all aspects of the mortgage transaction. This common structure ensures that from any viewpoint in the transaction, the name, meaning, and data type for information is consistent. Similarly, within the property- and valuation-related components, the data reference extends to all types of appraisal reporting, including property inspection and environmental reports. For instance, the data points for property value and effective date are always named and defined the same regardless of the report type they were used in.

Investor standards like those prescribed by Fannie Mae and Freddie Mac have played a primary role in instigating residential lending standards. In May 2010, under the direction of their regulator, the Federal Housing Finance Agency, Fannie Mae, and Freddie Mac introduced the Uniform Mortgage Data Program (UMDP) to provide common requirements for appraisal and loan delivery data. The result of a collaborative effort, this loan delivery data set has aligned Fannie Mae and Freddie Mac on the majority of the core data elements and their definitions. Where data differences persist, they are a matter of proprietary business or credit policy.[3] UMDP includes

- A Uniform Appraisal Dataset that standardizes key appraisal data elements to enhance data quality and promote consistency and is predicated on the MISMO Version 2.6 Valuation standard
- A Uniform Collateral Data Portal (UCDP) for the electronic collection of appraisal data to minimize impact on lenders
- A Uniform Loan Delivery Dataset that leverages the industry-recognized MISMO Version 3.0 standard

The decision to accept electronic reports using the MISMO data standards is a major tipping point for the industry, its key players, and valuation professionals. Specifically, it will enable MISMO to develop tools and metrics that will enhance collateral risk management by performing more sophisticated analysis and modeling and will ultimately improve decision-making capabilities.

---

3. "Uniform Mortgage Data Program Announced," https://www.efanniemae.com/sf/lqi/umdp/index.jsp/.

Commercial Lending

It is important to note that the MISMO Commercial Appraisal Work Group has begun to build momentum in its efforts to develop commercial standards. To that end, the work group's focus has been the incorporation and adoption of the Appraisal Institute's Commercial Data Storage Standards (CDSS), which has been underway since 2007. The CDSS project has resulted in approximately 2,000 distinct data elements for commercial properties, which are slated for adoption by both MISMO and OSCRE and which incorporated their efforts to promulgate data standards for commercial finance and general real estate uses.

Going forward, the primary goal of MISMO's Commercial Appraisal Work Group is the development of an "appraisal synopsis," i.e., a core third-party element of the asset summary report (ASR) that is widely used as a reporting mechanism and format within the commercial mortgage-backed securities (CMBS) marketplace. The Commercial Appraisal Work Group will focus on the initial metrics relevant to the credit decision made by a lender. Lenders currently have no means to effectively extract this information without costly rekeying of the essential data.

To assist in the adoption of this initial commercial standard, an incremental approach is underway, permitting users to experience the benefits of standards without having to change all of their internal processes simultaneously. MISMO Commercial is working to collect real world business cases of relevant example transactions that are currently in use throughout the industry to identify the critical data elements. This process serves two purposes:

1. To ensure stakeholder participation and involvement in the process (thus facilitating adoption and buy-in)
2. To promote understanding of current industry best practices in the preparation and usage of appraisal results and related business needs

MISMO efforts as a whole continue to work towards assembling a broad-based group of industry practitioners that includes lenders, appraisers, academics, and vendors. Appraisers will play an important role in the greater drive towards standards, which firmly establishes their role as stakeholders. In addition, there has been a greater effort towards identifying and reaching out to other stakeholders, including academic and assessment practitioners.

Open Standards Consortium for Real Estate (OSCRE)

The Open Standards Consortium for Real Estate (www.oscre.org) is a rapidly growing global organization that brings all of the data standards together from every sector of real estate. Appraisal is just one segment of the organization's collection of transactional standards, with other areas including construction, government, management, and brokerage. OSCRE is a 501(c)(6) not-for-profit consortium that drives the development, synthesis, and adoption of e-business standards that enable the real estate industry to function effectively and efficiently across various supply chains.

OSCRE represents all the players involved in the corporate real estate business, such as corporate real estate departments, owners, developers, property management firms, brokerage firms, and various service providers. Regardless of the industry sector, all real estate groups face the same needs. They require methodologies, systems, and tools to plan, buy, build, sell, lease, and manage real estate portfolios.

OSCRE's mission is to work with all real estate stakeholders to drive standards adoption–not just standards creation–throughout the industry. OSCRE is also different from MISMO in that MISMO's core focus is data standards specific to the universe of the mortgage transaction. OSCRE's focus is entirely real estate reporting information and processes, not exclusively related to a lending transaction.

Real Estate Standards Organization (RESO)

The Real Estate Standards Organization (www.reso.org) and its Real Estate Transaction Standard, or RETS, is a data standards effort that facilitates data transfer between partners in the real estate brokerage industry primarily for the sharing of listings of properties for sale. Creating and improving RETS is a collaborative effort of the National Association of Realtors (NAR) and MLS organizations to simplify moving real estate information from system to system and simplify development of services for NAR members. This effort has recently begun to take shape in a new initiative by NAR known as the Realtors Property Resource.

The development of a standardized data format allows brokers to more easily access, share, and propagate listings on the Internet, which is a primary shopping vehicle for buyers of real estate. These efforts were also a primary means of implementing the VOW (Virtual Office Web site) agreement with the Department of Justice in settling an antitrust case brought against NAR. The use of RETS has not only improved the accuracy and timeliness of the data being maintained on individual broker sites, but assists Realtors in VOW compliance.

The RETS community has focused primarily on developing a "transport" standard for listing and property information. Specifically, a data dictionary and core list of data elements have been identified, but the RETS approach relies heavily on *metadata* (i.e., data about data) to allow each MLS organization to have specific control over data definition, enumerations, and usage. NAR mandated that every participating MLS maintain a RETS server for listing transport by April 2009. In April 2010, RESO approved a set of Standard Names that provides for a higher degree of cooperation between participating companies on RETS but further engrains individual definitions communicated by metadata. Individual data translation on a per-MLS basis is necessary for any technology service or integrated access to an MLS system.

The goal of RETS is to create standards that allow brokers to expand access to their listing data using emerging technologies such as the Internet. The primary function of an MLS system is the marketing of properties for sale through a listing or offering. A single standard widely adopted across the industry enables real estate agents and brokers to enter listings once and deliver them when and where they want, including delivery to national property databases thereby exposing the listing to many more potential buyers.

Open Geospatial Consortium

Founded in 1994, the Open Geospatial Consortium is a 501(c)(6) nonprofit international voluntary consensus-based standards organization (www.opengeospatial.org) that is leading the development of standards for geospatial and location-based services. "Geospatial" online services provide data or information about events that are referenced to a specific geographic location. Geospatial data include

- A land use zone or flood zone
- The business occupying a specific floor in a building at a particular street address
- The position, orientation, and technical specifications of a surveillance camera

The OGC standards are important in multiple industries and are therefore considered horizontal standards. The real estate industry involves commerce related to geospatial features, i.e., land parcels and permanent man-made improvements and structures on the land. A direct result of the efforts of the OGC is the wide adoption of GIS-based services such as Google Earth, GPS tracking systems, electronic map plotting tools, and a whole host of geography-related, technology-based services.

A large number of OGC members are particularly interested in Building Information Models (standard schemas for recording and conveying virtually any information about a building), CAD/GIS integration (i.e., integration of computer-aided design and geographic information systems), and Open Location Services standards, all of which will benefit stakeholders in real property value chains. The OGC is involved because all of these types of standards have an essential geospatial or geographic location component.

Interoperability and Standards Convergence

Today, the primary appraisal reporting forms used by appraisers do not have fields for GPS or GIS coordinates. However, maps have long been required exhibits in the appraisal workfile. One major change in recent versions of the MISMO reference model is the recognition of a broader geospatial component beyond the latitude and longitude of the property. This additional information, aligned with OCG standards, enables appraisal data to be spatially defined in the context of a visual mapping format. This is significant given the importance of location to the appraisal process. The convergence of OGC and MISMO data enables appraisers to provide richer analysis methods to their clients. For example, the MISMO reference model accommodates the GPS coordinates of all of the properties that are referenced within the report.

GIS will continue to play a major role in facilitating a more exacting spatial element to valuation analysis. This information commonly includes a legal description, street address, property characteristics, property analysis, property tax data, and several hundred other fields.

Change on the Horizon

The Emergence of Consensus

The continuing cooperation of major standards organizations and their respective efforts are moving them closer to a collaborative and cooperative integration that will leverage literally dozens of organizations to achieve meaningful standards in all sectors of real estate. For example,

- MISMO is the standard for the mortgage lending industry in America. The Property Records Industry Association (PRIA) has maintained a strong alliance with MISMO in the pursuit of the fully electronic mortgage.
- OGC and OSCRE have recently signed memorandums of understanding to facilitate the development and acceptance of standards throughout real estate. Their efforts should have significant influence on the industry's acceptance of relevant and consistent data standards.
- The Appraisal Institute participates in these efforts through its own leadership involvement with OSCRE and MISMO, which brings a key stakeholder in the valuation arena to the table.

- Even The Appraisal Foundation, through its Industry Advisory Council (IAC), recognizes the importance of data standards efforts. The organization recently formed a Subcommittee on Data Standards to monitor and inform the IAC on matters pertaining to data standards such as outreach to appraisers, the need for training, and the manner in which this innovation could impact the industry.

The participation of all these significant players clearly underscores the importance that all stakeholders recognize as the push towards open data standardization in real estate continues.

Historically, competition and commerce have been key drivers for data standards in the private sector. But increasingly open data standards in the public sector are gaining ground. Open government at the local, state, and federal level is being driven by the implementation of data standards. Furthermore, the assessment community has tremendous opportunities to take advantage of technological advancements to keep pace with the community it serves because, ultimately, real estate is an amalgam of both private and public sector activity. Both are interdependent and synergistic, both rely on the underlying public records for land ownership and lien information, and both receive financial benefits from transactions in the private sector. In areas of taxation, there is a shared fundamental responsibility to ensure the accuracy of real estate data and the equality in taxation that uses that information to form its taxable values.

The Benefits to Valuation Professionals

It is becoming clear that more than a decade's worth of efforts for open data standards will have a future impact on appraisers. Appraisers deal with data–in fact, their entire business is about analyzing data. They should care and advocate for data standards because the easier and more efficient it is to acquire, analyze, and transmit data, the more efficient and profitable their business will be.

The ongoing mission of the data standards effort is to develop a common reference of data standards that creates a transparent, vendor-neutral environment. Promoting and achieving agreement between individual firms–often competitors–to use a common reference for data interchange is a core goal for the major standards organizations in the United States. This agreement does not threaten any competitive advantage. Rather, competitiveness is enhanced by reducing unnecessary complexity in an area common to all firms, allowing innovators to concentrate on their core competencies.

The industry generally recognizes that greater efficiencies are a critical benefit of the adoption of standards. Even greater structural changes on the horizon are likely to change the current paradigm of analysis and analytics. Both practical and theoretical ramifications must be considered and evaluated. The time is right for deeper insight into how standards could change the nature of the profession and ultimately drive aggregated data that can be used for research and analysis in more efficient ways.

Appraisers who are aware of the pace of change and the opportunities to take advantage of open data standardization will clearly have a competitive advantage in the marketplace. Access to data is one of the single biggest outcomes of the drive towards data standards. It will make access to data virtually effortless. Anything that allows appraisers to complete their jobs more effectively and efficiently will help their bottom line and allow them to serve the lending industry with more complete and transparent valuations. Ultimately, the winners will be valuation professionals who can focus on their core analytical expertise and their clients.

The appraisal community has a significant role to play and an opportunity to further the cause and participate in the promulgation of data standards throughout real estate. The opportunity to participate in meaningful dialogue for real estate standards will ensure that the input and requirements of valuation professionals are recognized and considered as standards continue to evolve. Ultimately, both public and private sectors should achieve greater efficiencies, cost savings, and opportunities for collaboration, trust, and understanding.

Conclusion

In many ways, technology has already revolutionized the way real estate information is retrieved, analyzed, transmitted, reported, and archived. It has also dramatically expanded who has access to the information. It is no longer clandestine information discreetly shared between a limited group of professionals operating in specific markets.

Data and process standards will flatten the terrain of valuation data and change significant portions of the legacy environment within which valuation professionals exist. Standards will ensure that real estate and valuation data flow to all parties in a transaction without friction, leading to a more efficient and transparent marketplace. And standardization will promote the use of tools to enhance the speed and accuracy of real estate data and permit the transformation of raw data into relevant knowledge.

REFERENCES

John Ross, former chief executive officer, Appraisal Institute

John Cirincione

Elizabeth Green

Geospatial Solutions, vol. 15, no. 6 (2005): 37.

Mortgage Industry Standards Maintenance Organization

Open Standards Consortium for Real Estate

Daniel Szparaga, MISMO/Mortgage Bankers Association

Andy Fuhrman, chief executive officer, OSCRE

Sam Bacharach, OGC

4 The Valuation Professional of the Future

Training for Innovation

Mark R. Linné, MAI, SRA

At its heart, appraisal is applied econometrics. In other words, the discipline of appraisal is the practical manifestation of analytics in real estate valuation. Even so, not that long ago the use of advanced analytics was unheard of in traditional appraisal unless the appraisal was to be used in court. In fact, statistical analysis was eschewed in common appraisal practice. Measures of central tendency such as the arithmetic mean and median were castigated for their potential for abuse, e.g., appraisers simply "averaging" value indications rather than "reconciling" them.

Previous chapters have examined the limited data and limited access to data that characterized the early years of the appraisal profession. This dearth of data spawned valuation techniques–such as paired sales analysis, developed in the 1970s–that attempted, as teaching tools, to convey how differences in property characteristics could be quantified. For example, in paired sales analysis, a "base" property could be compared to a secondary property with a difference in one key characteristic. The difference in sale price between the two properties serves as the pedagogical basis for illustrating the value of the difference in the selected property characteristic. E. T. Compere, MAI, an instructor for the American Institute of Real Estate Appraisers, was involved in the development of paired sales analysis as a teaching aid, but he would later relate his shock at finding out that the technique was being widely used in appraisal reports. In the classroom, Compere explained that the paired sales technique was never intended to be the primary analytic tool. Rather, it was intended as a demonstrative tool to explain the concept to students.

In contrast to the transformation of a pedagogical tool into a fundamental practical application, the adoption of analytics would pull appraisal methodology out of the past and into the future. But achieving the competence to deal with ever-greater amounts of data and to meet the Uniform Standards of Professional Appraisal Practice requirement for analyzing all the available data on top of the demands of clients for supportable, rigorous market analysis will require new ways of looking at how appraisers acquire and enhance their statistical skill set.

Understanding the Type of Education Required

What techniques should appraisers be using to meet the needs of an increasingly demanding clientele? No one can deny that more rigorous statistical analysis providing true insight into the dynamics of the market is a laudable goal and is increasingly needed by the market. The key question that arises is how this is effectively brought into the profession. Where does the training originate to achieve the goal of statistically based education? These questions are worth examining by the profession as a whole and, most critically, by those who provide training to valuation professionals. If the practice of real property valuation is indeed moving from an art to a science, are existing practitioners sufficiently empowered and knowledgeable to examine information in a meaningful way to produce the metrics that a data-hungry clientele demands?

In addition, what are the challenges posed by the need to retrain a profession of primarily adult learners in a challenging discipline that includes college-level analysis of data using tools such as regression analysis? What is the proper protocol for bringing advanced analytics education to the table with new and improved education for the 90,000 valuation professionals who make up the residential and commercial sectors of the appraisal profession? Adult learners are generally self-motivated but have higher expectations of the curriculum than traditional students at the college level. Experienced practitioners, and adults in general, learn differently than college-age students. Also, appraisal practitioners who are adult learners bring experience and knowledge to a classroom or online course. Sometimes educational expectations will be unrealistic or impossible, but they must be understood so that proper deployment of a new education curriculum is achieved.[1]

> **Why Should We Care About Statistics in Real Estate Appraisal?**
>
> There is a dichotomy when the scientific study of statistics is partitioned between its theoretical and applied components. It is critical to minimize the complexity of the theories underlying the applications that can be of practical use to appraisers. A better strategy would be to concentrate on providing instruction in those techniques that relate directly to an appraiser's need for a skill set that can provide an enhanced understanding of data and analysis.

The advanced education suggested in this chapter constitutes far more than a mere seminar to provide a sufficient understanding of fairly sophisticated techniques. Given the challenges in teaching adult learners, organizations such as the Appraisal Institute may be best suited to offering the more rigorous education to experienced practitioners that will be required.

Statistical Support for Appraisal Analysis

Recent court decisions, including the Guardian Pipeline decision of the U.S. Court of Appeals for the Seventh Circuit, clearly support a wider use of statistical techniques such as multiple regression analysis in valuation proceedings in place of the long-standing matched pair comparison.[2] Regression is moving towards becoming an accepted methodology in the courts, while matched pair analysis is still the most widely used technique in general appraisal practice.

Heller et al. note that there has been a clear progression from the "opinion" perspective to the "science" perspective for appraisers in recent years. Beginning with Daubert (*Daubert v. Merrell Dow Pharmaceuticals*), in which the U.S. Supreme Court ruled that a more defensible scientific basis for valuation was necessary, through Kumho (*Kumho Tire Co. v. Carmichael*) and Federal Rule of Evidence 702, and ending with the Guardian Pipeline decision, the courts are increasingly calling upon appraisers to use and implement a higher standard of valuation rigor in estimating values in a wide variety of valuation situations. The Seventh Circuit court of appeals, in its ruling, commented as follows:

> What puzzles us is why both sides were focused on pari-wise comparisons–that is, matching each subject parcel with a supposed "comparable" parcel that does not have a transmission-corridor easement. . . . A different approach would be to gather data about the actual selling prices of real estate

1. Amanda Kendle, "How to Teach Adults: Three Principles for Successful (and Fun) Classes," (2005), available online at http://www.googobits.com/articles/628-how-to-teach-adults—three-principles-for-successful-and-fun-classes.html/.
2. John A. Heller, Peter F. Colwell, and Joseph W. Trefzger, "Expert Testimony: Regression Analysis and Other Systematic Methodologies," *The Appraisal Journal* (Summer 2009): 253-262.

> with and without transmission-corridor easements and use these data to determine how much the easement reduces the value of real estate in real transactions. The law of large numbers would make up for the lack of closely matched comparison pairs. . . Using real transaction prices reduces the role of guesswork. Although no one suggested such an approach in this proceeding, litigants (and district judges) should keep it in mind for the future, as it has the potential to be faster, less expensive, and more accurate than a parade of witnesses offering estimates that cannot be verified.[3]

Heller et al. then comment on the practical ramifications of the court's decision:

> Judge Easterbrook's proposed new approach–questioning the use of typical adjustment factors derived from matched pair comparisons, and embracing multiple regression analysis as a faster, more accurate alternative–is so sensible, and long overdue, that is seems likely, eventually, to assume an important role in adjudicating such cases.

The article, in turn, includes a comment on the role of appraisers:

> Appraisers, in turn, will need to become more proficient in multiple regression methodology and, when testifying as expert witnesses in cases involving easement takings or other adjudicated valuation issues, be able to explain their analysis and data in a manner that is admissible in evidence.

The authors note in this article that the existing literature may not be sufficient:

> The Rubinfield "Reference Guide," to which the Guardian ruling renders, provides a concise summary of the theory underlying regression analysis and, in the appendix, a numerical example (involving gender discrimination in pay). But it is not likely to provide an appraiser new to the topic with sufficient grounding to apply multiple regression analysis in practice; so, other general references on statistics and econometrics, or other sources that focus specifically on regression analysis, may be needed.

While it may at first seem like a leap to expect that regression analysis will suddenly be demanded by clients for appraisal services, it is clear that the financial services sector is beginning to demand more than just an appraiser's opinion of value. In 2009, the 1004MC (for market conditions) became a requirement for appraisals completed for Fannie Mae and Freddie Mac. This format required a far more intensive consideration of market conditions influencing a given property than had ever been required by lenders before. Appraisers are clearly being called upon to provide more supportable valuations. If the courts also begin to demand a more scientific basis for valuation, then either this requirement will permeate to traditional appraisal services, or the market will see a bifurcation in which a greater sophistication may be required only for litigation services. The truth likely lies in between these two extremes, and appraisers in

3. Ibid., 254.

general would be well-advised to begin looking at ways that they can incorporate a greater scientific basis for their valuation conclusions.

Can current tool sets and levels of training and education meet the needs of a profession that until recently has not been called upon to deliver this level of detailed analysis to its clients? It is instructive to note that since 1982, a primary arrow in most appraiser's analytic quiver has been the HP-12C calculator, the only computational tool that is both simultaneously under glass in the Museum of Computer History and in the hands of practicing appraisers. An instrument that is both an exhibit in a history museum and a foundational analytical tool used by most of the nation's real estate appraisers is a serious oxymoron, or at the very least, presents issues for further consideration.

Even so, appraisers may have a greater understanding of the fundamental techniques and "real-world" practicality that belies what appears to be only intuitive or subjective insight in the valuation process. In the text *Real Estate Valuation Theory*, authors R. Kelley Pace, C. F. Sirmans, and V. Carlos Slawson Jr. discuss their thesis that appraisers already use techniques that mimic statistical methodology in the chapter "Are Appraisers Statisticians?"[4] The thesis of the chapter is that "statistically challenged (human) appraisers following ad hoc procedures" exhibit prediction errors substantially less than those exhibited by hedonic pricing models resting on a foundation of "statistical and economic theory."

Pace et al. note that rather than relying on the human equivalent of a statistical hedonic model, appraisers seem to focus on the comparable sales selection process. Furthermore, the authors develop the argument that real estate appraisers have gradually developed effective techniques for processing real estate information. The authors conclude that "the statistically naive practice of appraisers may have a more sophisticated basis than a casual examination would reveal." The conclusion by the authors is that "prediction in statistical models could be improved by incorporation of elements of what appraisers actually do."

Perhaps. But this perspective assumes that a stand-alone solution would be the only direction that this type of analytical structure could take. Looking at the process from another perspective, it makes sense to explore if appraisers could enhance the process. The (tongue-in-cheek) comment by the authors pertaining to "naive" appraisers is an interesting starting point for an analytic methodology that would take the best of analytics and meld it with the experience and local knowl-

4. Ko-Kang Wang and Marvin L. Wolverton, eds., *Real Estate Valuation Theory* (Boston: Kluwer Academic Publishing, 2002), 31-43.

edge of the appraiser. If models could be enhanced by the incorporation of appraiser emulation techniques, could not the opposite also be a probable solution? Is it logical to assume that the work of a valuation professional, augmented by more robust statistical analysis and techniques, could ultimately result in more supportable valuation that has the analytical support of sophisticated data analysis? Rather than attempting to build the best models by attempting to eliminate the appraiser component, it seems most appropriate to enhance the valuation professional with the mathematical rigor of appropriate statistics. Ultimately, it would seem that integrating the strengths of regression with the inherent skill set of appraisers would yield the best of both techniques, which should be the goal of statistical training for appraisers.

Guidelines in Teaching Statistics

A traditional statistics curriculum, in which a formula-oriented approach was taken to "teaching" the subject, has historically been the norm. That, however, is no longer the case. Analytics have been successfully embedded in a host of software applications, including Excel, SPSS, Minitab, and others. Memorizing formulae is no longer necessary. The focus of statistics education should be on studying the concepts and understanding the use of tools that perform the calculations. The focus should be to examine what the analysis means and how the data relates to solving the valuation problem at hand for valuation professionals.

Statistical competence is the result of a cumulative learning process, and the transference of knowledge must be accompanied by practical examples and case studies that reinforce the process. As previously noted, the predominant student population for continuing education in analytics for valuation will be current practitioners, who have used other techniques to arrive at the answers to valuation problems over the course of their careers. How can statistics be combined with practical knowledge to empower the intuitive process that appraisers have been using to solve valuation-related problems? Statistics is not learned successfully by just observing or watching. A hands-on regimen is necessary for advanced analytics.

Earlier chapters of this textbook have focused on the fact that significant amounts of data are now available (or soon will be) to practitioners in a manner that has never been possible before. It is also clear that the pace of data availability will only accelerate from this point on, demonstrating a need for education to make students comfortable with large amounts of data.

Characteristics of Successful Statistics Training and the Ideal Curriculum

In the past, statistics training within the existing coursework available to appraisers has been inconsistent at best. Historically, statistical training has been embedded within other courses or, worse, taught in small modules that fail to make the connection between theory and practical applications. Most courses on advanced statistical techniques that are available to appraisers today are essentially overview courses, generally structured as continuing education seminars or courses that attempt to use and extend some existing tools such as the HP-12C calculator or the functionality of software such as Excel.

The Appraiser Qualifications Board (AQB) of The Appraisal Foundation has current guidelines on the inclusion of a statistics component in the required core curriculum for certified residential real property appraisers, but the requirement is only part of a 15-hour course module focused on statistics, valuation modeling, and finance. The course is structured to conform to the 2008 Real Property Appraiser Qualification Criteria established by the AQB. The content of the Appraisal Institute version of the course intended to meet the AQB requirements starts with a discussion and examples explaining the importance of statistical analysis for real property appraisers. The course progresses through such topics as descriptive and inferential statistics, graphical analysis, measures of central tendency, and measures of dispersion. Then, mathematical and valuation modeling are demonstrated. In this same section, AVMs are also covered, along with linear regression and multiple regression analysis. The last section investigates the history of financing and the flow of funds. The role of the Federal Reserve System is explored, along with how mortgages work. There is an extensive analysis of mortgage types, terms, and calculations showing the effects of changes in terms of payment patterns. There is a required 60-minute exam at the end that must be taken in a proctored setting.

The total amount of time spent on statistics in this existing curriculum is slightly more than half of a 15-hour course, with the remainder dedicated to real estate finance. Eight hours of training is clearly not enough to develop any meaningful knowledge that can be successfully used in day-to-day practice. Expanding this initial introduction with a more technical and hands-on education experience is a key requirement for building an appraiser's skill set in this area of analysis. Even so, the existing curriculum does represent a starting point. The question is what is the next step the profession must take.

A prudent education strategy would provide a series of practical applications for real estate and appraisal professionals to truly

Exhibit 4.1 Course Objectives

1. Relate instances in which statistics can be helpful to appraisers
2. Distinguish between descriptive and inferential statistics
3. Describe graphic analysis
4. Perform basic functions on an HP-12C calculator
5. Calculate a mean
6. Calculate a median
7. Define a mode
8. Summarize the characteristics of each of the three measures
9. Describe sampling techniques
10. Define and illustrate a range
11. Calculate an average deviation
12. Calculate a standard deviation
13. Describe the characteristics of a normal distribution
14. Describe and illustrate inferential statistics
15. Calculate areas under a normal curve
16. Describe the use of a coefficient of determination
17. Demonstrate linear regression
18. Describe and illustrate mass appraisals
19. Define and illustrate AVMs

incorporate innovative best practices into their valuation work. Focusing on developing skills in the classroom that can immediately be employed in daily work efforts is ideal. An important perspective that must be instilled in the minds of valuation professionals is that appraisers in the future will have to use tools such as regression analysis to support their valuation analysis. Without a doubt, this is an accurate assessment of the future required skill sets of appraisers.

The majority of course material available to the valuation profession has been developed from an academic perspective, which is appropriate for basic and more advanced applications, but a practitioner's perspective is necessary to impart the statistical and analytic skills that practicing professionals require in their daily tasks. The concept of "modeling" property information is the critical difference between the theoretical and the practical. A practical application would involve taking raw data, applying statistical analysis to the data, and deriving meaningful outcomes that permit effective decisions to be made regarding adjustments, the underlying adequacy of the data, and supportable opinions of value. A hands-on approach that delivers practical applications is the only acceptable route to equipping appraisers with the skill sets that will be required for more rigorous analysis of data. Taking the theoretical and making it practical–by immediately taking data pertinent to a valuation assignment in the classroom and applying appropriate tools sets–will provide a best practices perspective that can demonstrate the costs and benefits of using advanced techniques effectively.

What Can Appraisers Learn from Assessment Professionals?

Since the early 1970s, the assessment profession has pioneered and perfected the use of statistical training with both a single-property and mass appraisal focus. Significant coursework has been developed that ensures competence and expertise in statistical analysis, including regression analysis, and most large assessment jurisdictions have developed training sufficient to ensure that their staffs have the expertise needed to value hundreds of thousands of properties with accuracy and an understanding of market trends and forces.

In examining the success of the assessment profession's educational efforts, a key consideration is a recognition and understanding that a delivery system for more advanced training has been effectively deployed in a manner that begins with a single-property focus but, once mastered, moves into fairly sophisticated real estate analytics that allows for portfolio-level analysis. The success of the educational program, most of which is provided by the International Association of Assessing Officers (IAAO), demonstrates that valuation professionals can be trained effectively to learn statistical techniques that will benefit them and proves the importance of holistic rather than piecemeal solutions to training. At the end of the coursework, students have a practical understanding sufficient to meet the needs of their discipline.

Overview of Potential Course Design

The focus of an effective curriculum for appraisers, especially those who are already experienced in their discipline, must stress the use of statistical software that can provide the basis for a meaningful analysis of real estate information. Students should be able to take real-world data, identify its strengths and weaknesses, and "unpack" the basic statistics that are relevant to understanding the data and the output of the analysis better than any traditional valuation methodology would otherwise provide. By using real-world data, students will already have a familiarity with the process and immediately see the benefits of the statitistical tools. Supporting adjustments, deriving depreciation from the market, and predicting market conditions are some of the concepts that students will need to become more familiar with.

To be effective, education programs must, at a minimum, include the following macro-level topic areas:

- Preliminary data analysis
- Statistics and real estate–what is really important

- Statistical measures of central tendency and dispersion
- Introduction to regression analysis
- Using regression analysis in the real world

Additional curricula can be developed that meets a broad and expanded range of practitioner needs. An ideal curriculum could include a full range of training that begins with helping appraisers in the use of statistical programs, beginning valuation modeling, intermediate modeling, modeling commercial and industrial properties, and modeling vacant land.

Suggested Body of Knowledge

Valuation professionals should be familiar with and capable of applying the following elements in the body of knowledge:

1. Know the data collection methods and understand how to categorize data.
2. Construct and interpret a scatter plot, joint frequency table, Pareto chart, frequency histogram, and various types of bar charts.
3. Compute and understand the mean, median, mode, and variance for a set of data.
4. Compute the coefficient of variation, the coefficient of dispersion, and z scores, and understand how they are applied in decision-making situations.
5. Apply the common rules of probability and identify the types of processes that are presented by discrete probability distributions.
6. Understand the main approaches to assessing probabilities and determine probabilities associated with binomial and Poisson distributions.
7. Discuss the important properties of the normal probability distribution, calculate probabilities using the normal distribution table, and apply the normal distribution in appropriate business situations.
8. Understand the concept of sampling error and the importance of the Central Limit Theorem.
9. Determine the mean and standard deviation of the sampling distribution of the population mean and proportion.
10. Construct and interpret a confidence interval estimate for a single population mean using both the standard normal distribution and t distribution.
11. Determine the required sample size for estimating a single population mean.

12. Establish and interpret a confidence interval estimate for a single population proportion.
13. Formulate null and alternative hypotheses for applications involving a single population mean, proportion, or variance.
14. Know how to use the test statistic, critical value, and *p* value approach to test the null hypothesis.
15. Calculate and interpret the simple correlation between two variables.
16. Understand the basic assumptions behind single and multiple regression models.
17. Recognize some potential problems if regression analysis is used incorrectly.

Observations on Academic Involvement

Substantial benefits could result from the academic community becoming more involved in the effort to take advantage of the amount of data currently available. Better-quality data will enable more robust analysis and enhance research opportunities–the type of research efforts that are necessary to determine the analytic techniques and metrics of a data-rich real estate environment. Furthermore, mundane tasks relating to the standardization of data sets are easily taken care of.

In exploring what a standardized world looks like, the role of academia must not just touch the future, but take an active role in building it. Open data standards are developed by active volunteer participation, and academia is well positioned to contribute its time, efforts, and resources. Such resources have been available and have been channeled to other current standards efforts, most notably through organizations such as the Open Geospatial Consortium (OGC), which has developed relationships with a number of universities that have contributed time and labor to the effort of standardizing geographic information.

Exhibit 4.2 Core Curriculum

Specific core elements in an effective statistics curriculum should include the following specific topics:

- The nature of statistics
- Types of measurement
- Quantitative and qualitative data
- Organization of data
- Measures of location
- Measures of variability
- Summary statistics
- Numerical/quantitative scales
- Categorical/qualitative variables
- Descriptive statistics
- Displaying data
- Frequencies and histograms
- Correlation analysis
- Tables and charts
- Concepts of probability
- Probability distributions
- Sampling distributions
- Statistical inference
- Identifying characteristics about populations
- Differences between populations
- Fundamentals of hypothesis testing
- Analyzing significance
- Clearly communicating results
- Simple regression analysis
- Multiple regression analysis
- Model building
- Presenting statistical results

What is needed at present is greater academic participation in other aspects of the standards effort. The academic community is clearly a stakeholder in this process, and getting involved at the front end of the process is of vital importance. The curriculum of the future must be considered and designed today in a manner that will drive research, theoretical applications, and practical applications, and ultimately develop graduating students who will be more employable. The goal would ideally be the development of skill sets that are appropriate to the real estate world as it will be in five to 10 years, not as it exists today.

It is as stakeholders that the participation and influence of academia is most critically needed. Academia and the broader valuation profession will both benefit as the future is explored and defined from a collaborative perspective that challenges participants to imagine a future made all the more relevant, interesting, and ultimately prosperous.

Conclusions

Valuation professionals have a compelling need for more comprehensive training in statistical analysis than is presently available. Complicating this situation is the fact that the majority of those who have a need for more information are adult learners whose professional experience has been focused on the traditional techniques that grew from an environment of limited data availability. The new techniques that will be required in the future cannot be learned in a simple seminar or two-day course. There is a compelling reason to support a redefinition of the type of education that is most effective in helping appraisers meet the challenges of data analysis. Rather than being purely reactive, the valuation profession must codify in more exact terms the expectations, nature, and implementation of an analytics framework that will serve the needs of a rapidly evolving landscape.

5 Introduction to Geographic Information Systems

Michelle M. Thompson, PhD, and Milton Ospina

- GIS is a collection of computer hardware, software, and geographic data for capturing, managing, analyzing, and displaying all forms of geographically referenced information.

  –ESRI (http://www.gis.com/whatisgis/index.html/)

- In the strictest sense, a GIS is a computer system capable of assembling, storing, manipulating, and displaying geographically referenced information, i.e., data identified according to their locations. Practitioners also regard the total GIS as including operating personnel and the data that go into the system.

  –USGS (http://webgis.wr.usgs.gov/globalgis/tutorials/what_is_gis.htm/)

- A geographic information system (GIS) is a computer-based tool for mapping and analyzing things that exist and events that happen on earth. GIS technology integrates common database operations such as query and statistical analysis with the unique visualization and geographic analysis benefits offered by maps.

  –ESRI (http://www.gis.com/whatisgis/index.html/)

- GIS is an integrated system of computer hardware, software, and trained personnel linking topographic, demographic, utility, facility, image and other resource data that is geographically referenced.

  –NASA (http://science.larc.nasa.gov/biomass_burn/glossary.html#G/)

Overview

This chapter provides a summary of the origins of geographic information systems (GIS) and a look at major changes in the technology to date and what significant changes have yet to emerge in the ongoing evolution of GIS in research and practice.

Geographic information systems are made up of a set of specific tools and methods that allow the visualization and analysis of a plethora of data that spans from social to economic to empirical data. However, as illustrated in probably the most familiar summary of the components of GIS (Exhibit 5.1), the most important piece of the puzzle is **people**. In order to formulate an appropriate research question and response as well as create a dynamic system for analyzing output, the users of GIS systems must develop patterns and methods of critical thinking that will influence the whole of the GIS.

The collection of **data**–from using a handheld GPS to accessing regional digital orthophotographs–is now readily accomplished through communal downloads for use by both public and private entities. The ability to synthesize data formerly considered proprietary (e.g., a credit report) and merge this with housing transactions and market segmentation (e.g., buyer behavior) has increased knowledge shares of "social capital" that affect public policy and private investment.

Exhibit 5.1 Components of GIS

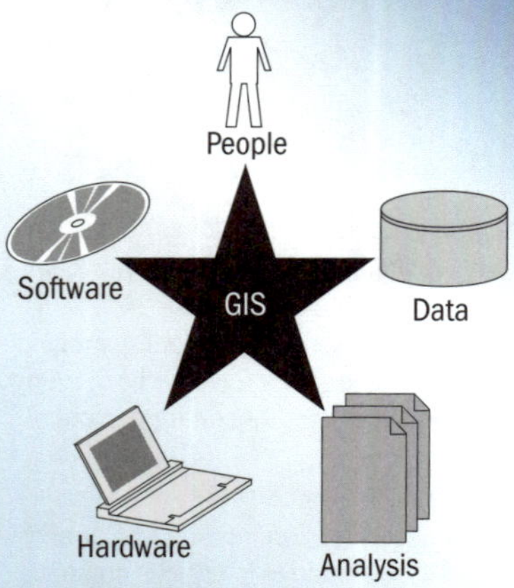

Significant changes in the collection and use of data, the type of software available, the hardware necessary to run programs, and the types of geospatial analysis that can be conducted have influenced the business environment.

Source: ESRI

Over the course of its history, GIS **software** has changed to accommodate the less expensive, yet more powerful and user-friendly **hardware** such as laptops and smart phones that have allowed room-sized, tape-driven data processors to become pin-sized microprocessors. GIS software was formerly encapsulated or encrypted, but with the advent of desktop publishing GIS software is more readily accessible to less sophisticated users of geographic data. A shift from command line GIS to Windows-based GIS as well as Web-based GIS and mapping has increased the ease of use of GIS, which also increased general GIS and geographic knowledge by making the tools accessible to more users along with complementary training, both formal and informal.

During the 1980s, advances in computer science propelled the world of GIS to new heights. Specifically, *relational databases*–databases with a GIS data structure in which GIS attribute tables are logically associated with each other by shared common fields such as in Oracle, MS SQL, and MySQL–allowed data to be manipulated into components, which previously were not easy to merge using ASCII tables.

The *cost* of a geographic information system previously depended on the user's ability to access software and hardware, both in the setup and maintenance. As with the sophistication of the software, the evolution in the cost of desktop computing revolutionized the ability of companies of all sizes to customize GIS for their business practices. In the discipline's early years, setting up a GIS systems was quite expensive, and the technology was only available to large corporations or government agencies. As software became easier to use and desktop computing became ubiquitous, businesses, universities, and nonprofit organizations were able to take advantage of GIS for their organizations' business goals.

GIS allows for the execution of scalable spatial **analysis**–the final component–within the same system. The ability to develop models that allow for historic and prospective market analysis is only limited by the extent and validity of the data. Multiple linear regression analysis is still not immediately integrated within most geographic information systems, but the expansion of specific models, such as inverse distance weighting, are now more common as options for advanced spatial analysis, and companies such as ESRI have developed GIS extensions for geospatial statistical analysis (e.g., ArcGIS Geostatistical Analyst) that allow for statistical analysis of surfaces such as grids, contours, filled contours, and hillshades or any combination of these renderings. Conversely, many statistical programs, such as SPSS and Geoda, now include a mapping element after statistical analysis.

In all cases, the expansion of *mapping applications* on the World Wide Web has undoubtedly had the most profound impact on what *contemporary GIS* is and what it will become in the near future. Internet mapping services now allow the distribution of data services, not just data sets. These services also allow for the use of basic GIS functionality such as mapping a property and doing basic real estate comparative analysis (e.g., www.zillow.com) or conducting basic economic development research for locating or relocating businesses in a community (e.g., http://availableproperty.seda.org/).[1]

Contemporary GIS

The term *GIS* has become synonymous with any study, analysis, or reporting related to the collection, transformation, visualization, and analysis of geospatial data. GIS is a complex system of tools, techniques, and theoretical applications of how objects that were previously identified in two dimensions have now been transformed into three dimensions with the increasing availability of GIS layers (see Exhibit 5.2) that capture *Z* values for roads as well as building structures (3D GIS). In the future, GIS will also take into account how the fourth dimension–time–will be considered in models to reflect and predict behaviors that can also leverage all dimensions.

Exhibit 5.2 Data Layers

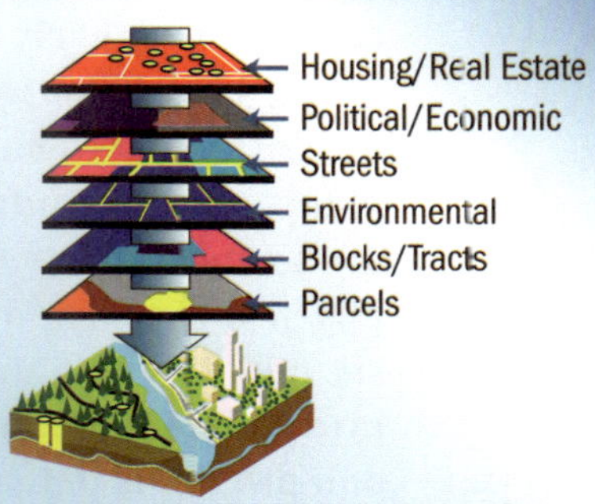

Source: ESRI

In the past, temporal analysis has been possible using, for example, combined historic data sets of population growth or crime instances over a specific period of time. However, today the ability to draft analysis that reflects "real-time" phenomena such as the spread of disease due to food contamination can be monitored as it is happening, for example, by the U.S. Centers for Disease Control (CDC) as pharmacies across the country report real-time sales of medication for controlling diarrhea.

1. Chapter 7 will provide a look at the shifts in how GIS is being used in real estate and business analysis with a case study of GIS applications in economic and business development.

The key to an effective GIS is an accurate and up-to-date GIS database (the old computer science adage "garbage in, garbage out" applies here as well) within a GIS organizational structure (i.e., a GIS program) that promotes enterprise-wide use and prevents duplication of effort. (Exhibit 5.3 illustrates common GIS input and output.) This GIS program can help to promote a more comprehensive and holistic approach to government and public service where costs are reduced and customer service is more readily available. Applying an enterprise approach to a GIS program helps ensure that databases are not duplicated, many business objectives are served by a single GIS database, and software and hardware costs are minimized by leveraging centralized approaches to GIS storage and application development. In an effort to create consistent methods and to advance the use of geospatial analysis beyond GIS, the field of *GIScience* has emerged.

Exhibit 5.3 GIS Input and Output

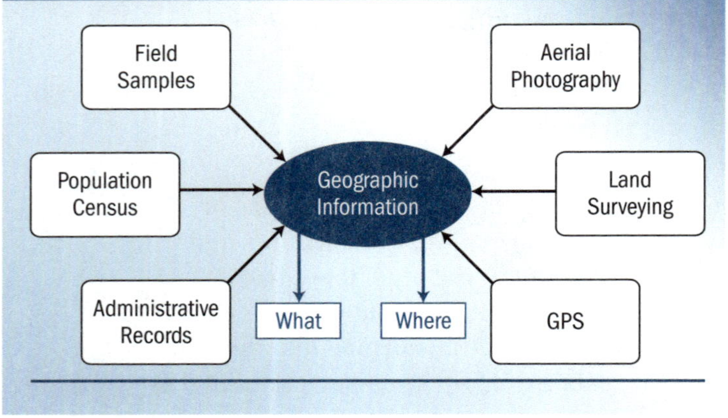

GIScience

The National Center for Geographic Information and Analysis (NCGIA) based at the University of Buffalo was one of the first organizations to develop an early history of GIS project called "History of GIS." It focused on the outgrowth of GIS from computer applications to the expansion of *GIScience* as a subarea of study that has confluence with many areas of study and theoretical models.[2] GIScience is study of the science, art, technology, institutions, and socioeconomic and biophysical dimensions involved in the collection, management, analysis, and dissemination of geographically referenced data in order to provide planning and decision-making solutions to societal demands and problems that are usually ill-defined.[3]

---

2. http://www.ncgia.buffalo.edu/gishhist/.
3. Michael F. Goodchild, "Geographic Information Science," *International Journal of Geographic Information Science*, vol. 6, no. 1 (1992): 31-45; P. A. Longley, M. F. Goodchild, D. J. Maguire, and D. W. Rhind, *Geographic Information Systems and Science* (John Wiley and Sons, 2001), especially pages 21-26.

GIS is deeply rooted in science, from the the algorithms required to process simple mapping displays to the high-level mathematics required for complex 3D, raster, routing, logistics, sales territory planning, site evaluation, market penetration, customer prospecting, and other GIS modeling applications. Today, GIS technology companies make the application of GIS look easy. However, GIS still requires a strong technical background to interpret the results generated by clicking on a few menus and buttons in the GIS software.

GIS companies and their users realize that while increasingly simple to use, GIS requires a great deal of knowledge to design valid models and interpret their results, which is why private sector vendors like ESRI have dedicated a great deal of human and financial resources to supporting GIS education and research in collaboration with the traditional providers of GIS education.[4] According to Angela Lee, ESRI libraries marketing manager, in the United States 220 four-year higher education institutions offer a degree in GIS, 128 of which have graduate programs. In 2008, 1,184 two-year colleges offered GIS courses, of which 48 offered certificates and degrees and 68 offered AA or AS degrees.[5]

The University Consortium for Geographic Information Science (UCGIS) counts about 60 member universities and is dedicated to advanced the use of GIS "as an effective, unified voice for the geographic information science research community and to foster multidisciplinary research and education." Among other things, UCGIS (www.ucgis.org) focuses on developing science-focused curricula that all higher education institutions can use to teach with GIS applications rather than simply to teach GIS skills and theory in isolation.

Summary

The evolution of geographic information systems from a technology-driven program to a decision support system has been accomplished in a limited period of time. Even with recent advances in the discipline, the basic elements of a geographic information system remain the people, software, hardware, and the data. Configuration of these components and reliance on standards and systems that will render the GIS analysis accurate and reliable are all dependent upon the user.

At its core, GIS is considered a program, but it is also considered a methodology on which GIScience is based. There are systematic means to develop projects, as well as conduct and report analysis, but

---

4. http://www.esri.com/industries/education/index.html/.
5. Much of this information is gathered from the ESRI's Online Database of Academic GIS Programs available at http://edcommunity.esri.com/universityprograms/onlinedb.cfm/.

open source and customized programming provide great flexibility in the application of GIS. The endless opportunities for using data to address micro and macro public, business, policy issues from the local to the regional scale have allowed GIS to go beyond programming and evolve into the realm of science.

6 Integrating Geographic Information and Valuation Modeling for Real Estate

Mark R. Linné, MAI, SRA, and John Cirincione, SRA

Introduction

A *geographic information system (GIS)* is "a computer system for capturing, storing, checking, integrating, manipulating, analyzing, and displaying data related to positions on the Earth's surface," according to the Association for Geographic Information, as cited in the Open Geospatial Consortium Glossary.[1]

Put simply, the use of GIS in real estate enables the user to capture, store, analyze, and visualize real property information data points of interest in a map or geospatial context. GIS makes possible solutions that are impossible to achieve viewing the same data set in a spreadsheet format or without geospatial context. Adding a spatial component to data provides the user with an opportunity to gain greater insight into the data and see new patterns that might be less noticeable in a purely tabular or narrative representation of data.

"GIS brings information together; it unifies and integrates that information," says Jack Dangermond, cofounder of Environmental Systems Research Institute (ESRI). "It makes available information to which no one had access before, and places old information in a new context."[2]

1. www.geo.ed.ac.uk/agidict/welcome.html/.
2. Jack Dangermond, "Software Components Commonly Used in Geographic Information Systems" (1989).

Geographic information systems have, for the most part, been available to real estate practitioners for more than 20 years, with vendor platforms integrating data into various layers. However, the use of GIS by appraisal professionals is still relatively limited in comparison to the advanced valuation applications in tax assessment.

In recent years, some basic forms of GIS technology and techniques have become more common, especially aerial images and the mapping of data with applications such as Google Earth, Bing, and Yahoo Maps. These applications facilitate the geographic map presentation of comparable sales, for example, against a Google Earth overlay. Often termed *mashups*, these integrations have been very useful in presenting data effectively from a spatial or GIS perspective.

While these mapping functions have certainly provided a window into the market, they lack more sophisticated mapping functions embraced by other professions such as tax assessors. Applications in tax assessment have been much wider in scale, but many of these applications still do not provide a direct link between GIS and mass appraisal valuation. However, the trend to embrace or leverage new tool sets that resulted in the use of automated valuation models (AVMs) is expected to spur the transfer of GIS technology and techniques in the private sector as well. Assessors were among the first real estate professionals to realize the benefits and potential of the enabling technologies of GIS, focusing on the spatial display of data, and among the first to embrace the technology, as much out of necessity as desire. The use of GIS with public record data became the mechanism for bringing a spatial perspective to data for assessors, which became an analytic tool the assessment community used to manage large quantities of data.

The benefits of embracing enabling technologies from mass appraisal techniques in the private sector have never been more obvious. Current uses range from instantaneous property value indicators that could be reconciled with other approaches to value for lending purposes to portfolio valuations for asset management. Other examples in this book demonstrate that the use of CAMA (*computer assisted mass appraisal*, the assessment community's

> There has long existed a dichotomy within the appraisal profession, essentially a partitioning of the appraisal world-view between mass appraisal and single property appraisal. Both disciplines, to some extent, have differing views with respect to the techniques and perspectives of the other. In many ways, this partitioning inhibits the transference of knowledge between the two. Assessment practitioners, charged with the simultaneous valuation of thousands of commercial and residential properties, have had to develop perspectives and techniques that enable them to process myriad data points into a reasonable and defensible framework. Conversely, fee appraisers have traditionally focused on fewer data points to develop their estimates of value. There is clear evidence that the core elements of the two disciplines are rapidly converging with the integration of statistical techniques and data availability.
>
> – Mark R. Linné, M. Steven Kane, and George Dell, *A Guide to Appraisal Valuation Modeling* (Chicago: Appraisal Institute, 2000)

term for mass or multiple property appraisal) technology is beginning to permeate all sectors of the valuation industry, from the agricultural sector to the interactive tools used in the residential sector.

There is great potential for synergism between the available computing power and the application of this power to the available database. Practioners can take advantage of statistical techniques and geographic information-based technology and presentation by understanding how these tools function in tandem.

Analyzing all relevant market trends, which is a presumption of the Uniform Standards of Professional Appraisal Practice (USPAP), has never been truly practical due to the constraints on an appraiser's ability to efficiently process and understand the large quantities of data available in healthy, active markets. An appraiser using both GIS and statistical valuation modeling can leverage these tools to help master the data and make more accurate and timely valuations.

AVM, IVM, and CAMA Development

Within the last 10 years, both the assessment and fee appraisal communities have employed GIS to improve accuracy in property value determinations. A GIS can identify, through mapping and imaging, external geographic influences on a subject property such as the proximity of traffic corridors, i.e., influences that can and do affect property value. For example, a subdivision bounded by both a busy highway and a golf course will experience different effects from these external influences across the entire range of properties within the subdivision. An interactive valuation model (IVM) that incorporates spatial attributes can assist in the valuation of the effect of these influences for all properties in the subdivision.

Geographic information systems can also be used in single-property appraisal. For example, if the subject property is vacant land with a wetland issue, a GIS can help identify soil limitations and provide insight into the development potential for a highest and best use analysis. Exhibit 6.1 illustrates how a GIS layer of the subject property boundary lines is placed over a GIS layer of a map showing wetlands areas using Maptitude GIS software. The areas of the subject property affected by wetlands can be seen clearly.

Public and Private Sector Source Development

Unlike a traditional automated valuation model system, IVMs allow an appraiser to refine and validate modeling output and neighborhood level analytics, making the final value estimate a truly synergistic collaboration between the appraiser and a valuation model.

Exhibit 6.1 **Wetlands Data and Property Boundary**

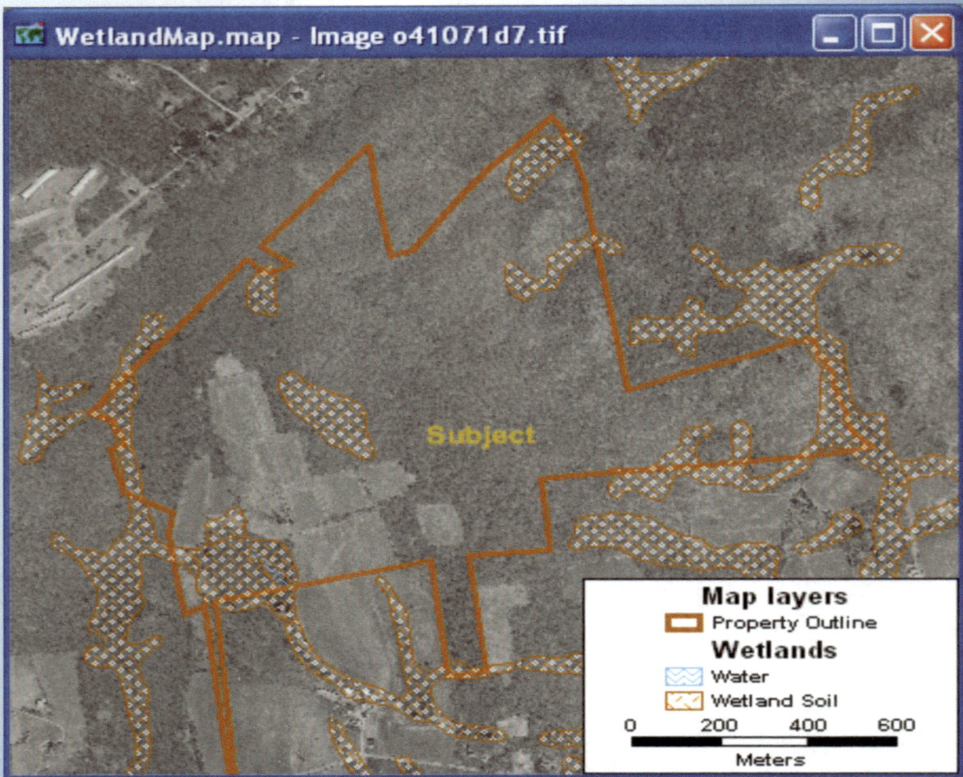

Source: Chris Miner, MAI, SRA, Greenfield Advisors

While regression-based modeling has been in use for many years on the assessment side of the profession, today both the public and private sectors have IVM/CAMA valuation players. GIS today is primarily used in the public sector as an illustrative, consumer search, or adjustment tool. Combining IVM technology and GIS has been less common. An example of a GIS-based valuation system is the AgriStat system currently used by Farm Credit Services of America, which will be discussed in more detail in a later chapter. While the public sector has led the private sector in terms of GIS-only use, it is possible that the private sector will actually solve the "riddle" of creating an IVM/GIS valuation system first because there is a clear need for better analysis tools at the field appraiser level. Combining current appraisal practice with the integrated use of mapping, GIS, Global Positioning System (GPS), and other available technologies could provide a solution for maximum appraisal productivity.

All mortgage transactions link a geographic location to related information necessary to move the loan process forward, and the appraiser is often the independent party who gathers and evaluates information relating to the property. This information commonly includes the legal description, street address, property characteristics, property taxes, and several hundred other data fields. These data points provide vital information of interest to the GIS community.

Today many geographic information systems consist of simple (or complex) layering functions that literally stack visual data sets on top of one another. These "stacks" can be linked, providing interactive functions. For example, a user can draw a shape on one GIS layer, and the system in turn provides data from other layers for the area covered by that shape. Exhibits 6.2 through 6.5 on the following pages demonstrate this functionality and additional analysis using the application Spatialest.

The combination of the layers of data becomes information to the user. This transformation of real estate data into real estate information may provide the key to the linkage of modeling data with spatial analysis through GIS. Most geographic information systems today do not allow for any transfer (in a numeric format) between the GIS screen and the file data. Although some GIS vendors have coded some analytic ability into their systems, the classic model is a two-way pipeline between the regression-based valuation system and the geographic information system.

For example, assume that an IVM is developed for a neighborhood that has a golf course located in its center. The IVM could demonstrably account for all significant valuation attributes concerning the location and physical characteristics of each home, with the important distinction of golf course proximity. Appraisal theory provides indications that golf courses do affect the value of homes with direct frontage on fairways. The appraiser therefore has only a few options concerning use of the IVM in this neighborhood:

1. Do not use the IVM and perform traditional appraisal procedures.
2. Use the IVM and provide a manual adjustment for the golf course influence based on appraisal judgment.
3. Use the IVM system to develop a base value, and then use the GIS map to provide support for the value adjustment developed manually.
4. Use the IVM model and add a golf course variable to the regression model to account for golf course frontage; manually code the variable based on information from the GIS map; and rerun the IVM with the golf course variable.

Exhibit 6.2 Plotting a Property by Location

Find a Property

By address:

State: CA

County: Arcadia

Zip: 91006

Address: 5623 Lenore Ave

From a map

spatialest®

www.spatialest.com

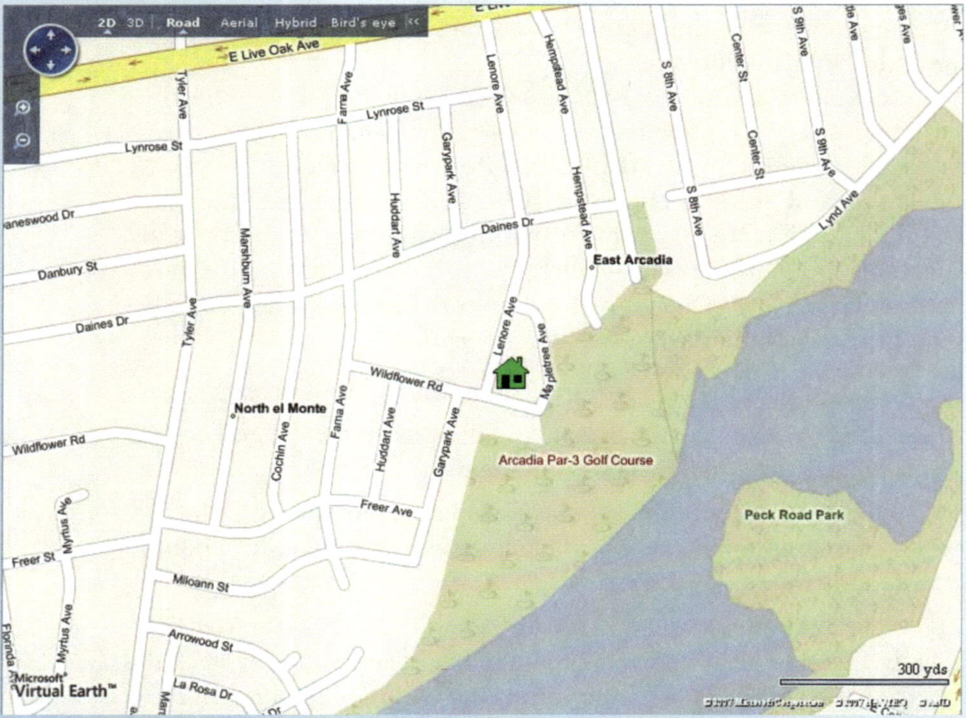

Exhibit 6.3 Adding Aerial Map Imaging and Recent Sales Information

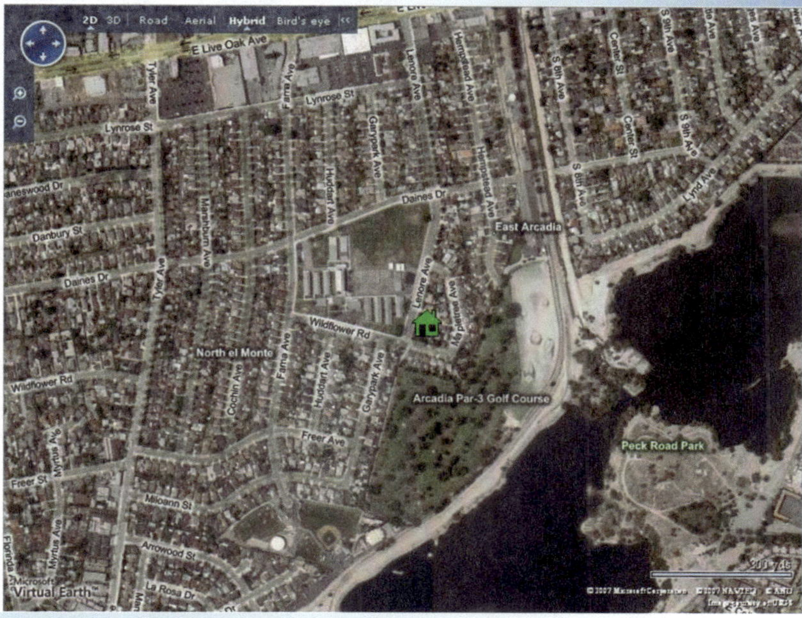

spatialest®

www.spatialest.com

Exhibit 6.4 Adding Foreclosure Sales and Other Data

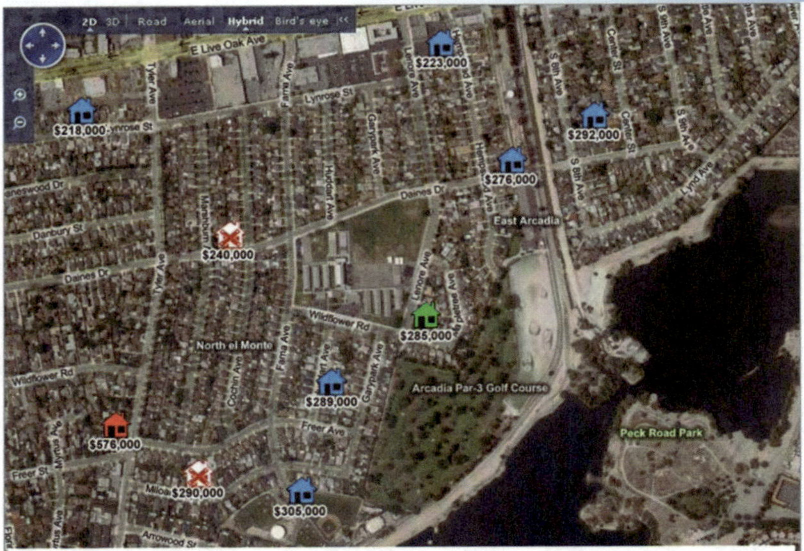

Apply GIS layer of foreclosure sales

Add properties from appraiser files

spatialest®

www.spatialest.com

Exhibit 6.5 **Running Statistical Analysis**

5. Use the IVM system and the GIS link to the data in the regression model; click or draw a polygon around properties so that the GIS automatically creates and codes a variable that is then transported into the regression modeling system; and rerun the IVM with the golf course variable.

It is important to note that all five options are integrated appraisal-GIS solutions. Options 4 and 5, however, are integrated IVM-GIS valuation solutions. While the latter options are similar in functionality and outcome, the issue of scale becomes problematic. Option 4 can be used for smaller scale IVMs but quickly becomes impractical for larger scale models with many properties and many GIS influences. Only Option 5 provides a true and practical solution to IVM-GIS integration.

Other paths to GIS integration depend on the scope of work of the assignment and the experience and knowledge of the practitioner. The following sections of this chapter describe integration efforts in other real estate valuation areas.

Other Private Sector GIS Applications

Outside of valuation modeling, GIS coupled with a real estate sales or listings database can assist a buyer or seller in understanding the influences of supply and demand by showing properties that are available and those that have sold. Consumer uses of GIS tools have been popularized by online applications such as Zillow and Trulia, among others. With newer tools such as Google Earth and other aerial applications, users can understand the context of how, in a macro sense, location affects their properties or properties that they may have an interest in.

Government Uses of GIS and Valuation

GIS technology continues to offer many tools to assist the valuation function of tax assessors. While not a replacement for a computer-assisted mass appraisal program, GIS, if carefully developed, can be integrated with a CAMA program and enhance numerous aspects of an assessment practice. Ultimately, the linkage between data and GIS ensures that the interests of the community are served through equitable taxation.

In areas with a large number of property parcels, it is not unusual to find many outliers or exceptions that require extra attention to detail and additional research and verification. A graphical, visual method of displaying the properties makes it easier for assessors to identify potential problems in the appraisal process when dealing with a potential outlier.

GIS also provides useful tools for the assessment process by integrating a variety of factors into the valuation process based on the characteristics of a particular property parcel such as the following:

- Soils
- Comparable sales within a given distance
- Zoning
- Proximity to bodies of water and floodplains
- Area or size
- Transportation access and frontage
- Land use patterns

While assessment is considered here, the transference of this same functionality to the private sector can clearly provide significant benefits to fee appraisers.

Using GIS to Improve the Modeling Process

Knowing exactly where a property is located is the most basic GIS application in the appraisal process. Moreover, GIS can be used effectively in the appraisal process to handle large data sets, which lend themselves to use in statistical modeling as well. Mass appraisal has already defined which statistical methods are credible tools in the valuation process, and integrating statistical modeling tools within a GIS-centric environment can make GIS practical and useful to appraisers.

Building a modeling process that fully integrates GIS and mass appraisal requires that location determinations become an explicit part of the modeling process. The GIS software need not be directly linked to the modeling software, but such linkages can streamline the process.

Using GIS to help determine or define market areas can start with the examination of assessor neighborhood maps. Assessors usually build their own internal models using neighborhoods and subdivisions, which normally involve one or more periods of construction, similar property characteristics, one or more builders, and, of course, location.

More sophisticated GIS applications may in fact aggregate many neighborhoods into a few large modeling areas, though it is generally clear that such large-scale applications are less accurate than smaller-market modeling systems. The property grouping process is two-sided: neighborhoods are first clustered by attributes such as age, size, and housing style, and the properties are then grouped geographically by market area.

The clustering process allows the appraiser-modeler to group similar neighborhoods to increase the number of sales. The clustering process also allows smaller neighborhoods with few or no sales to be modeled with larger neighborhoods that have a greater number of sales, in effect mimicking the appraisal process used in traditional single-property appraisal.

Neighborhood-level adjustments occur within the boundaries of the modeling area and the neighborhoods that make up the modeling area. Standard mass appraisal models typically cannot account for locational influences within or adjacent to the modeling area. For example, if a row of homes is located next to a highway corridor, any deleterious valuation impact from this location will not be considered in the valuation model unless a variable representing the location effect is explicitly present in the model. Such a variable would allow the modeler to test for value differences within or adjacent to the modeling area. This analysis process assumes, of course, that sufficient sales exist within the influence area to test against sales not located within the influence area. GIS applications allow for several levels of adjustments, from simple lump-sum adjustments to layered adjustments where the valuation impact follows a gradient away from the location impact. The advantage of GIS is that these location effects can be identified and tested in the valuation model through a spatial or visual examination of the data plotted on a map.

The full integration of GIS into mass appraisal modeling requires that GIS-based data be included in and enhance the market sales data set. Ideally, modelers should be able to test and retest for location factors as part of the modeling process. This interaction allows for the complete accountability of location inputs. The GIS information, in effect, refines the modeling data and process. Moving such information from the GIS screen into the modeling data, however, is not easy with many GIS software applications, but with a greater focus on enhancing real estate data, this sort of functionality will likely be available in the near future.

Conclusion

In the future, the integration of GIS information into the mass appraisal modeling process could allow for a complete visualization of the valuation process. Currently GIS data is geographically oriented, allowing for the layering of data and the manipulation of the information stored within those layers. This scenario simply represents one facet of the visualization process. Using GIS fully in the valuation process is likely to result in a flood of new applications and valuation innovations in the not-too-distant future. The Farm Credit application mentioned earlier is one example of the potential for applications that leverage the functionality of GIS with data to provide better and more meaningful analysis for users.

It is clear that a comprehensive valuation system must include at least a location component that is linked to a basic geographic information system. Part of the challenge will be to effectively add GIS to the valuation ingredient list at every point in the valuation process. Mass appraisal models can be built without a formal GIS tool, but without GIS such models have difficulty accounting for significant location effects and phenomena. Adding GIS to this process will have to be accomplished if mass appraisal valuation is to advance beyond regression modeling. Another result of the integration will be the creation of formidable valuation systems that outperform current mass appraisal systems and traditional appraisals because local appraisers, knowledgeable about the market and the valuation process, will be involved in the analysis.

Needless to say, understanding the issues of data are of the utmost importance. As GIS expert Grant Thrall, professor at the University of Florida, states, "It can be said that GIS is 90% data."[3] Through the incorporation of data with GIS functionality, appraisers can discover much about the markets they are analyzing and provide their clients with a more granular view of the marketplace.

3. John Cirincione, Daniel Szparaga, and Grant Ian Thrall, "Mortgage Data Standards," Geospatial Solutions (June 1, 2005), available online at http://www.gpsworld.com/gis/.

7 GIS for Business and Economic Development

From Data to Analysis

Michelle M. Thompson, PhD, and Milton Ospina

How do community-based organizations (CBOs) use geographic information systems (GIS) at the local level for business and economic development efforts? Most CBOs focus on small geographic areas and their impact is very much local and limited to their geographical area of interest. CBOs working with GIS to enhance the economic vitality of their communities tend to support the local real estate market by preventing, or reducing, phenomena that negatively impact real estate market prices such as blight, unemployment, safety, or community environmental health. In most cases, CBOs support government efforts to improve the quality of life in their communities (a direct benefit of improving economic conditions at the local level), but in many cases they implement those same government programs because they are the only ones who can. As CBOs help improve local economic, housing, social, and environmental health conditions in their communities, these communities become more stable, and therefore the housing stock (real estate) becomes also more stable thus improving housing market values and real estate assessments.

Access to data and the tools to analyze that data play a large part in the success of the activities of CBOs. The ability to merge data types and sources and to integrate tools within GIS continues to evolve. Business uses have expanded to address the integrated approach to using economic data to make business decisions. This chapter will explore how geographic information systems (GIS) have influenced private industry and how decisions, by communities or private businesses, are made for economic development using those tools.

GIS for Business and Economic Development

Not long ago–20 years, really–Milton Ospina was analyzing Home Mortgage Disclosure Act (HMDA) mortgage data for the City of Albuquerque, New Mexico. Thousands of records of HMDA data were available only in paper format, and Milton hired his younger brother to spend hundreds of hours entering the data into a statistics program. Once the statistical analysis was done, "Zip-A-Tone" pattern strips were cut into the appropriate census boundary shapes and then pasted onto paper maps to make thematic maps by hand showing the geographical distribution of the statistical results. Desktop GIS software did not exist at the time, and the cost of a copy of high-end GIS software was tens of thousands of dollars, too expensive for an individual researcher or most community-based organizations back then. Furthermore, hardly anyone, especially a community-based organization, could afford so much time and effort doing data entry, while at the same time advancing the organization's goals.

Today, economic data, HMDA reports, and all types of business and demographic data are available for download from the Internet, in most cases at no cost. Free or low-cost online mapping services can map, query, and analyze such data in a matter of minutes and still produce good-quality maps for presentations and reports. Many of these mapping services are packaged with business, demographic, and economic data for ease of use. Information technology, particularly the Internet, has facilitated the access and analysis of economic and demographic data in such an effective way that individuals, government agencies, business groups, and CBOs can now focus more on interpreting the data and on setting economic development policy to improve the quality of life of their communities and neighborhoods.

The power and availability of GIS has become ubiquitous, and the availability of data is sometimes overwhelming. Analysts have, in a short span of years, gone from little access to GIS technology and ready-to-use data to a situation where they are data-rich but sometimes information-poor. In other words, business and demographic data is widely available, mostly in a raw format, but difficult to make sense of without the proper training or education. CBOs working to advance economic development projects at the local level need to have clear goals as to what they want to get out of their GIS and data, with the understanding that the goals to be achieved should help advance their organization's mission statement.

The American Economic Development Council defines *economic development* as

> [T]he process of creating wealth through the mobilization of human, financial, capital, physical, and natural resources to generate marketable goods and services. The economic developer's role is to influence the process for the benefit of the community through expanding job opportunities and the tax base.[1]

As is probably clear, the use of GIS technology alone will not improve the economic situation of the communities served by CBOs. What GIS can do is facilitate the economic development process by shortening the amount of time it takes to conduct business and demographic analysis. It can also save CBOs time in mapping transportation routes, infrastructure availability, and the location of existing businesses. And it can certainly improve the process by which communities recruit and retain companies and small businesses and help them grow. Being able to easily map business data for economic analysis leaves more time to actually focus on the analysis of the data and what that analysis means for developing creative and realistic economic development efforts and also to strategize how these efforts may impact the community.

Using GIS to Build Successful Partnerships with the Local Business and Economic Development Community

It takes time for CBOs to be accepted by the business community, and it also takes time for CBOs to integrate GIS into their business processes. A great number of public and private organizations use GIS at different levels, but many underuse the technology. For example, a CBO may use GIS technology only for simple tasks such as producing maps for presentations or updating enterprise zone boundaries or land use maps. However, to really make GIS an integral part of their business processes, CBOs can leverage GIS to analyze appropriate enterprise zone boundaries, explore the criteria behind these boundaries, and identify other areas that could also be considered enterprise zones or other types of economic development zones so those areas can receive special tax incentives for development or redevelopment.

As an example of fruitful public/private partnerships, some CBOs work closely with local governments to carry out housing development programs. Local governments and banks are likely providing

1. American Economic Development Council, "Economic Development Today: A Report to the Profession" (1984), available online at http://www.economicdevelopment.net/tourism/papers/index.htm#what/.

the funding while the CBOs may be working with the developers, managing construction, and eventually screening and recruiting new homeowners. It's also likely that all parties involved are using GIS in one form or another to identify possible housing development sites and to analyze the demographics and the effect the development would have on the transportation networks, school enrollment, police/fire resources, local taxes, and proximity to employment sites. Often individual CBOs are using GIS to perform portions of these analyses, but they are not sharing their data, technical capabilities, or GIS applications to support each other and produce the greatest benefit to their community. Cooperation is key in these situations. If these partnerships share and analyze their respective data sets–"sharing their maps," so to speak–they may find common ground more quickly, or they may identify better alternatives for proposed or future development. This does require that they all have access to and an understanding of how GIS can help with their respective goals.

CBOs can be sure of one thing–the businesses and local governments they work with will likely have access to GIS and to the business and demographic data needed to make decisions. It is imperative CBOs also have access to the same data and tools as their partners to have a more equitable business relationship.

Leveling the Playing Field by Using the Same Tools Businesses Use

Businesses vie to compete, make money, and endure as viable, successful for-profit entities in their communities. Understanding how businesses think and act is sometimes complicated for CBOs, who do not generally see themselves as businesses that must turn a financial profit, but rather as organizations that exist to facilitate a greater good for the entire community. These two perspectives do not have to be contradictory, but they must be reconciled in order to increase the economic vitality of the communities served by the CBOs.

CBOs are a great asset to community development, but they do not necessarily sit at the business development meetings involving public/private partnerships. This does not have to be so. To level the playing field, CBOs will have to be able to not only use "business jargon," but they must also have a clear understanding of the processes, business cycles, and plans by which organizations determine what businesses will generate the best long-term financial performance. CBOs need not become businesses themselves, but rather demonstrate that they understand what businesses must do to be successful and competitive. Community land trusts, for example, make numerous business decisions on a daily basis. The staff of such an organization may include lawyers and business managers who have the knowledge and expertise necessary to expedite the type of projects that have potential for success.

Data that Drives Business

As discussed in Chapter 5, a geographic information system (GIS) is composed of hardware and software, applications, people, and, of course, data. Data can be both spatial and attribute data. The combination of spatial data–e.g.,business locations–with attribute data–e.g.,business revenue, type of business, taxes paid, business owner information, North American Industry Classification System (NAICS) codes (formerly known as *SIC codes*)–creates intelligent data that can be studied, analyzed, and mapped. These groupings of information are called maps, or GIS layers, and they may also include redevelopment areas or enterprise zone boundaries, land use, zoning, transportation networks, and, of course, business and demographic data (see Exhibit 7.1). Other examples of GIS layers include boundaries of telecom broadband coverage areas, water and sewer line networks, parks, recreational facilities, transportation networks, and locations of educational institutions.

CBOs can access essential data in many ways. In some cases, local government agencies may be willing to share their spatial data, and most business and demographic data will be available from government sources such as the U.S. Census Bureau, Department of Labor, Small Business Administration, and local planning and economic departments. Last but not least, CBOs can obtain this data from private data providers, albeit at a potentially high cost.

Once a CBO has these data sets, it can carry out a number of economic development and business analysis such as

- **Community profile.** The community profile showcases the strengths and weaknesses of the community. These may be in the form of indicators such as level of education, tax incentive programs, demographics, crime and schools statistics, market profile, business, segmentation, spending data, and other data. (See Exhibit 7.2.)
- **Site selection.** Site selection comes down to the old real estate adage, "Location, location, location." Business and economic development truly is all about the "right"

Exhibit 7.1 **GIS Data Layers**

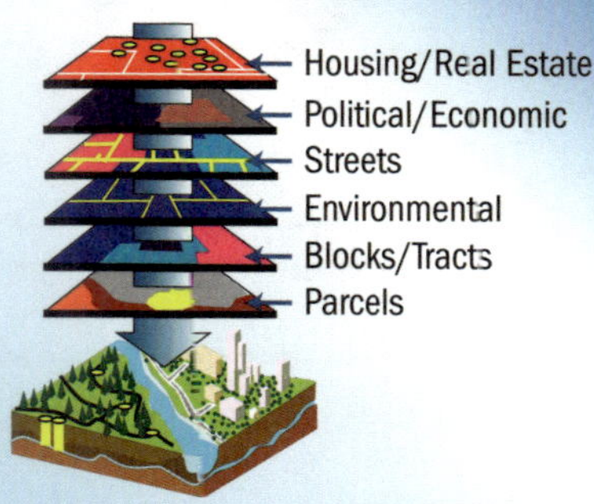

GIS layers are combined to create a computerized geographical representation of the community.

Source: ESRI

Exhibit 7.2 Community Profile

Households

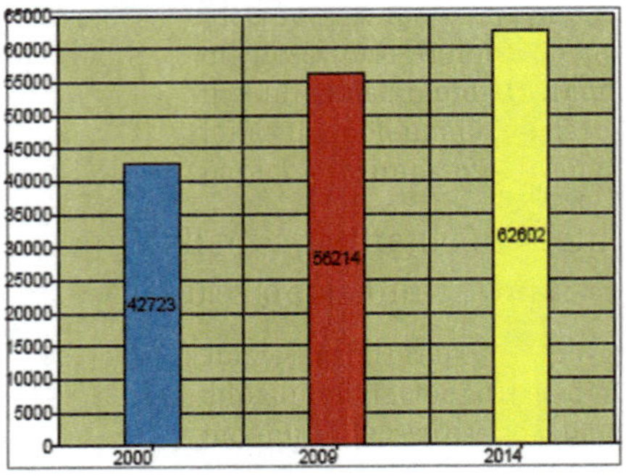

2009 Households by Income

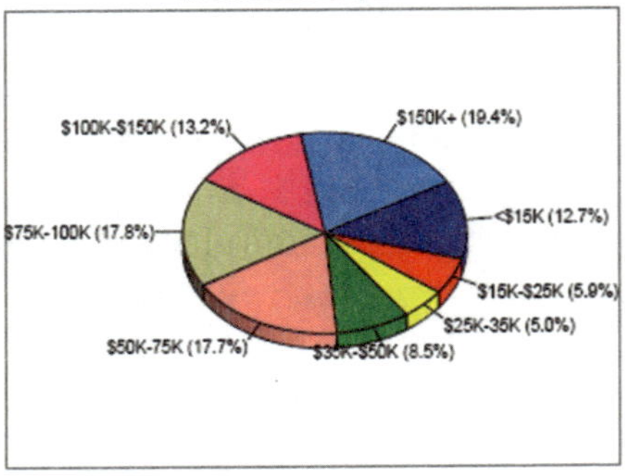

2009 Population by Age

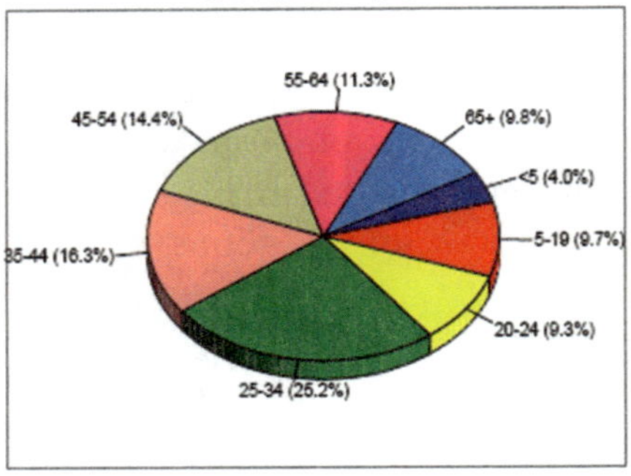

2009 Owner Occupied HUs by Value

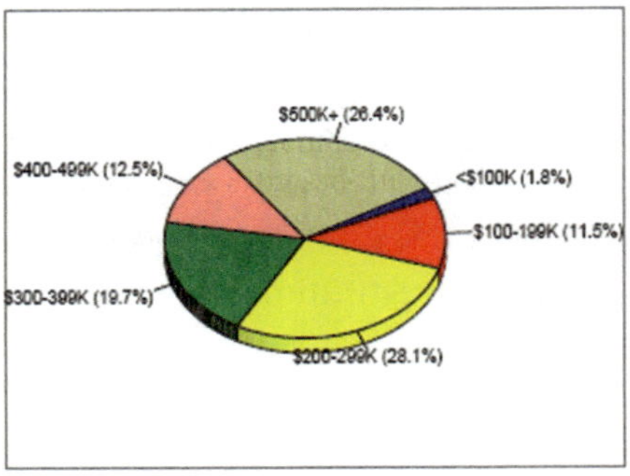

2009 Employed 16+ by Occupation

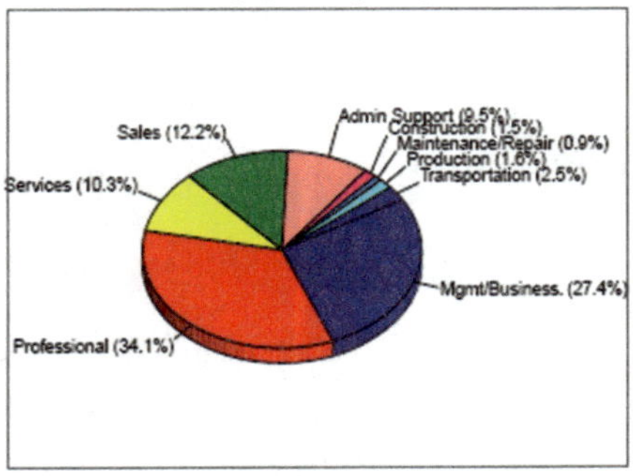

2009 Population by Race

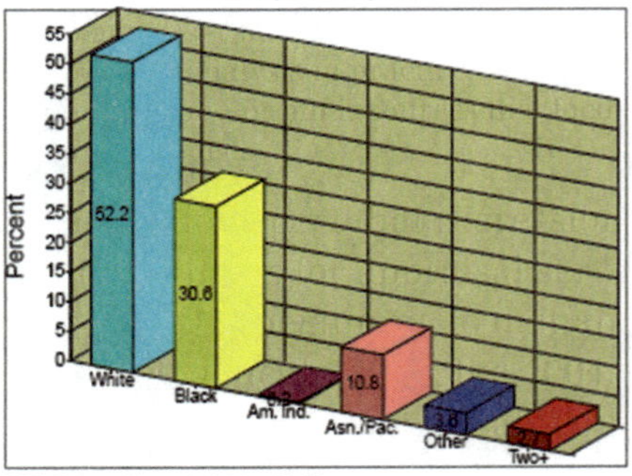

2009 Percent Hispanic Origin: 10.1%

Source: U.S. Bureau of the Census, 2000 Census of Population and Housing. ESRI forecasts for 2009 and 2014.

location for the "right" type of business. Site selection is the process by which a business identifies ideal locations to build stores, factories, or offices. GIS can assist businesses, economic development agencies, and CBOs in selecting the appropriate site for the next desired business in the community, be it a retail store, child-care center, restaurant, warehouse, senior retirement center, factory, or gas station. GIS allows an analyst to combine community indicators (i.e., the community profile) with financial data, consumer/customer surveys, and local community knowledge to visualize potential market penetration, market share, and business cluster, and in doing so select the "right" location for a new business. (See Exhibit 7.3.)

- **Competitive market analysis.** Finding the right location is only part of the equation. Helping potential businesses identify the competition is part of the market analysis CBOs can offer to the business community and is a type of analysis that GIS illustrates effectively. (See Exhibit 7.4.) Knowing who the customers are and what they like will help identify the types of products and services that will attract these customers to a business. If too many businesses are offering the same products or services to too few customers for all these businesses, this imbalance of supply and demand will lead to constant saturation in a trade area. The competitive analysis can help find the right balance of businesses for the community.

As CBOs learn "how to speak the language of business" using the accepted tools of the trade like GIS, they can gain respect from businesses by having a complete understanding of the market forces at work in their communities. Knowing how to conduct the applications above will be a valuable asset CBOs can bring to the table as they become a player in the economic well-being of their communities.

Quality of Business and Demographic Data and Data Ethics when Using GIS

Business and demographic data providers include professional statisticians and demographers. Their data collection and tabulation methodology is often available for inspection and review by anyone interested in licensing their data. Local users of the data will know fairly quickly if the data is accurate enough for in-depth business and economic development analysis. Local knowledge plays a key role in determining the usability of the data and the amount of trust to be placed in the data. And that's where CBOs have a local advantage by being quite knowledgeable of the economic forces affecting their communities.

That said, it's not uncommon for local entities to choose not to share certain variables or data sets with the public or business com-

Exhibit 7.3 Site Selection

123 N Wacker Dr
Chicago, IL 60606

Site Type: Rings

Latitude: 41.88384
Longitude: -87.63683
Rings 3, 5, 7 Miles

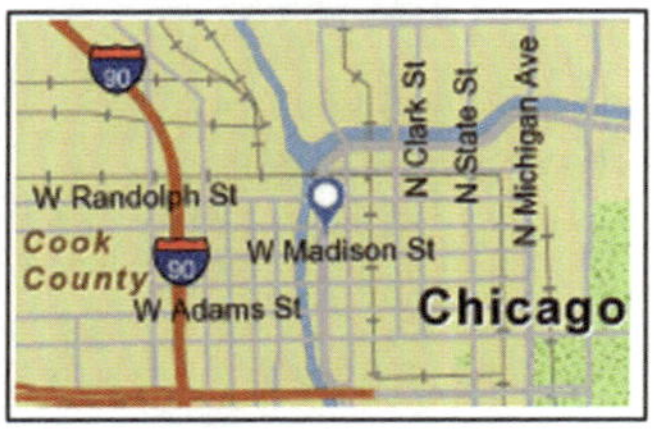

Source: ©i-cubed

 On-demand reports and maps from Business Analyst Online. Order at www.esri.com/bao or call 800-292-2224 Page 1 of 1

Exhibit 7.4 **Competitive Market Analysis**

Miami Location
530 SW 24th Rd
Miami, FL 33129-1951

Site Type: Rings

Latitude: 25.75941
Longitude: -80.20915
Rings 1, 3, 5 Miles

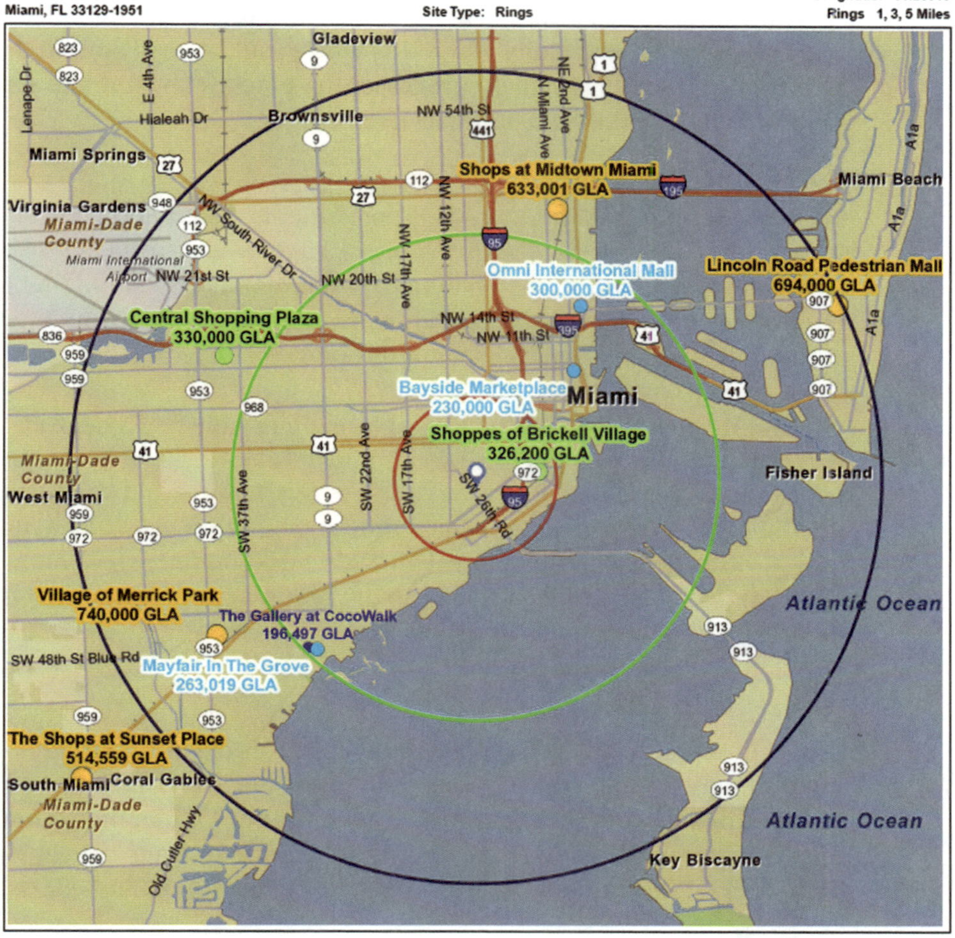

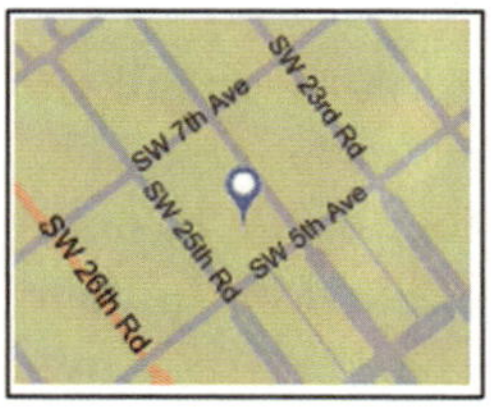

Gross Leasable Area
- Less than 200,000 sq ft
- 200,001 - 300,000
- 300,001 - 500,000
- 500,001 - 800,000
- More than 800,000

Source: Directory of Major Malls, Inc.

Directory of Major Malls, Inc.

On-demand reports and maps from Business Analyst Online. Order at www.esri.com/bao or call 800-292-2224 10/05/2009 Page 1 of 1

munity as they may not reflect a "good" or "appropriate" business environment. The community profile may in fact show a community in distress. Hiding the bad aspects of the community from potential businesses interested in relocating to the community can do more harm than good, especially when this reality can be easily found on the Web or from the business processes the potential businesses will follow as they make relocation decisions.

Case Studies

In Puget Sound, Washington, NWProperty.net is a commercial property and lease locator tool that provides 14 Puget Sound–area cities with a new level of international, national, and regional exposure. The Web site is unique in that it provides seamless coverage for member cities and leverages the Web to combine several disparate date sets for a specific service portal (Exhibit 7.5).

In Lincoln, Nebraska, the City of Lincoln established a centralized partnership to manage all economic development activity and consolidate individual efforts by the Lincoln Chamber of Commerce, Lincoln Electric Systems (the local utility), and other agencies within city government. The Lincoln Partnership for Economic Development (LPED) turned to GIS technology as the best way to present Lincoln to prospective businesses (Exhibit 7.6).

Opportunities

CBOs have many different opportunities to promote local development initiatives that can generate real estate and sales taxes while generating jobs and helping improve the quality of life in the community. Each one of the following opportunities can leverage the capabilities of GIS technology to assist in economic development efforts.

Brownfields

Brownfields, grayfields, and portfields (abandoned ports or rail yards) are vacant or underused parcels of land or properties that may or may not be contaminated. These properties number in the hundreds of thousands in the United States, and they can sit idled for quite a long time after they have fulfilled their original use. The U.S. Environmental Protection Agency (EPA) Brownfields program provides grants to communities and nonprofit organizations to assist in cleaning up brownfields and similar contaminated sites and eventually putting them back on the tax rolls where they generate tax revenues and create local jobs.

According to the U.S. EPA, since 1985 the Brownfields program has attracted over $14

To learn more about brownfields, visit http://www.epa.gov/brownfields/about.htm/.

Exhibit 7.5 NWProperty.net

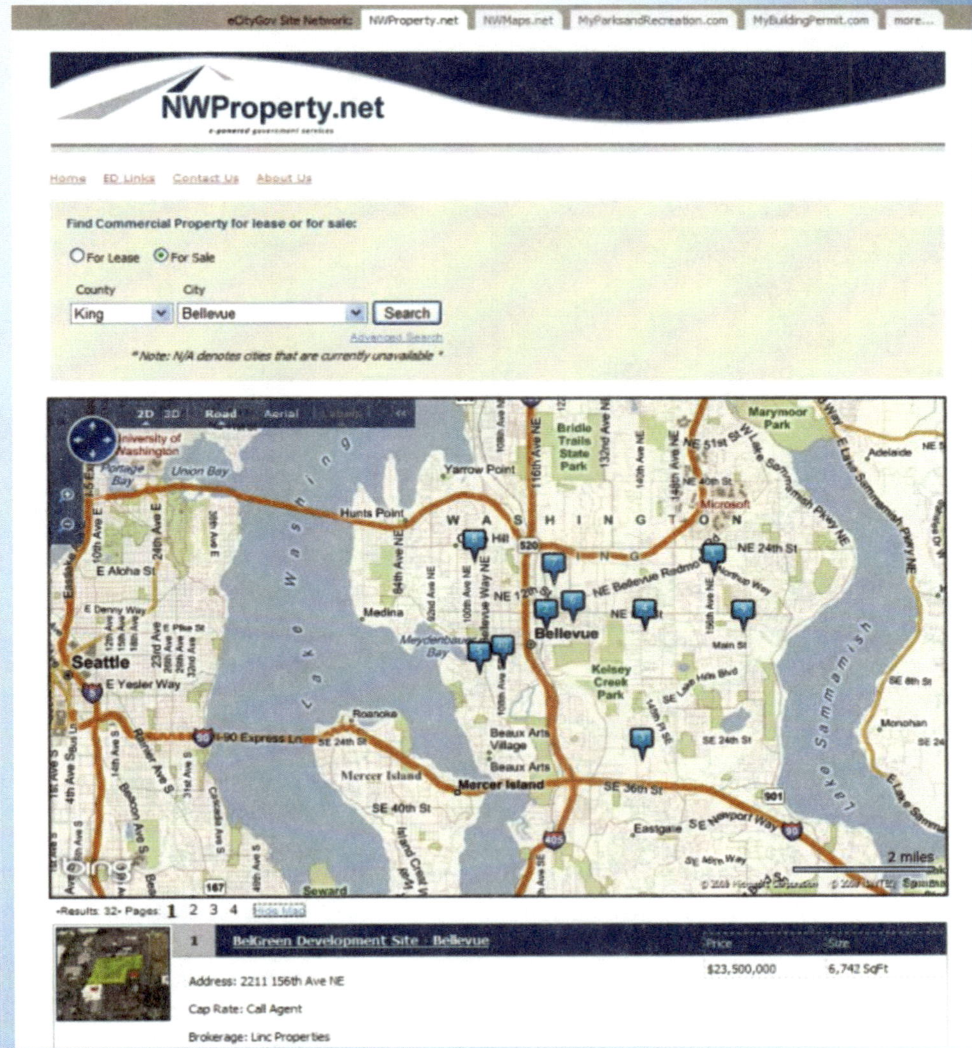

The NWProperty data sets include local GIS data, up-to-date commercial property data, county assessor data, and detailed business demographic data.

billion in investment for cleanup and redevelopment of contaminated sites. As an example of using GIS for brownfield redevelopment, the City of Milwaukee, Wisconsin, manages a remedial excavation with GIS to maintain accurate data during the remediation process.[2]

2. http://www.esri.com/news/arcnews/winter0607articles/maintaining-accurate.html/.

Exhibit 7.6 Locate Lincoln

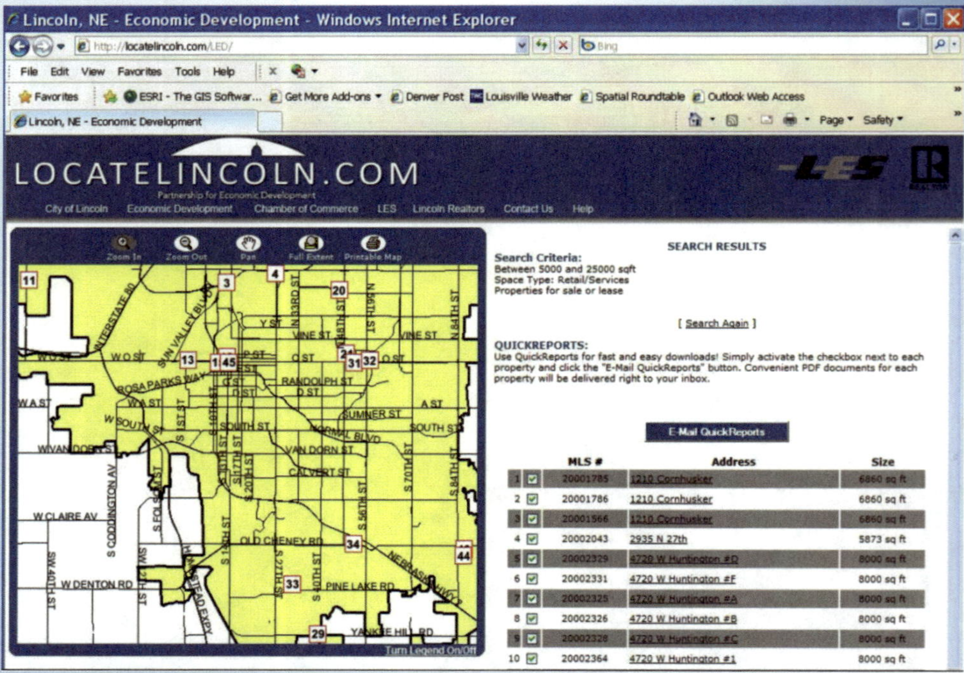

Locate Lincoln allows for quick search on commercial property availability and serves as a gateway to the city's economic development-related agencies.

Vacant Parcels

The National Vacant Properties Campaign defines *vacant properties* as

> [V]acant residential, commercial, and industrial buildings and lots that exhibit one or both of the following traits:
>
> - The site poses a threat to public safety (meeting the definition of a public nuisance), or
> - The owners or managers neglect the fundamental duties of property ownership (e.g., they fail to pay taxes or utility bills, default on mortgages, or carry liens against the property).

The exact number of vacant parcels across the United States is not known, but it's estimated to be in the hundreds of thousands, just like brownfields. One of the goals of the National Vacant Properties Campaign is to identify and catalog all vacant properties and to work with public and private partnerships, including CBOs, to revitalize these properties and put them to productive use. Facing an increase in foreclosures, like many communitites across the country in recent years, the City of Chula

To learn more about the National Vacant Properties Campaign, visit http://www.vacantproperties.org/.

Vista, California, built a GIS to monitor the status of abandoned properties in the metropolitan area.[3]

BRAC

Since 1985, the Base Realignment and Closure (BRAC) Commission has been tasked with evaluating the human and economic impact that the recommendations by the Department of Defense (DOD) will have on communities with regards to what military installations are to be closed or realigned. While many communities suffer economically by the loss of jobs due the hundreds of facilities closed, they can also take advantage of the opportunities presented by base closures and to redevelop these installations as housing, commercial, or recreation areas that can generate new jobs and help communities in diversifying their economic base. CBOs can join forces in public and private partnerships to make a claim and a positive impact in the revitalization of these facilities.

> To learn more about BRAC, visit http://www.brac.gov/, and to see how communities and CBOs can get involved, visit http://www.icma.org/.

Case Study 1: Oakland's Urban Strategies Council

The Urban Strategies Council, based in Oakland, Calif., is a community-based organization focused on improving the quality of life for children and their families in the areas of redevelopment, health, education, and economic development. In the council's own words, its mission is "to eliminate persistent poverty by working with partners to transform low-income neighborhoods into vibrant, healthy communities."

The Urban Strategies Council is well aware of the intricate relationships between economic opportunity and health and crime and education issues, and that's why the council's economic development efforts is key to achieving the goals of the council's suite of program areas, which are

- Economic opportunity
- Community safety and justice
- Education excellence
- Community capacity building
- Research and technology

The council's GIS program is located in the research and technology program area.

3. http://www.esri.com/news/arcuser/0609/repo.html/.

GIS plays an important role in making all Urban Strategies Council program areas successful. It's easy to see the role of GIS and mapping across the programs as maps, some interactive and some in PDF format, are found throughout each program area's Web pages, from maps related to crime analysis to home foreclosures, demographics, public health, and, of course, economic development. Examples of maps are shown in Exhibits 7.7, 7.8, and 7.9.

Over the years, the Urban Strategies Council has developed such strong GIS skills, capability, and knowledge that the organization currently supports other CBOs in developing their own GIS skill sets or developing interactive mapping Web sites for other CBOs so that those organizations can advance their own missions. Under the council's research and technology program area, the council promotes its GIS Services, which go beyond simply mapping demographic data or community indicators to include complex and sophisticated spatial analysis and GIS implementation services. The council offers these services on a fee-for-service basis to other CBOs and stakeholders including public agencies. The council does make it

Exhibit 7.7 **Urban Strategies Council's GIS Tools**

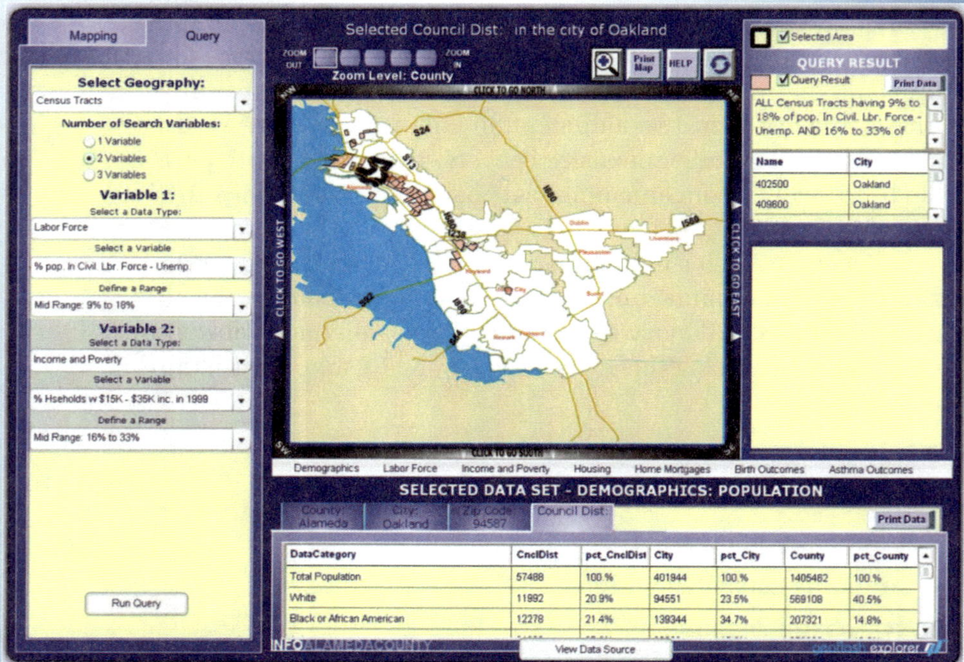

The Urban Strategies Council's Web-based GIS tools allow citizens and businesses to explore numerous demographic variables for the city of Oakland.

Exhibit 7.8 Notice of Trustee Sales with Redevelopment Project Areas

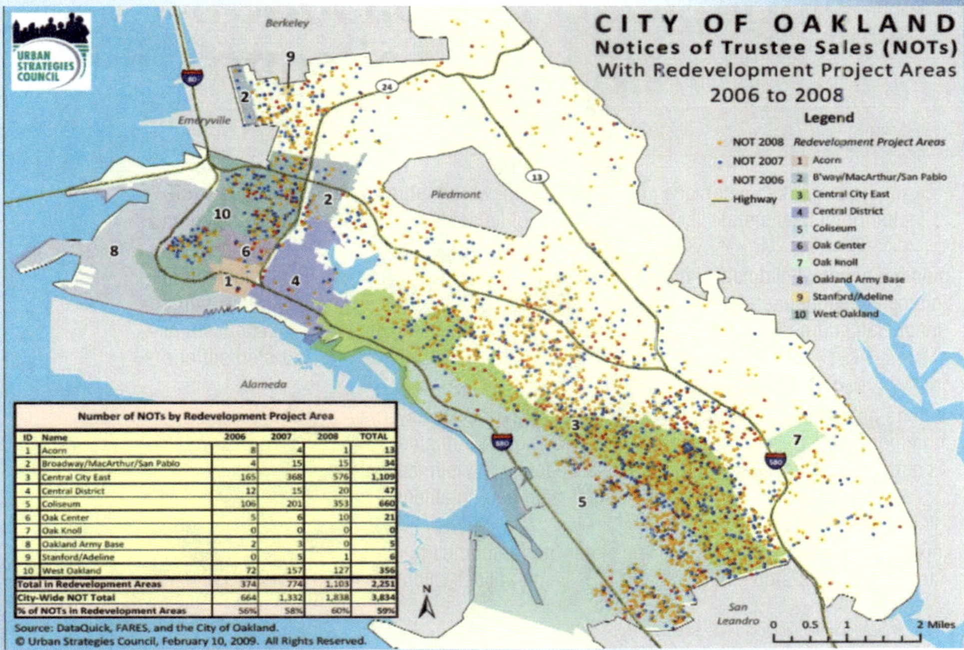

Number of NOTs by Redevelopment Project Area					
ID	Name	2006	2007	2008	TOTAL
1	Acorn	8	4	1	13
2	Broadway/MacArthur/San Pablo	4	15	15	34
3	Central City East	165	368	576	1,109
4	Central District	12	15	20	47
5	Coliseum	106	201	353	660
6	Oak Center	5	6	10	21
7	Oak Knoll	0	0	0	0
8	Oakland Army Base	2	3	0	5
9	Stanford/Adeline	0	5	1	6
10	West Oakland	72	157	127	356
Total in Redevelopment Areas		374	774	1,103	2,251
City-Wide NOT Total		664	1,332	1,838	3,834
% of NOTs in Redevelopment Areas		56%	58%	60%	59%

Source: DataQuick, FARES, and the City of Oakland.

clear that the organization provides these services to other organizations "who align with our mission and operating philosophies."

In addition to supporting the mission of the council and its internal program areas, the GIS Services area focuses on five main services:

- GIS planning and implementation including training and database development
- Planning for and development of external or intranet online mapping systems and services
- Custom desktop mapping services for CBOs and other organizations
- Full spatial analysis and research projects
- Education in terms of speaking engagements

To learn more about the Urban Strategies Council, visit http://www.urbanstrategies.org/index.php/.

Case Study 2: New Orleans Redevelopment Efforts

In the aftermath of Hurricanes Katrina and Rita, the Beacon of Hope Resource Center provided a refuge for residents to begin the process of recovery in New Orleans. Having access to technology–as simple

Exhibit 7.9 Median Household Income by Tracts and Toxic Sites

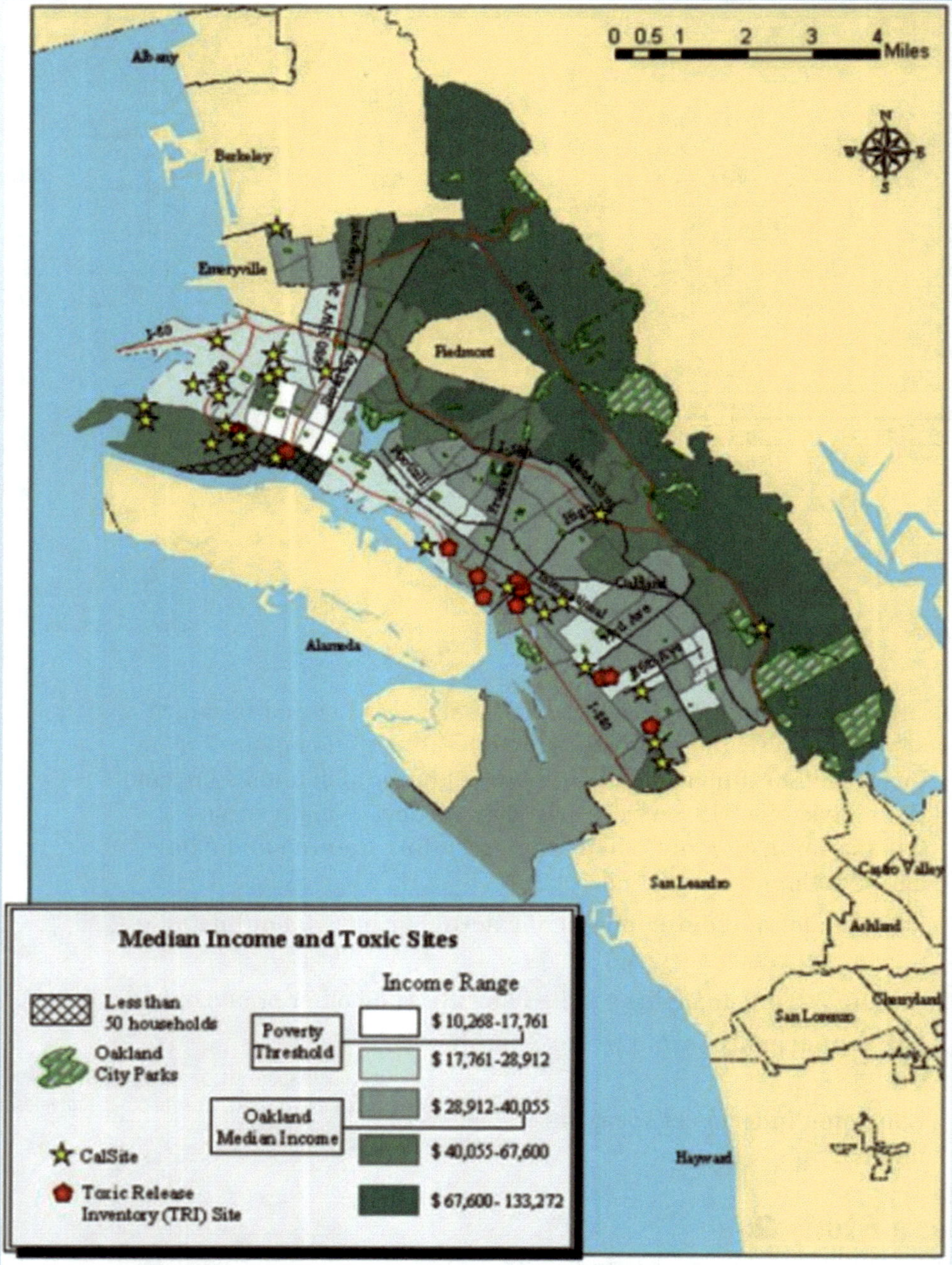

Sources: 2000 US Census,
US Environmental Protection Agency (EPA-1998),
Cal. Dept. of Toxic Substances Control (DTSC-2003)

Produced by: www.InfoOakland.org

as a phone or a fax–proved significant to residents and businesses lacking current information on the storm's impact on the community. Residents and business owners yearned to make informed decisions about how to determine the best course of short-term and long-term action for reinvestment in the city.

In an effort to "inform and empower" the community through the use of technology, the Beacon of Hope Resource Center and the public participation geographic information systems (PPGIS) project of the University of New Orleans Department of Planning and Urban Studies responded to the post-hurricane needs of the community, showing how community data can improve municipal data and aid in economic redevelopment. In this case, the crisis inspired residents who traditionally were closed out of the community planning process to develop creative solutions to their immediate needs for relevant, accurate, and current recovery information. The PPGIS project focuses on the how nonprofit organizations incorporate technology to collect, manage, and evaluate data that typically has been summarized by public organizations and fed, in limited quantities, to the public due to the top-down, "expert-driven" approach to traditional neighborhood planning.

The difference between planning in a vacuum and planning in a crisis–or even post-crisis–depends on having the ability to customize the way in which organizations merge their ideas and resources to make for a better, collaborative system. The Beacon of Hope/University of New Orleans Community Recovery Program (BUCRP) employs standard approaches to working as a community-municipality-university project that other PPGIS projects nationwide can use as a model. What remains different is the ability to collaborate given a diminishing level of resources from public, private, and academic sources when these resources should be increased, expanded, and supported locally and nationally.

To learn more about the Beacon of Hope Resource Center, visit the Web site at http://www.lakewoodbeacon.org/. To learn more about the Beacon of Hope/University of New Orleans Community Recovery Program (BUCRP), visit http://planning.uno.edu/BUCRP/.

Valuation in New Orleans Pre- and Post-Katrina

The public participates in the community planning process in many ways. In the case of post-Katrina New Orleans, citizens were effectively closed out of shared public resources due to lack of adequate access, lack of coordinated data standards, and a general lack of reliable data provided by public officials and related departments. The process of gathering public information, such as assessment data, to make informed business decisions was considered infeasible and unreliable. A *Times Picayune* series on "questionable assessment" practices helped support the recent realignment of the City of New Orleans Assessing Department.[4] In 2010, citizens will vote to eliminate the seven assessors and hire a single assessor to improve services and assessments. The hope is for standardization and stabilization of a methodology of ad valorem taxation that is not influenced by power or money.

Establishing building values after a catastrophic disaster requires that assessors and appraisers revisit the methods used to establish values using the three approaches to value. The definition of *market value* and how it is applied locally must be modified to reflect the current state of the market. In this situation, market values are not based on what is known but what can be predicted based upon limited and faulty data. Real estate appraisers have been trying to estimate land and improvement data in New Orleans, but the value estimates do not reflect a market that was undervalued pre-Katrina and remains at less than 100% valuation after. The issue remains that if policy prevents undervaluation, then determining true "market value" for properties that have been significantly damaged creates a level of error that is almost immeasurable.

State and quasi-governmental organizations such as the Louisiana Recovery Authority (LRA) and the New Orleans Redevelopment Authority (NORA) are influencing market value en masse. As part of the Road Home Option 1 program, homeowners of more than 39,000 properties were given up to $150,000 toward rebuilding their homes. The first 2,308 properties have rehabilitation contracts that are set to expire, and the LRA is evaluating the progress, or lack thereof, in a citywide survey. The basis for the amount of monies provided to homeowners was the pre-Katrina assessed values that, as stated, were artificially low. The conflict between the actual repair needs and the funds available from the Road Home program or insurance companies remains a constant battle for current and former residents.

For the Road Home Option 2 or 3 properties, the New Orleans Redevelopment Authority has developed two programs to turn formerly abandoned properties back into commercially viable properties. NORA is trying to sell approximately 5,000 adjacent properties to homeowners in their "Lot Next Door" program. In the spring of 2010, properties not sold to owners were made available to the public for purchase at "fair market value." Many of the sales are considered "below market" because NORA wants to primarily clear any tax liens and expedite ownership back to willing buyers. Evaluating the *new* New Orleans market is ongoing, but many program administrators

4. Gordon Russell, "Dubious Value: Why New Orleans Property Taxes are Unfair," *Times Picayune* (April 05, 2004), available online at http://www.nola.com/speced/dubiousvalue/.

consider the fluidity of values and the methodology used by appraisers to be inconsistent and, in some cases, unsupported.

To learn more about the Lot Next Door program, visit http://www.noraworks.org/Lotnext.htm/.

Using Data for Community Recovery

NORA, civic associations, and Beacon of Hope Resource Center (BOH) are working together to share reinvestment strategies and identify recovery area priorities. The "NORA Blight Committee" has allowed community members to report nuisance properties that may be privately owned or publicly managed through ongoing working groups. Code enforcement has received results from condition surveys that previously were summarized in Excel spreadsheets with street address and a condition rating. This information, in turn, is shared with other city offices such as assessment and planning in order to update neighborhood monitoring and analysis.

The Beacon of Hope Resource Center began using resident volunteers in Lakeview to conduct block-by-block, parcel-by-parcel conditions surveys in 2006. The focus was on collecting property conditions and on reoccupancy. This information was provided to business owners and helped to reestablish the Harrison Avenue Marketplace. In the summer of 2008, BOH began to evaluate the use of GIS technology to create property condition maps (Exhibit 7.10). From 2008 to 2009, a pilot mapping program standardized and automated the survey process using ArcGIS 9.2, the "gold standard" GIS used by the City of New Orleans and high-end private organizations.

Prior to BOH's use of GIS technology, the organization used paper maps and Excel spreadsheets in a manner that was effective to quantify change. However, using GIS allows BOH to visualize and analyze the changes in a more immediate, sustainable manner as well as to integrate state and local data that was otherwise inaccessible. Specifically, the data that influences reinvestment (Lot Next Door with NORA) or discourages reinvestment (Code Enforcement Blight properties) is now contained in the BOH GIS with results that are accessible to the public in form of maps and reports (Exhibit 7.11).

To learn more about the Harrison Avenue Marketplace, visit the Web site at http://www.harrisonavenuemarketplace.org/.

Conclusion

Exhibit 7.12 show ways communities are maximizing the power of GIS and the Internet. Regardless of what approach CBOs take to promote economic development at the local and regional level and what role GIS plays in facilitating that process, CBOs must ask themselves if such efforts are in fact aiding in the growth of the economy. Are those efforts increasing the property and sales tax base and creating

Exhibit 7.10 Option 1 Map

Exhibit 7.11 Neighborhood Map: Lakeview Property Conditions

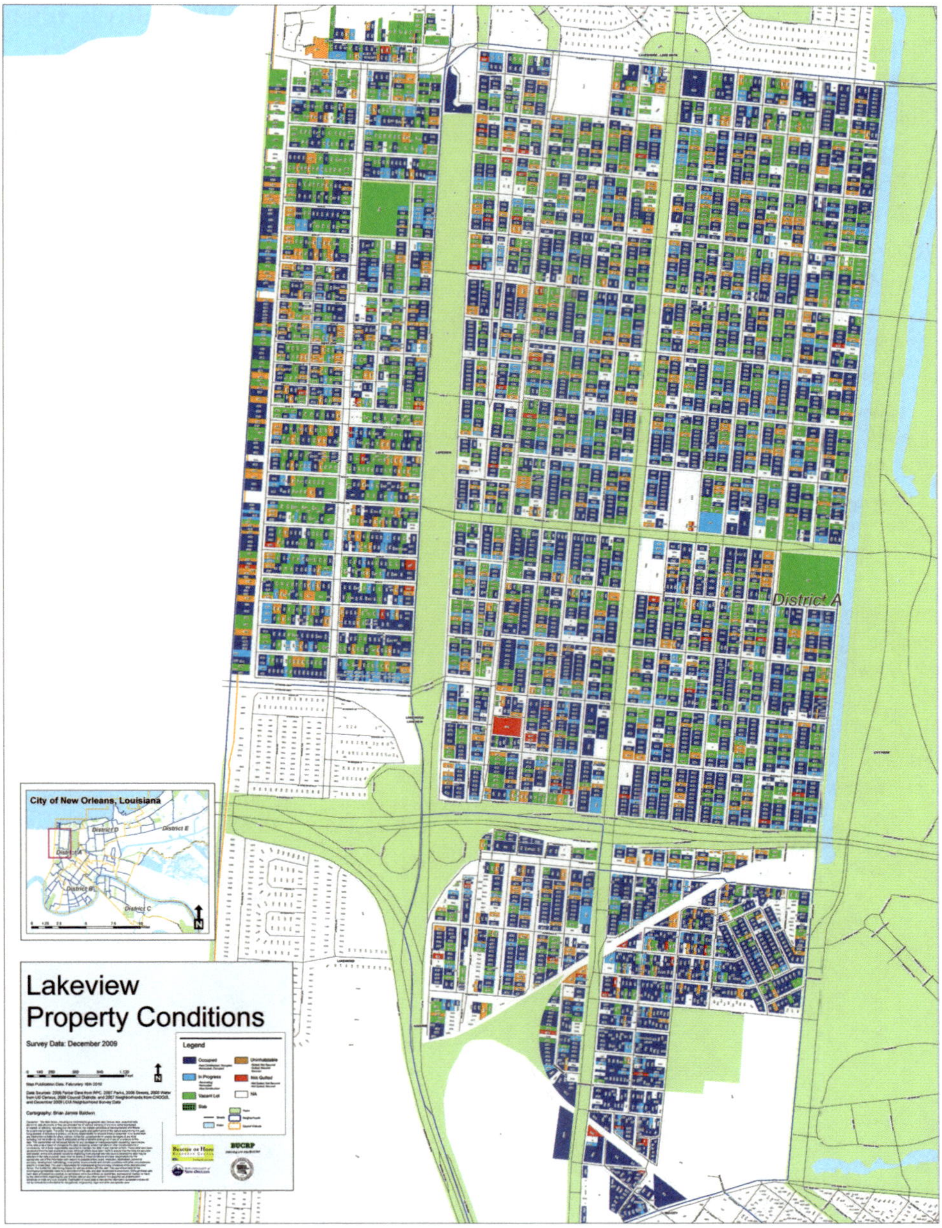

Exhibit 7.12 Uses of GIS by CBOs

- Virtual brochure of community
  - Increase visibility
  - Highlight community assets
  - Travel and tourism
- Extends marketing opportunities
  - Business attraction
  - Business expansion
  - Business retention
- Improved economic development services
  - Financial and tax incentives
  - Stimulus spending tracking and allocation reporting
- Workforce development
  - Employment location finder
  - Economic gardening
- Extends opportunity to interact with real estate community
  - Commercial
  - Residential

new jobs for locals while guaranteeing the quality of life for a great majority of the members of the community? If the answer is "yes," then the organization is making a real difference in strengthening the economic base of communities while at the same time improving the attractiveness of the local real estate market.

RESOURCES

International City/County Management Association (ICMA)
http://www.icma.org/

National Association of Counties (NACo)
http://www.naco.org/

Smart Growth America
http://www.smartgrowthamerica.org/

ESRI Economic Development
http://www.esri.com/econdev/

SECTION II

Implementing Innovation: Practical Applications of Advanced Technology

8 GIS and GPS Usage in Land Appraisal

Christopher Miner, MAI, SRA

OVERVIEW

Topics covered in this chapter include

- Using a GPS device in the appraisal workflow
- Basic and common mapping on the Internet and using a GIS
- Using GIS in the appraisal of acreage in transition areas
- Using GIS in the appraisal of agricultural properties
- Using GIS in the appraisal of forestry properties

This chapter will also examine the geographic data most commonly used for appraising land:

- United States Geographic Survey (USGS) topographic maps
- Aerial photographs
- United States Department of Agriculture (USDA) soil types from the Natural Resource Conservation Service of the U.S. Department of Agriculture

Introduction

Land appraisers now have a tremendous advantage over the old days of paper maps and copy machines. County geographic information systems (GIS) are available at kiosks or over the Internet, and Internet-based aerial imagery is commonplace. The goal of this chapter is to show how having an in-house GIS can take a land appraiser's practice to the next level through enhanced analytics and reporting. It also shows how a GIS and global positioning system (GPS) device can share data so that using a GPS is even more meaningful.

The chapter discusses choosing GIS software, accessing governmental data sources, and using a GPS device while inspecting land. It discusses how a GIS can be used to understand land capabilities and limitations and to communicate with basic maps: topographic, aerial photo, flood zone, and soil maps with property boundary overlays. The chapter gives insight into how USGS soil databases can be used within a GIS so that soil capabilities can be visualized instantly. Workflow processes and automation are also discussed. Three land uses are considered (land in transition to residential development, farm land, and forest land) along with the different GIS functionality that can be used in the analysis of each.

This chapter assumes that you are familiar with the fundamental features of GIS and land appraisal. If you encounter GIS terms that are unfamiliar, review the definitions in the dictionary on the Web site of Environmental Systems Research Institute at http://www.esri.com/. Follow the links to Support, and then search for GIS Dictionary.

After implementing the applicable concepts found in this chapter, land appraisers will find themselves implementing highly effective workflow methods for the analysis of land and for the creation of professional-looking maps that enhance communication with report readers.

In this chapter I share some of my experiences using a geographic information system (GIS) and Global Positioning System (GPS) receiver in land appraisal. This chapter contains excerpts from the Appraisal Institute's online GIS curriculum that I wrote.

GIS Software

My laptop has the following GIS software loaded on it:

- ArcGIS by ESRI (ArcGIS is a component of ArcView)[1]
- Maptitude by Caliper Corporation[2]
- Google Earth[3]
- Microsoft's MapPoint[4]

I have used ArcView for over a decade and have written many scripts to automate some processes in my appraisal and consulting practice. All the maps in this chapter were created using ArcView. I also use Maptitude because it has some capabilities that are superior to ArcView. (Because of its functionality, attractive price point, and the data that comes with the software, Maptitude was selected as the GIS software used in the case study seminars in the Appraisal Institute's online GIS curriculum.) These geographic information systems are quite capable for most tasks needed for land appraisal. Exhibit 8.1 compares the differences between Maptitude and ArcView (as of this writing) that are relavent to land appraisal.

Choosing a geographic information system depends on both the minimal functional requirements of the system and the intended application. Be sure to identify the features and capabilities of the GIS software that are important for your practice, the capabilities of the computer(s) you intend to use it on, and a network plan if you intend to load the software on multiple computers in your organization.

1. ArcView, ArcMap, ArcIMS, and ArcGIS are registered trademarks of ESRI (www.esri.com).
2. Maptitude is a registered trademark of Caliper Corporation (www.caliper.com).
3. Google Earth is a trademark of Google (www.google.com).
4. MapPoint is a trademark of Microsoft (www.microsoft.com).

Exhibit 8.1 Software Comparison	
Maptitude	**ArcView**
2010 price of about $500 (20% less when purchased for an Appraisal Institute GIS seminar)	2010 price of about $1,500 for the base product
Can calculate the slope of land	Requires the Spatial Analyst Extension to calculate slope ($2,500)
Comparatively small user base—not many available software utilities	Has largest installed user base—users and commercial developers have created software for use with the GIS
Connects to Microsoft's TerraServer's* black-and-white images from the USGS and Google Earth's images, which are subject to license restrictions regarding commercial usage	Connects to ArcIMS Web Servers (by ESRI) and other map image servers and Web feature servers, which allow data layers to be served over the Internet by the organizations that maintain them and allows them to open in an ArcGIS map on my computer
Images must be north-up before georegistration (also known as *registration*)	Rotates images during the georegistering process

* TerraServer is a trademark of Microsoft (www.microsoft.com).

Prepared GIS Data

Every GIS requires geographic data to produce the maps that appraisers use. Today's full-featured GIS software will readily integrate projected and unprojected data into the same map display "on the fly."[5] This capability allows data from different sources to be viewed in the same map, which is a vast improvement over older systems for which users often had to convert data to a common coordinate system before the data from different sources could be viewed in the same map.

When assembling data, local sources should be examined first because locally produced GIS data is usually the most current and most detailed. Tax assessors generally drive the production of GIS data (aerial photography and parcel and building layers), and utilities, elevation, and zoning data is generally produced for the departments that need the data (e.g., public works, planning and zoning) and integrated into municipal geographic information systems. Accordingly, appraisers will usually acquire data from the county or town. If the municipality does not have GIS data, appraisers must look to the state and, if all else fails, a national data source.

Raster data (images such as aerial photos and topographic maps) and vector data (e.g., roads, state, county, and town boundaries, flood hazard areas, soil map unit boundaries) are often downloadable for free from the municipal, county, and state departments responsible for cataloging this data. The names of appropriate departments at the

5. Unprojected data is what you would find on the three-dimensional surface of the near-spherical earth and is generally described using latitude and longitude similar to locations reported with a GPS. The process of projecting data generally results in a somewhat distorted two-dimensional depiction of three-dimensional data for use with a flat sheet of paper or a computer monitor.

state level tend to vary from state to state and have abbreviations like *DNR*, *DEQ*, *DEP*, and *DOE*, to name a few. In some states, universities host servers that permit downloading.

Acquiring data from in-state sources increases the likelihood that the data will be in the same datum, projection, and coordinate system and will use the same map units. For example, the "State Plane" coordinate systems present familiar projections for individual states (large states may have multiple projections) and use feet as the horizontal and vertical units from a point that is common to all the state's data sets.

The Internet provides access to data posted by private companies and government sources. As of this writing, the key national sources applicable to land appraisal are

- **United States Geographic Survey (USGS)** for geographic data and images from federal- and state-funded sources. The USGS Web site (http://seamless.usgs.gov/) is rather slow but has an incredible amount of data that can be downloaded in small file sizes.
- **United States Department of Agriculture (USDA) Geospatial Gateway.** The USDA Web site (http://datagateway.nrcs.usda.gov/) offers a lot of the same data as the USGS site but allows data to be downloaded in very large files from an FTP site.
- **USDA Soil Data Mart.** Soil data is most readily available from the Soil Data Mart Web site (http://soildatamart.nrcs.usda.gov/). The databases include both geographic data and tabular data.
- **GeoCommunity** is a quasi-public Web site that aggregates data from many different sources and offers the data to the public. This site (http://www.geocomm.com/) offers data in convenient chunks. Downloading some data requires payment of a nominal fee, which is money well spent.

Accelerate Basic Mapping Using GIS

Someone who is already familiar with a property can look at a map and mentally estimate where the property is on that map without having to draw boundaries. However, maps without property boundaries do not help the reader of an appraisal report understand the property's precise location. An outline of a property is critical to understanding a subject or comparable property in land appraisal. A basic map series for land appraisal displays the outline of a property over the following:

1. Topographic map
2. Aerial photo
3. Flood zone

4. Soil map

Appraisers acquire map data from many sources and at different scales when preparing to map a property. Without a GIS, some land appraisers draw freehand property boundaries on paper or scanned maps for each map they make. Map series can be made more efficiently with a GIS once a property boundary has been integrated into the GIS. Using an outline in a GIS is usually quicker and more accurate. Furthermore, GIS output in an appraisal report looks more professional than marked-up photocopies of paper maps.

When obtaining a property outline, the best data would be a Shapefile[6] provided by a surveyor that lines up with other GIS data. The second-best is a parcel outline made for the tax assessor. The third-best data would be an outline made with a GIS from a paper survey or map either by drawing freehand or by tracing a property outline from a georeferenced image.

To trace an outline from a georeferenced image of a survey (or map), start by obtaining a digital image (TIFF is the usual format) of the survey. The image can be e-mailed from the surveyor, or a paper version can be scanned at a blueprint shop. TIFF images can be opened in a GIS. (In Maptitude, the image must be rotated so that north is at the top of the view window and then saved.) To align the image with the features in the map, the user must georeference the image. Georeferencing, or *referencing*, is a process by which the image is resized, moved, and possibly rotated so the survey lines shown in the image are correctly placed within the map. The last step is to create a new area (polygon) geographic file, like a Shapefile, and add a new geographic feature to this file by tracing the boundaries of the property shown in the georegistered image of the survey.

The amount of time this process takes depends on the complexity of the property boundaries. Using geographic features on aerial photos (stone wall intersections, corners of buildings, and edges of fields) will increase accuracy while georegistering an image. Geographic information systems have tools for creating parcel boundaries reasonably accurately, including

- Zoom in
- Zoom out

6. A Shapefile is a set of digital files with the same name but different file extensions in the same file folder on one computer or network. Shapefiles, initially developed by ESRI, are possibly the most commonly used file format for transferring geographic data from GIS data. The Shapefile contains the information needed to locate geographic features on a map in a GIS and tabular data relating to each feature. Each geographic feature is associated with one row in the database table. Optionally, a Shapefile's set of files may contain files that tell the GIS the projection, coordinate system, and datum of the data, how to symbolize the data, and other information.

- Pan at a zoomed level
- Magnifying window

Using a GIS, annotation can be added once for all the maps in a series. A basic map series can be produced by having the parcel layer symbolized to show only the parcel boundaries, with the inside of the parcel transparent, and then turning underlying layers on and off. The next two illustrations (Exhibits 8.2 and 8.3) demonstrate this capability.

Exhibit 8.2 displays the topography of a property with annotation pointing out the right of way. The parcel layer was drawn on top of an image of a topographic map so both layers can be seen simultaneously. This map can give the appraiser and report reader an understanding of the property's topography.

Exhibit 8.3 shows an aerial image of the same property and its environment. To show more of the neighborhood surrounding the property, the map was zoomed out to a smaller scale. Note that the same parcel outline and annotation were used. The only differences between Exhibit 8.2 and Exhibit 8.3 are the image that is displayed under the parcel layer and the scale.

Exhibit 8.2 Topography

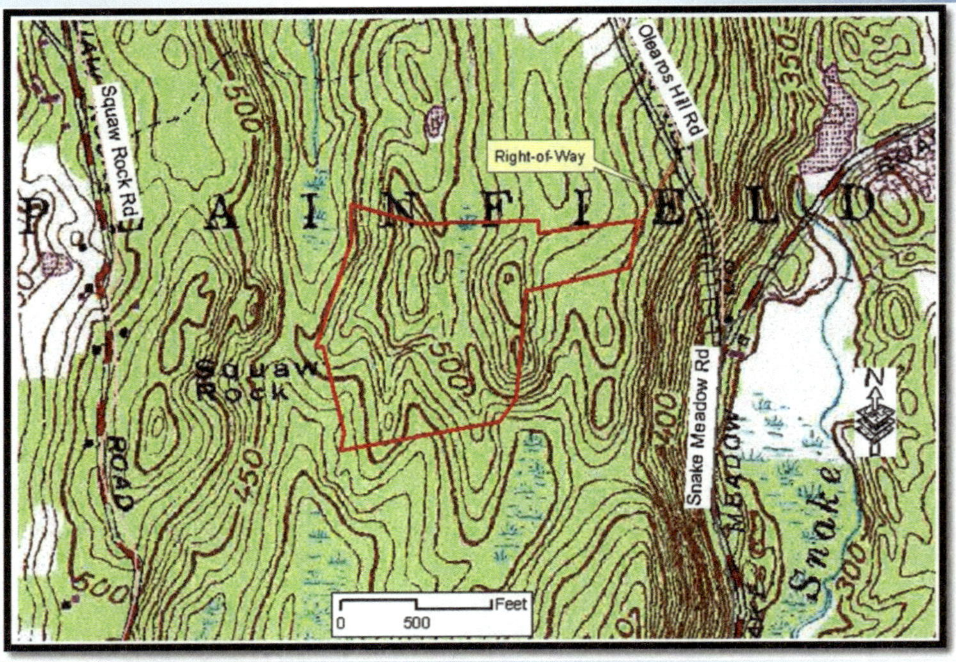

Exhibit 8.3 **Aerial Photograph**

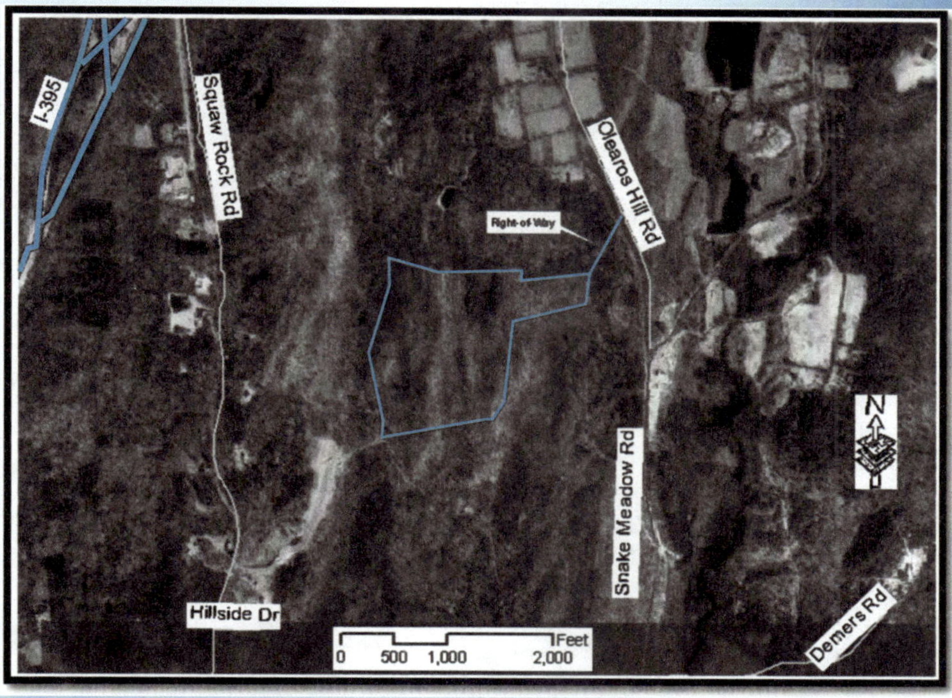

USDA Soil Database

The USDA maintains soils databases for the United States with both national interpretations and, when available, state interpretations that supersede national interpretations.[7] Exhibit 8.4 is an example of how the USDA describes a soil mapping unit, which is a generally consistent grouping of major and minor soil types.

Soil databases can be downloaded for a whole state or part of a state at http://soildatamart.nrcs.usda.gov/. The databases are in Microsoft Access format, so an appropriate version of Access or the Access runtime engine is needed to use the database.[8] The databases generate PDF reports that are easy to print. The reports are presented by soil mapping unit.

---

7. States may have different interpretations about bounds used for categorizing different soils and the capabilities of those soils. For example, the USDA may find that some soils are capable of being used for septic systems while a state may decide the capability should be listed as limited. As another example, the USDA does not regulate wetlands but notes the hydric properties of a soil type, while a state may define wetland soils as a list of soil types that may not exactly correlate with only hydric soils.
8. Access is part of the Microsoft Office suite of products and may be purchased without the rest of the suite. Both are trademarks of Microsoft.

Exhibit 8.4 **Description of Soil Mapping Unit**

Map unit: 32A—Haven and Enfield soils, 0 to 3 percent slopes

Nontechnical Description

This map unit is in the Connecticut Valley New England and Eastern New York Upland, Southern Part Major Land Resource Area. The mean annual precipitation is 40 to 50 inches (1016 to 1270 millimeters) and the average annual air temperature is 45 to 55 degrees F. (7 to 13 degrees C.) This map unit is 60 percent Haven soils, 25 percent Enfield soils. 15 percent minor components.

Haven soils

This component occurs on valley outwash plain and terrace landforms. The parent material consists of eolian deposits over glaciofluvial deposits derived from schist, granite, and gneiss. The slope ranges from 0 to 3 percent and the runoff class is low. The depth to a restrictive feature is greater than 60 inches. The drainage class is well drained. The slowest permeability within 60 inches is about 0.57 in./hr. (moderate), with about 5.1 inches (moderate) available water capacity. The weighted average shrink-swell potential in 10 to 60 inches is about 1.5 LEP (low). The flooding frequency for this component is none. The ponding hazard is none. The minimum depth to a seasonal water table, when present, is greater than 6 feet. The maximum calcium carbonate within 40 inches is none. The maximum amount of salinity in any layer is about 0 mmhos/cm (nonsaline). The Nonirrigated Land Capability Class is 1.

Typical Profile:
0 to 7 inches; silt loam
7 to 14 inches; silt loam
14 to 20 inches; silt loam
20 to 24 inches; fine sandy loam
24 to 60 inches; stratified very gravelly sand to gravelly fine sand

Enfield soils

This component occurs on valley outwash plain and terrace landforms. The parent material consists of eolian deposits over glaciofluvial deposits derived from schist, granite, and gneiss. The slope ranges from 0 to 3 percent and the runoff class is low. The depth to a restrictive feature is greater than 60 inches. The drainage class is well drained. The slowest permeability within 60 inches is about a 0.57 in./hr. (moderate), with about 6.8 inches (high) available water capacity. The weighted average shrink-swell potential in 10 to 60 inches is about 1.5 LEP (low). The flooding frequency for this component is none. The ponding hazard is none. The minimum depth to a seasonal water table, when present, is greater than 6 feet. The maximum calcium carbonate within 40 inches is none. The maximum amount of salinity in any layer is about 0 mmhos/cm (nonsaline). The Nonirrigated Land Capability Class is 1.

Typical Profile:
0 to 3 inches; slightly decomposed plant material
3 to 4 inches; moderately decomposed plant material
4 to 12 inches; silt loam
12 to 20 inches; silt loam
20 to 26 inches; silt loam
26 to 30 inches; silt loam
30 to 37 inches; stratified coarse sand to very gravelly loamy sand
37 to 65 inches; stratified very gravelly loamy sand to coarse sand

The USDA's Web Soil Survey application (http://websoilsurvey.nrcs.usda.gov/) uses the same database as the agency's state soils databases. The Web Soil Survey can be used to quickly approximate boundaries for a simple property with a drawing tool, to calculate areas, and to produce soil reports. Because the drawing tool is so primitive, the accuracy of the property boundaries and the resultant area calculations decreases with the size and complexity of the property.

Compared to this drawing tool, a GIS locates a parcel's boundaries and calculates its size more precisely.

Geographic information systems have programming capabilities that can automate common, tedious tasks. ArcGIS has a programming tool called Model Builder. A model is a series of inputs, actions, and results (outputs), i.e., it is a programming tool. When building a model, inputs are identified and operations are defined. The operations act on the inputs to produce results that can be displayed on a map or that can be used as inputs into additional processes. Exhibit 8.5 presents the model that was used to identify soil mapping units and their areas and to create the map shown in Exhibit 8.6. The program "clips" the USDA soils map layer to the size of the selected parcel, calculates the areas of each soil mapping unit, creates a label field, and then populates the field with text that contains the number of acres of each soil mapping unit, its map unit symbol, and map unit name. The layer is then added to the map, and the map is nearly ready to be output. The program takes a couple minutes to run, which is quicker than doing the same operation by hand or even approximating by eye.

Exhibit 8.6 shows the different soil mapping units that are present in the property. This map results from running the model shown in Exhibit 8.5 and then adding some finishing touches by hand, such as symbolizing the different soil types. Once the soil map is generated and the soils present on the property are identified, soils reports are run in the USDA soil database to provide information on the character, capabilities, and limitations of the soils.

Exhibit 8.5 **USDA Soils Flow Chart**

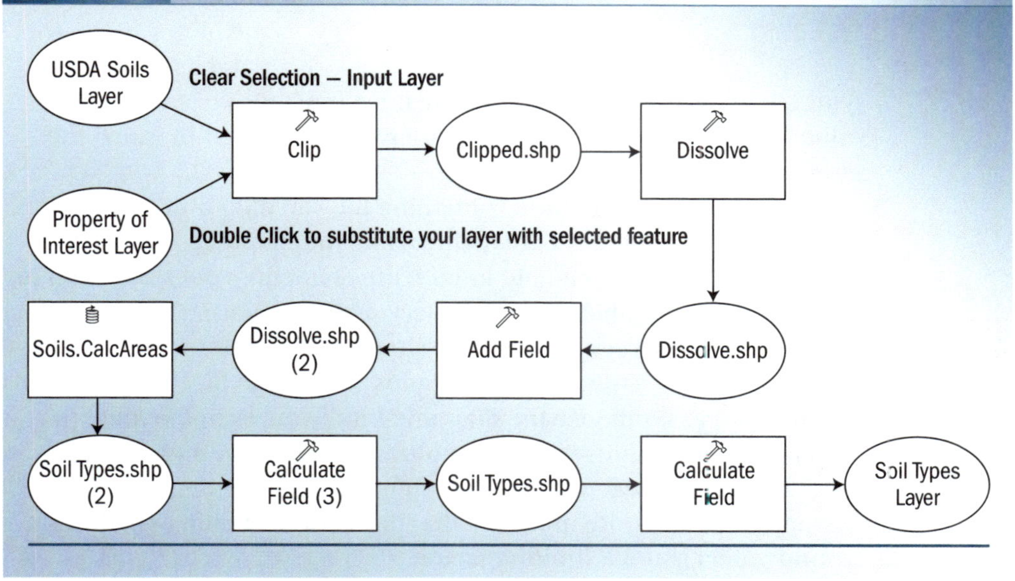

Exhibit 8.6 **Soils Map**

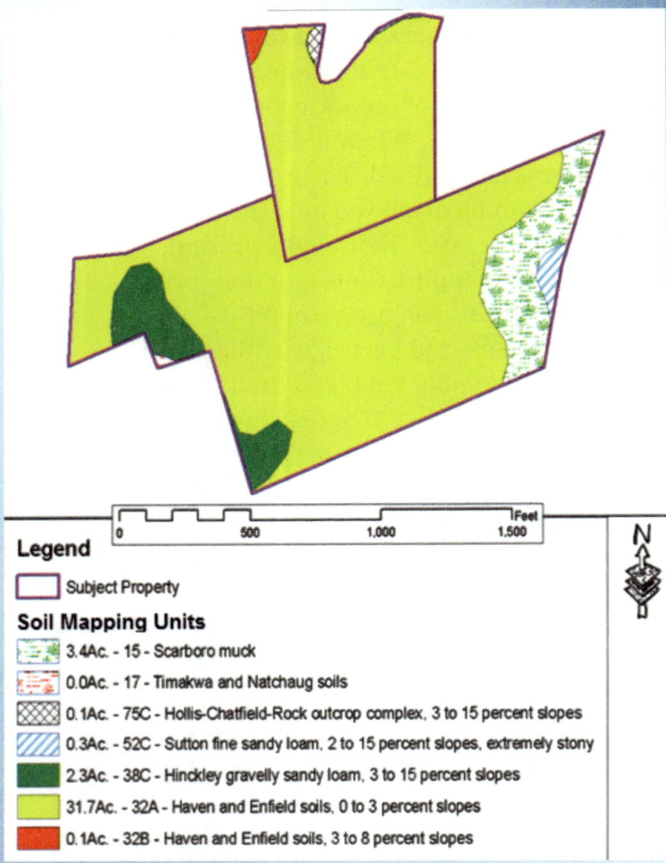

Evaluation of Development Limitations

Colored thematic layers make evaluating soil capabilities quicker by pointing out where development limitations are located. As an example, the three most meaningful limitations of development potential in Connecticut are

- Depth to bedrock
- Depth to seasonal high water table
- Slope

The map series shown in Exhibits 8.7, 8.8, and 8.9 was created by joining the soil layer's Shapefile to one of the tables in the soil database and then symbolizing every mapping unit's feature (polygon) based upon the value in the appropriate field. The maps were created in much the same way as the map in Figure 8.6.

The table in the database containing the soil map unit's characteristics was joined to the soil Shapefile using the mapping unit symbol because that value is in a field in both the Shapefile's data table and in the soil database's tables. This, in effect, appends more columns of data to the data table associated with the shapefile. Therefore, more data is available on each feature in the mapping unit shape file. This data can now be used to symbolize the mapping unit features in the map.

The mapping units were symbolized based upon their capabilities. The title and legend of each map explain the contents. Depicting map units as white indicates that depth to the feature (bedrock, groundwater) is not a limiting factor.

Exhibit 8.7 Soil Type and Typical Depth to Bedrock

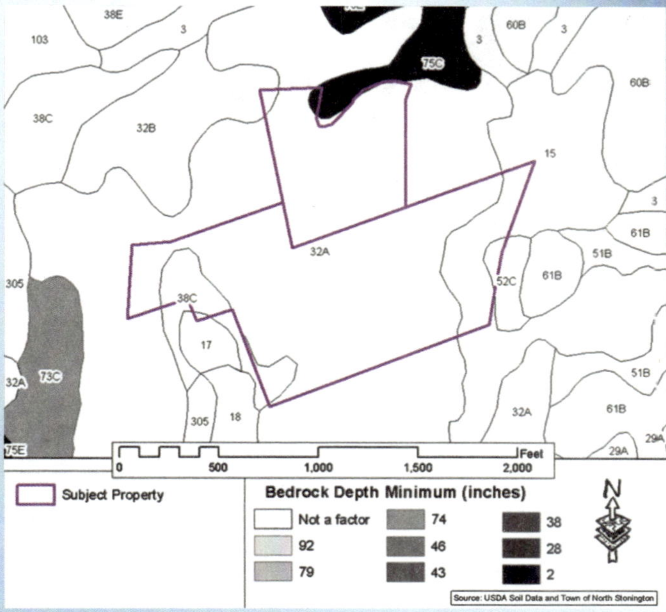

Exhibit 8.8 Soil Type's Depth to Seasonal High Water Table

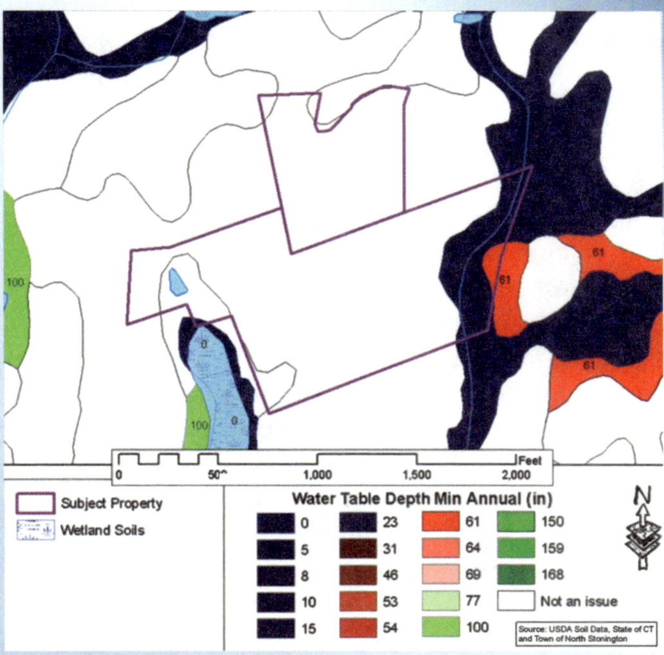

Exhibit 8.9 **Soil Type and Slope Gradient**

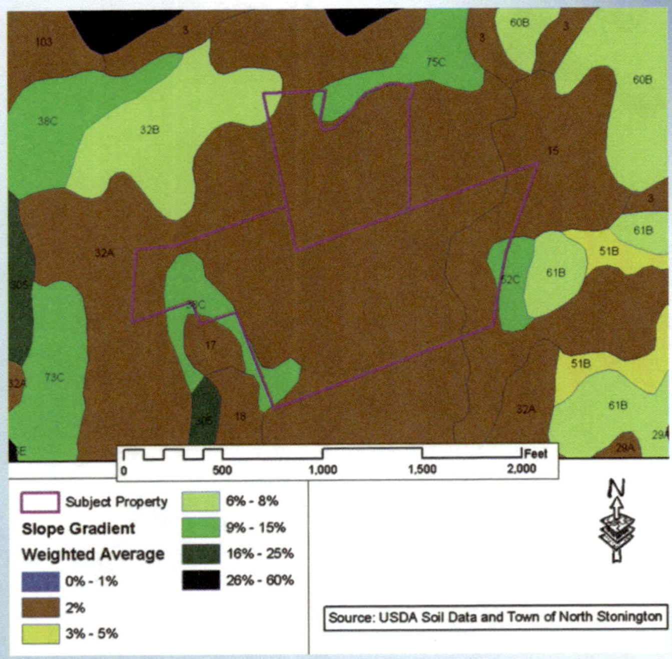

Subdivision Analysis

On occasion appraisers may be called upon to form an opinion of the development potential of a property. This process can be tedious and is not for the timid. Some appraisers may consider this activity outside the scope of their capabilities or too speculative in nature. While engineering expertise would be preferred, an appraiser's ability to render a highest and best use opinion based upon readily available data (public records, zoning regulations, and data from a site inspection) improves the quality of analysis. A GIS can help the appraiser form an opinion of the number of lots and infrastructure quantities as a prelude to understanding a property's financial potential.

Sometimes, the highest and best use of a property is not clear. Using a GIS to analyze the soil mapping and geometry of the property while reviewing development regulations can help an appraiser form an opinion of the highest and best use of a property. As an example, Exhibit 8.10 shows a concept plan for the development of a parcel of land. The town planner provided me with the parameters that would probably be approved by the wetland and planning commissions. Professional planning and engineering assistance was not available. Yet, I still had to determine the highest and best use of the property.

In this case, the analysis resulted in a development scenario for the maximum number of lots that the commissions would allow on a cul-de-sac and infrastructure quantities needed to develop that many lots. The analysis took about a day to develop as different road locations had to be tried before my concept plan was finalized. The planning was complex as the lots had to be sized correctly for the frontage,

Exhibit 8.10 Subdivision Plan

Lot	SQ. FT.	Acres	Total
Concept Lots			
1	179,925	4.13	
2	136,240	3.13	
3	135,249	3.10	
4	294,075	6.75	
5	145,580	3.34	
6	228,110	5.24	
7	131,575	3.02	
8	171,325	3.93	
9	229,079	5.26	
10	369,703	8.49	
11	526,522	12.09	
12	409,625	9.40	
13	261,144	6.00	
14	268,079	6.15	
15	697,200	16.01	
16 (CCF)	21,483	0.49	96.53
Open Space			
OS 1	32,754	0.75	
OS 2	59,941	1.38	
OS 3	2,302,146	52.85	54.98
Road			
Road	151,849	3.49	3.49
Total All			155.00

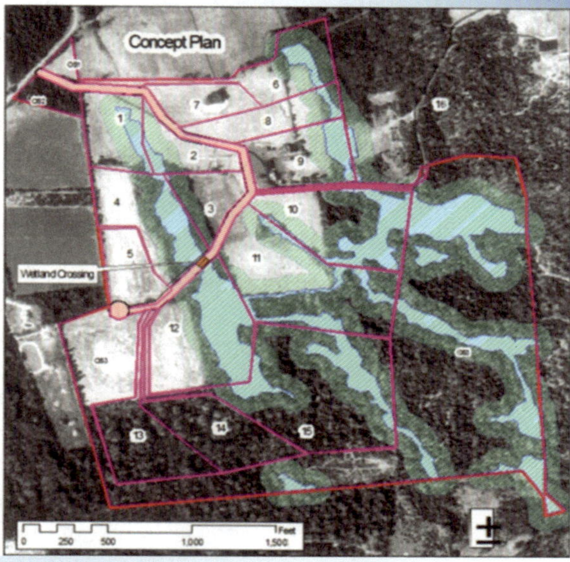

minimum area, wetland area (the town allowed a variable portion of the wetland area to be counted toward the minimum lot area based upon the wetlands' percentage of total lot area), road length, and other characteristics. The GIS allowed rapid calculations of the areas and lengths of these features. The GIS also made short work of calculating and drawing a nondisturbance buffer around the wetland areas and a 50-foot-wide road as a buffer around a hand-placed road centerline.

Having identified the number of lots and the infrastructure needed to support the lots, I was then able to compare the value indicated by this development to other uses and finalize my opinion of the property's highest and best use. In a situation like this, inclusion of the map in the appraisal report is optional. In this assignment, the map was not included in the report because a golf course, not subdivision of the parcel, was ultimately determined to be the highest and best use.

Agricultural Land Appraisal

For the appraisal of properties with agricultural use as the highest and best use, the USDA database provides specialized information about soil mapping units, which can be useful in analyzing properties. The USDA estimates a yield per acre for a typical year for grazing productivity (number of animals) and for farmland (irrigated and nonirrigated) productivity (bushels, tons) for select crops (corn, grass hay, grass-clover, improved Bermuda grass, Irish potatoes, peanuts, soybeans, tall fescue, tobacco, tomatoes, watermelons, and others) that are important in the local soil database.

Appraisers of agricultural lands can use a GIS and the USDA soil database to calculate quantities of a common crop that can be grown at the subject and comparable properties during a typical year as a basis for comparison. While the land's productivity may be a very important factor, distance to market and silos, access to water, and numerous other factors also influence the value of agricultural land.

In the near future, appraisers may be able to use Crop Productivity Indices in nearly the same manner as will be described in the next section on the appraisal of forestry land.

Forestry Appraisal

Forestry appraisal is similar to agricultural appraisal but has some other twists. The different species of wood on a property have different values based on market conditions. Growing times will vary depending on the market for different wood products from pulp to boards to poles. The USDA provides forest productivity estimates for select species of trees. The USDA describes these estimates on the basis of a volume number and a site index:

> The "volume of wood fiber," a number, is the yield likely to be produced by the most important tree species. This number, expressed as cubic feet per acre per year and calculated at the age of culmination of the mean annual increment (CMAI), indicates the amount of fiber produced in a fully stocked, even-aged, unmanaged stand.[9]

> This section [Forest Site Productivity] lists the minimum and maximum site index and annual productivity of the major tree species. The annual productivity per acre per year in cubic feet at culmination of mean annual increment (CMAI) is listed for each identified species. Annual productivity per acre per year in other common units of measurement, (boardfoot-doyle, i.e.), may also be listed for one or more of the identified species.[10]

The tree site index (per acre) is provided by major tree type by mapping unit. The USDA describes this index as

> The "site index" is the average height, in feet, that dominant and codominant trees of a given species attain in a specified number of years. The site index applies to fully stocked, even-aged, unmanaged stands.[11]

> The site index is correlated to each soil and may be further interpreted in terms of cubic meters per hectare.[12]

9. USDA Natural Resources Conservation Service Soil Survey database.
10. ftp://ftp-fc.sc.egov.usda.gov/NSSC/National_Forestry_Manual/2002_nfm_complete.pdf/.
11. USDA Natural Resources Conservation Service Soil Survey database.
12. http://soils.usda.gov/technical/manual/contents/chapter6.html/.

Either of these statistics (volumes or indices) could be a basis for adjustment along with the analysis of the existing plantation, past harvesting and reseeding practices, and stumpage. As with agricultural land, a GIS can be used to calculate the areas of the different soil map units and calculate weighted average site indices for the whole subject property and for comparable properties. However, a growth index does not tell the whole story for forestry lands. The value of timberlands is determined both by their potential for growing and the ease with which trees can be harvested. Let's examine how harvesting issues can be easily handled by a GIS.

Slope

When appraising forestry land, the slope of the land influences how the logs can be extracted from the forest. Land-based extraction methods do not work well when the slope of the land exceeds 45 degrees. Logs must be removed by cable or helicopter, which, of course, affects the net profit. The acreage of areas with slopes over 45 degrees can be calculated using Maptitude or ArcGIS (with the Spatial Analyst Extension). Polygons of steep areas can be created using readily available elevation data known as Digital Elevation Models (DEM) from LiDAR (Light Detection and Ranging).[13] The raw data consists of a collection of points that have the elevation as an attribute. The points are distributed in a grid. Distances between points on the grid can differ based on the scale at which the data was collected.

The value loss for slope is handled in a GIS by attributing a factor reflecting lower values to those lands that fall within steep areas and full values in non-steep areas. In this example, the steep areas are 40% of the value of non-steep areas (100%).

Streams and Rivers

In addition to federal protection of designated wild and scenic rivers, some states require riparian management zones that limit changes to the forest line within defined distances of designated streams and rivers. The distance varies with the classification of the tributary. When states prohibit harvesting within certain distances of a watercourse, the state frequently provides a hydrology geographic data layer that includes the classification of the tributary in the associated data table.

13. LiDAR uses lasers to determine the distance to an object or surface. LiDAR is similar to radar, but it measures the time delay between transmission and reflection of laser pulses rather than sound waves. Industries such as atmospheric physics, geology, forestry, and oceanography use LiDar. In geography, it can be used to create three-dimensional topographic maps. Source: http://clear.uconn.edu/data/index.htm/ accessed on September 7, 2009.

A GIS can easily create buffers of varying widths around line features (tributaries) of varying classifications.

The value loss for streams and rivers is handled in a GIS by attributing a value diminution (as a percentage) to those lands that fall within the buffer areas surrounding designated streams and rivers. Ignoring any recreational value, in this example, the appraiser would say the area within the streams and river buffer zones would have 0% of the value of lands outside these zones (100%).

Clip and Overlay

USDA soil mapping unit features can be *clipped* by other geographic files such as a parcel feature in the parcel layer to produce an output data layer that consists only of those portions of features in the soils layer that lie within the parcel's boundaries. This is much like pressing a cookie cutter into marbled cookie dough (see Exhibit 8.11). The marbling remains but the veins are cut into smaller pieces depending upon which side of the cutter blade they fall. Exhibit 8.6 was created using clipping.

Geographic information systems also allow clipping with multiple clip features in a single operation. Imagine a donut cutter cutting two concentric circles, one for the hole in the center and one for the perimeter of the donut. A soil mapping unit could be clipped into multiple parts depending upon where the donut cutter blades fell across it.

An *overlay* can be defined as "a process in geographic data processing that involves registering and combining layers of thematic data to form a new composite layer."[14] As an example, if an appraiser were to overlay the "steep areas" layer with the "streams and rivers buffer" layer, the result would be a new layer containing more features than either of the two layers as a result of them clipping each other.

For example, suppose a river runs through a steep area into a non-steep area. The buffers surrounding the river would be in both steep and non-steep areas, and the steep areas would be both in and out of the buffer. The overlay pro-

Exhibit 8.11 Clipping Soils Maps

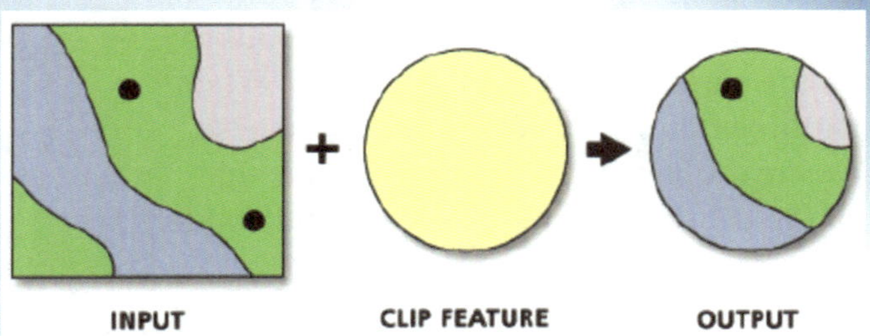

Source: ESRI, ArcGIS 9 Geoprocessing Commands Quick Reference Guide (2008)

14. http://www.prenhallgeo.com/Lo/glossary.htm/, accessed on September 7, 2009.

cess would generate a new layer (the result of clipping each layer at its boundaries with the other) and combine the tabular attributes of both layers in the data table of the clipping output data table. The data table would contain the data represented in Exhibit 8.12. The percentages for the steep areas are a weighting percentage for the higher cost of extraction in steep areas relative to non-steep areas. The percentages in the streams and rivers buffers reflect the percentage of timber that can be harvested from those areas relative to areas outside the streams and rivers buffers.

Exhibit 8.12 Clipping Output Data

Feature	Steep Area		S&R Buffer		Result
1	In	40%	In	0%	0%
2	Out	100%	In	0%	0%
3	In	40%	Out	100%	40%
4	Out	100%	Out	100%	100%

The Result column in Exhibit 8.12 is calculated after the overlay operation. It is the product of the percentages a feature might have for being in or out of a steep area and for being in or out of the streams and rivers buffer (S&R Buffer) zone. This product is multiplied by a productivity unit of measure (either the wood volume number or the site index) for the soil mapping units. The product represents a productivity unit of measure weighted for the unique characteristics of the property being analyzed.

The next step would vary depending upon how the analysis is conducted. If volumes are used, each weighted volume would be multiplied by the acreage with that value. The sum of these products would be an approximation of volume for the whole property. If the site index is used, each weighted site index would be multiplied by the acreage with that value, and then these products would be summed and that total divided by the total number of acres in the property. This quotient would be the weighted average site index for this property.

Access and Proximity

The location of sawmills and the type of wood they will accept may be stored in a geographic file for use in calculating the distance to the saw mills that buy different species and types of wood. Also important is the quality of the road network (on and off the site). Dangerous mountain roads and a lack of bridges make transporting logs more difficult. Roads are best inspected during a timber cruise or appraisal inspection of the property, but the character of the roads can also be estimated by looking at aerial photographs draped over a three-dimensional simulation of the changes in elevation (terrain). Terrain layers that show mountains, hills, valleys, and plains are based upon the elevation data discussed earlier.

Because these access and proximity (A&P) factors vary from time to time, market to market, and property to property, they may be handled in a ranking or qualitative fashion. For example, an appropriate analysis of access and proximity would include statements like the following:

- At the time of sale, the property was within *X* miles of Sawmill *X* that was buying *X* product and *Y* miles from Sawmill *Y* that was buying *Y* product.
- Interior access roads had appropriately placed bridges across rivers.
- Exterior roads in easement areas were passable gravel roads, and county roads to the sawmills were paved.
- Overall, comparatively good access.

The GPS

I use a Garmin GPSMAP 76CS-series GPS receiver, a handheld mapping GPS unit with a color map display, ample memory, compass, altimeter, and both a high-speed USB interface (for uploading and downloading data) and a serial NMEA interface for streaming location information to other devices like a laptop. The GPS unit is rugged, waterproof, and floats, and it can be used for navigating in a car with optional street data, on a boat using optional NOAA charts, and on foot while walking land and for displaying USGS topographic maps overlaid with parcel outlines.

Garmin has a large user base that has written software utilities that interface with the unit, such as the Minnesota Department of Natural Resource (MNDNR) software[15] that lets the unit exchange lines and polygons as track files–the equivalent of a breadcrumb trail–and point files as waypoints with ArcGIS. While these were the priority uses I identified, you should consider how a GPS device will be used and develop minimum specifications before shopping for a GPS unit.

Using a GPS device

I like to upload the outline of the property to be inspected and a topographic base map into the handheld GPS device prior to inspecting a parcel of land. At a minimum, the outline shows me where I am in relation to the property boundaries. If I want to look at specific features (cabins, waterfalls, mines, artifacts, etc.) while inspecting the property, I upload these features into the GPS as waypoints so I can find them while inspecting the property.

15. http://www.dnr.state.mn.us/mis/gis/tools/arcview/extensions/DNRGarmin/DNRGarmin.html/.

Loading a Property Outline into the GPS Receiver

My company collects GIS data for the different municipalities where we work. With the parcel layer turned on in ESRI's ArcView, I select the parcel (or parcels) I will be inspecting. The MNDNR utility program mentioned earlier then uploads the perimeter of the parcels as track points into my GPS device. After this process is complete, the screen of my GPS unit shows the boundaries of the property (or properties).

A subject property outline (two parcels, really) are shown as tracks in Exhibit 8.13. The subject property was acquired in two parcels, which is how they were entered into the municipality's GIS database by the GIS technician. In ArcView, both were selected and uploaded into the GPS as one track file without eliminating the boundary line between the two. This GIS-to-GPS conversion resulted in an extra line being drawn to connect the ending point of the second parcel to the beginning point of the first parcel. This additional line runs through the interior of the two parcels and can safely be ignored. (See Exhibit 8.6 for an image of the properties in the GIS.) From start to finish, this process took a few minutes.

Exhibit 8.13 GPS Display

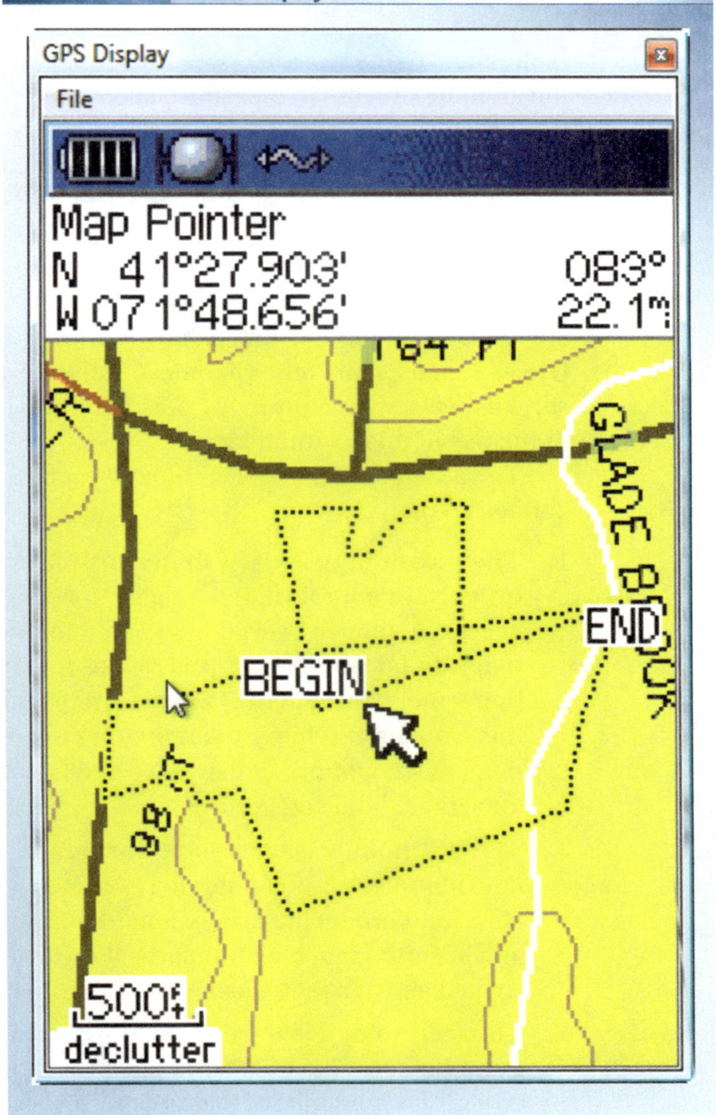

With this property boundary data in the GPS, I can find the corners and boundaries of the property and see where I'm standing in relation to the features in the USGS topographic map, which is the background base map.

Conclusion

The foundation of appraisal theory relies upon how appraisal methods imitate the "market." My first appraisal class drilled this into me when explaining how the principle of substitution was at the heart of each of the three approaches. What you read in this chapter is how I have used GIS to imitate the pricing decisions made by those who make their living working with the land. I have not tried to duplicate the analytic process of decision makers, but rather to build systems that reach similar pricing conclusions, systems that are credible because they consider those major factors that drive pricing decisions.

Consider that while GIS is decades-old technology, it has only been available on desktop-caliber computers, like appraisers use, since the mid-1990s. After writing the first draft of this chapter, I changed companies. My new company has had a full-time GIS technician for years, which is hardly typical. Our work depends upon GIS as much as any other database software and almost as much as spreadsheets, which probably puts our company on the leading edge in the use of this technology.

The adoption of new technology by a user group follows well-defined steps:

1. **The technology is used to perform familiar functions more quickly.** In appraisal, this relates to viewing municipal data, accessing Internet aerial imagery, and making comparable sale maps. In land appraisal, this relates to creating property outlines then quickly producing the basic maps that were discussed in this chapter. It relates to automating measurement and calculations that could previously have been accomplished only through amazingly laborious processes.
2. **The technology is used to perform familiar functions better.** In land appraisal, this equates to breaking away from the copy machine and producing professional-looking maps that can be digitally inserted into digital reports. It means measuring irregularly shaped acreage more accurately. It means automating calculations.
3. **The technology is used to do things that could not have been done before.** This chapter does not venture into this area for the GIS profession, but individual appraisers may feel this stage applies to their practice or them personally as they implement some

of the techniques and applications shown in this book.

4. **The technology is pervasive.** For over a decade commercial companies have produced GIS systems that appraisers have adopted. The low-hanging fruit of GIS applications is now available in kiosks in municipal offices, on the Internet, and even in GPS systems on the dashboards of our cars. Aerial images, GPS, and routing and location-based services in telephones are in daily use by appraisers.

If you have additional interest in using GIS in your practice, the Appraisal Institute has an entire online GIS curriculum for real estate appraisers. The curriculum includes an array of offerings from a nontechnical "Executive Overview" of GIS applications in real estate appraisal and consulting to hands-on case studies, using Maptitude GIS. The case studies start at the novice level for the first-time user and prepare you to use Maptitude in your practice.

Stepping into the next level of GIS adoption will be more difficult than using Internet mapping systems. Implementation of GIS in your practice will require an investment in time, software, and data. If integrating GIS into your practice differentiates your practice from the other appraisers in your market and if your clients find this valuable, you could gain greater market share and possibly earn higher rates.

9 Developing Enterprise-Level Valuation Applications

Incorporating Appraisers in Innovation

John Connelley, Kirk Manker, Mark R. Linné, MAI, SRA, and James Gresham

In 2004, a diverse team of geographic information and valuation professionals were assembled to assess the feasibility of providing an integrated solution incorporating robust GIS and interactive valuation modeling (IVM) technology to Farm Credit Services of America (FCSA), currently a $13 billion entity and one of America's largest agricultural lenders. The initial project scope of work and the outcome and results of this 4½-year effort are detailed in this chapter and the next.

In 2004 FCSA first requested the development team to consider the possibility of developing an agricultural land AVM that had a strong GIS or spatial component. During an initial site visit by the development team, this idea developed into a much clearer vision of the need for a GIS-based AVM solution with specific goals and functional elements within the FCSA appraisal department. The development team provided an initial feasibility analysis and solution proposal in 2005. (Chapter 10 details the results of the implementation of the solution.)

The development team's work began with interviewing key organizational leaders in the appraisal and lending departments of FCSA. The initial study provided a high-level overview of the components necessary for the successful implementation of a GIS-based agricultural AVM system for use by FCSA. The system specifications were designed around the customization and use of the core functions of existing and proposed valuation software products. The new system would leverage FCSA's internal ValueSource system, an automated valuation system that FCSA had been working on for several years that was nearing completion at the time.

FCSA was looking for a system that could augment and enhance the skill set of appraisers and improve appraisal productivity and accuracy as well. Such a system would be implemented within the framework of ValueSource, extending the software's already significant capabilities. The FCSA Appraisal Leadership Team (ALT) had a clear vision of what the organization's needs were and, in addition, had completed sufficient research to understand that their needs could not fully be accommodated with available off-the-shelf products.

Goals

After all key personnel were interviewed, the key objective of the project quickly resolved itself as an initiative to develop and deliver an integrated "enterprise-level" (i.e., meeting the needs of the entire organization or enterprise) automated valuation and geographic information solution to FCSA that aggregated, refined, analyzed, and contextualized all relevant valuation and property data and significantly enhanced FCSA's Credit Modernization program. Specific project attributes were identified as the following:

- Extend the functionality of a valuation system throughout the FCSA organization.
- Provide FCSA staff access to market and valuation data in visual and tabular formats.
- Integrate property and valuation data with a GIS system.
- Enhance appraiser and financial officer productivity through rapid property identification and value estimation.
- Help FCSA better serve existing customers.
- Incorporate value and market trend data into corporate intelligence.
- Maximize benefits from existing in-house data through data mining.
- Provide better understanding of existing customer base and competitors' customers.

Background

Typical production farmland in the United States is valued using the sales comparison approach. Variables for this property type include scale, production, and attribute data. The unit of comparison is typically the price per acre, given that land is bought and sold on that basis in most markets. Agricultural AVMs–to the limited extent that they exist–are similar to residential AVMs in that they require appraisal guidance in the framing and model specification. Coefficients generated and found to be significant must follow established valuation principles. For example, the price per acre of gravity-fed irrigated land should fall between the price per acre of nearby center-pivot irrigated land and of non-irrigated land based on the relative utility of each type of land. More intuitively, land designated with a subjective quality label of *above average* should have higher value than otherwise similar land designated as *average* or *below average.* Valuation parameters of agricultural valuation in many ways resemble residential valuation with respect to model development and testing.

However, agricultural modeling involves other considerations not found in residential modelling. Most importantly, soil productivity is a key factor and must be accounted for in any market value model. Soil problems, known as *hazards,* must be considered. Examples of hazards include

- Sandy soils
- Rocky soils
- Poor drainage
- Erosion

Tract configuration and topography must be considered. Issues such as rainfall and climate are mostly localized factors that may not vary significantly in a market area and may be accounted for by the presence or absence of irrigation. Urban encroachment and highest and best use issues must also be considered. An appraiser should consider all of these elements when developing a valuation model, whether it is traditional or statistical in nature. As always, appraisers must know their market and the nature and availability of data before proceeding with valuation modeling. Finally, sales of agricultural properties may or may not occur at regular auction sites or times, and thus the nature of the sales marketplace must be fully understood.

The use of GIS in agricultural valuation must also be modified. Given the scale of the parcels of land and number of agricultural properties in a given market, mapping and layering must typically be considered from a slightly less granular perspective, covering a broader area than what is typically seen in residential analysis. Agricultural properties are larger than residential. (Farm size is typically ¼ square mile–160 acres–or more.) So location factors may affect a property in either a positive or negative direction. There may be larger-scale influences to consider, such as broad floodplains and urban development, that affect a portion of the modeling area. Transportation linkages are also critical, especially in western Nebraska and South Dakota or Wyoming, where distance to market increases and road quality can decrease. GIS can effectively map and analyze routes to and from urban distribution centers.

General Overview

The FCSA service territory encompasses all of four states: Iowa, Nebraska, South Dakota and Wyoming (Exhibit 9.1). It is the largest agricultural lender in this region. Geographically, these states comprise about 8% of the U.S. land area, but only about 2% of the U.S. population. More people reside in the single county of Los Angeles. However, these four states produce about one-third of the U.S. corn crop (Exhibit 9.2), one-fourth of the U.S. soybean crop (Exhibit 9.3), and one-fifth of U.S. cattle production. The great majority of that production comes from Iowa and the eastern portions of Nebraska and South Dakota. This is a large and important component of the U.S. agricultural heartland, and the modeling effort was to be significant.

Geography

The four FCSA states cover a fairly narrow range, north to south, of only a few hundred miles. East to west, however, the distance is 1,100 miles, beginning at the Mississippi River and ending at the Continental Divide. At the eastern end, in Iowa, soils tend to be rich and water abundant. Corn and soybeans predominate and are produced without the need of irrigation on farms of typically about 140 acres.

Exhibit 9.1 Farm Credit Services of America Territory

Exhibit 9.2 **Relative U.S. Corn Production**

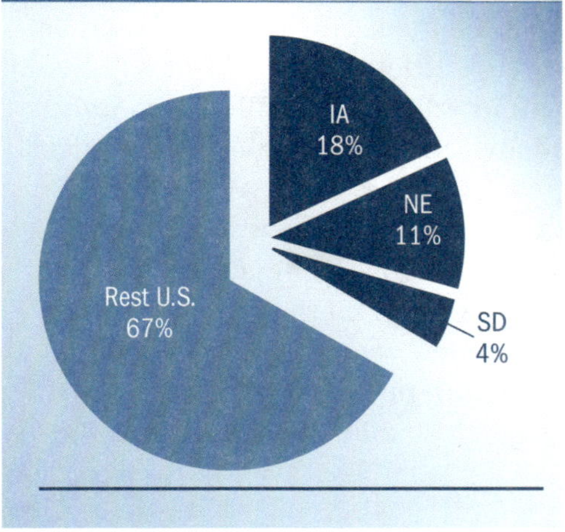

Exhibit 9.3 **Relative U.S. Soybean Production**

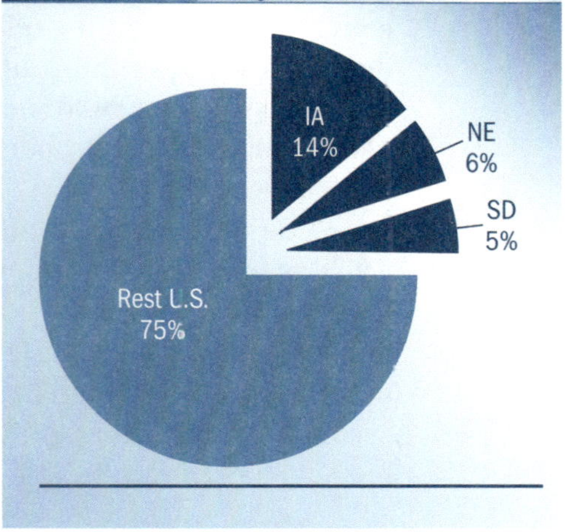

Moving west through Nebraska and South Dakota, elevations gradually rise and the climate becomes dryer. Agricultural productivity problems tend to be associated with too little water and less productive soils in more westerly locations. Farm or ranch size tends to be greater–around 380 acres is typical–and more heterogeneous. Distance to market begins to become an issue. Crops like wheat and sunflowers become relatively more important.

Finally, in Wyoming, climate issues, non-agricultural influences, recreational uses, and reliance on public lands for grazing begin to play a great role. The growing season may be less than 100 days annually in many areas and can limit crops to alfalfa and hay. With a decrease in land quality, typical farm or ranch size can be well in excess of 1,000 acres.

Proposed GIS-Based AVM Structure

After months of discussion and evaluation of what software FCSA needed, a proposed system was offered that functioned as an integrated, enterprise-level valuation and geographic information environment. The proposed system was designed to be modular and flexible with the capability of adding new components within a seamless user interface, all based around the development team's GIS/AVM engines and FCSA's ValueSource application and associated appraisal and sales database.

The proposed system would enhance the valuation and data infrastructure of FCSA by extending the existing knowledge base at an enterprise-wide level. The development team's proposal combined the latest GIS and AVM technologies to provide primary FCSA users such as the field appraiser and financial officer with the necessary tools to estimate a market value for properties and identify suitable comparable sales in the office or in the field. The analysis would be delivered via the ValueSource product developed by FCSA. Additional delivery systems would include an interface to be called *AgriStat* used to provide key property and appraisal information to appraiser and client users.

FCSA and the development team wanted the new system to help redeploy current personnel into data enhancement or, even further, to allow the agricultural lending unit to expand market share without increasing staff size. The development team recommended that the solution be implemented with a core multifunctional GIS/AVM system that was geographically specific to all FCSA markets in the four-state region.

The GIS/AVM Engine

The core analytic component of the solution was constructed as a GIS-based automated valuation system that had the capacity to estimate market value for fields, tracts, or complete farms on either an individual or mass (e.g., county) basis. The analytical framework used product technologies relying on both multiple regression analysis and spatial analysis as part of the sales comparison approach to valuation. The analysis was delivered using AgriStat, the product interface developed as an end-product of this process. In addition, the GIS/AVM engine could be run in the background as a core component of the ValueSource system.

Benefits and Enhancements

All of the stakeholders believed that the solution that FCSA sought was far more than mere software. The AgriStat solution was expected to provide enhancements to the appraiser tool set and knowledge base, as well as increase FCSA's understanding of its changing business environment. FCSA could thus leverage its two greatest assets–its employees and the data that was painstakingly collected and verified from previously performed appraisals.

Industry-Oriented Benefits

Development team members had experience in constructing USPAP-compliant interactive valuation modeling solutions for the lending marketplace that had gained acceptance from such industry leaders as Freddie Mac as alternative valuation products. This industry experience was a key element in FCSA's belief that such a product must ultimately have secondary market acceptance. An IVM, linking a data source to sophisticated valuation technology and directed and controlled by the appraiser, was the preferred outcome of the AVM/GIS solution.

The importance of an interactive valuation modeling process was important to FCSA precisely because they required a system that met compliance efforts and represented a best practices perspective. In addition, the application/solution had to meet all Uniform Standards of Professional Appraisal Practice (USPAP) and current industry standards requirements as well as future requirements. FCSA needed to ensure that the product continued to meet both the regulatory and market demands going forward.

FCSA saw a strong opportunity to leverage the expertise of its development team to make AgriStat/ValueSource the first interactive valuation modeling product serving the agricultural community. Through proper consideration of industry initiatives throughout the greater real estate community, FCSA could ensure that ValueSource became the de facto standard for agricultural valuation standards, data standards, and best practices on a national and perhaps an international basis.

GIS/AVM Benefits to FCSA

The solution that was agreed upon between the developers and FCSA was not a traditional software application, due to the elements of the solution that integrated GIS expertise and valuation modeling expertise. The aggregation of these elements represented, in effect, an expert system that integrated sophisticated technology and allowed interaction with an appraiser-user. Specifically, the benefits of the solution provided were identified as

- Increased appraiser productivity
- Increased appraisal accuracy
- Increased information timeliness
- Improved understanding of market dynamics
- Better understanding of market trends
- Development of true forecasting models
- The ability to track all FCSA property value trends
- Decreased turn times (e.g., deadlines in days, not weeks)
- Ability to visualize data
- Ability to identify significant determinants of value from market data
- Ability to simulate valuation environment

- Ability to value one property using multiple AVM models simultaneously
- Ability to demonstrate valuation veracity more concisely
- Use of the intellectual property already developed in ValueSource to its fullest extent
- Ability to capture and use the FCSA appraisal intelligence that had been built up over the past decade
- Improved market data quality and quantity
- Improved appraisal and data consistency

Ultimately, the system would serve the needs of the entire enterprise and position FCSA as the leader in agricultural lending.

AgriStat Solution Structure

The design of the solution assumed three primary types of end users:

- Appraiser
- Financial officer (FO)
- Management

Two operating modes were assumed:

- Connected mode
- Disconnected mode

The appraiser or financial officer was assumed to operate either in the office (connected mode) or in the field (disconnected mode). Field operations could run in connected mode if Internet connectivity exists, but the development team followed the more restrictive assumption of disconnected operation. Management was assumed to operate in connected mode only.

Connected Mode

In the connected mode, the appraiser or financial officer would be connected to the FCSA network and servers and have access to all functionality based on permissions level. More importantly, given the vast increase in data due to implementation of a GIS, a connected user would have potential access to all ValueSource data. The user in this mode could be in the office with a hard network connection or in the field with a wireless connection.

Disconnected Mode

In the disconnected mode, the user is not connected to the FCSA network and experiences some functional limitations. Disconnected mode limitations arise from the massive amounts of data in the fully developed system, especially the GIS component. A key assumption of the development process was that anyone operating in disconnected mode will have an opportunity in-office (i.e., in connected mode) to prepare for a field trip, with knowledge of the location and some details of the description of the parcels to be examined.

ValueSource runs a background synchronization program that keeps the remote data set synchronized with the central database anytime a user is operating in connected mode. That is, data newly entered on the remote computer is uploaded to the central database, while data new to the central database is downloaded. Local data sets are typically determined by a subscription process. For an appraiser working in the field in a subscribed county, no special arrangements need to be made. If an appraiser, for example, is planning on being in Hamilton County, Nebraska, the next day, and that county is not part of the appraiser's regular territory, the relevant ValueSource, AgriStat, and GIS data could be preloaded on a laptop. In most cases, the field user would never notice that functionality is limited in disconnected mode.

Appraisers and Financial Officers in Connected Mode

As the first step in any agricultural property appraisal, the appraiser will typically identify the subject tract. How a tract is identified can be a matter of user preference. AgriStat allows map-based or textual identification. Map-based identification uses the GIS interface to zoom in for selection. The more geographic information initially provided by the user, the quicker the zoom. To locate a tract textually, a user would first select a state from a dropdown tool or checklist. The next step would be to select a county from the state county list (see Exhibit 9.4). When the county is known, a list of available range and township combinations can be offered. Lastly, the available sections can be shown, usually identified as sections 1-36. At any point in this textual drill-down process, the user will have the option of switching to map view. It may, for example, be easier to locate a township from within a county or a parcel from within a section by examining an aerial view (Exhibit 9.5).

A requirement of this project was the ability to scan legal descriptions and locate a tract after passing the legal description through character-scanning software. Because of the many format inconsistencies and complexities in legal descriptions, this feature needed to be interactive with the user. That is, the system can suggest a possible combination of meridian, range, township, or section, but the user should sign off on the accuracy and correct any mistakes before the system accepts the results obtained.

Exhibit 9.4 **Textual Tract Search**

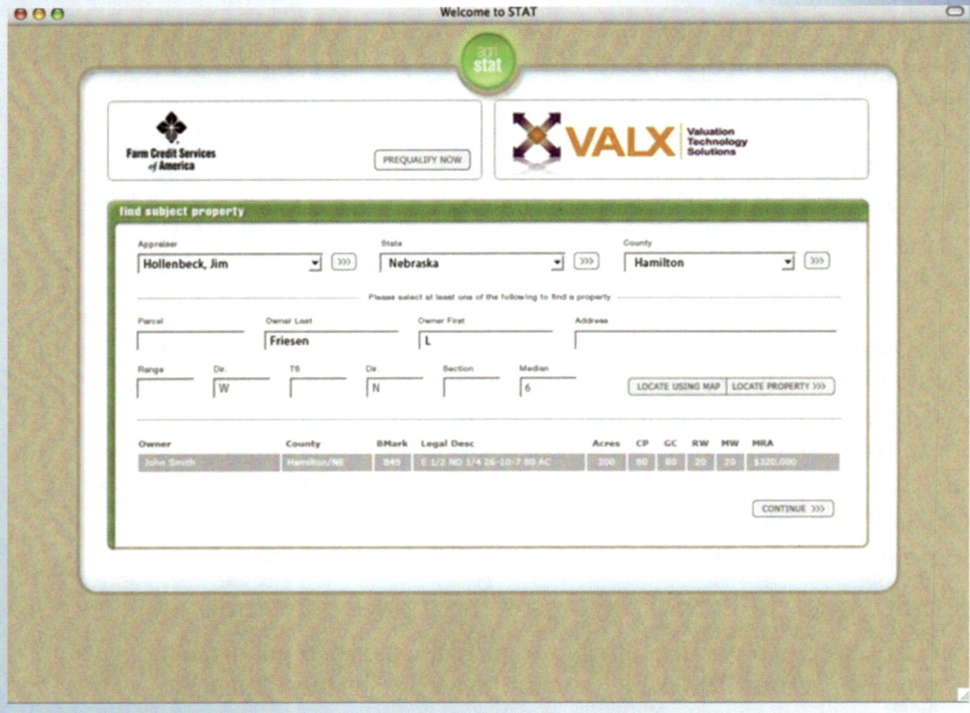

The software would offer sufficient linkages into ValueSource that a property could first be located in ValueSource. Then the user would click on an AgriStat link. If the subject tract was known to already exist in ValueSource, AgriStat could allow alternative methods of identification based on the data previously captured. Owner name and street address are more common possibilities, although there are others such as general property type or nearest town.

ValueSource, and by extension AgriStat, would not use county-assigned parcel numbers as the primary identification for tracts. It is entirely possible for a farm owner to split a farm into fields for financing purposes, obtaining a loan on one piece and leaving the rest unencumbered. Both portions could be contained within a single county parcel designation. Similarly, multiple tracts, each a separate parcel, could be combined within a single loan. An FCSA tract can therefore be a portion of a parcel, an entire parcel, or multiple parcels. Prior to implementation of a GIS, duplicate entries could be a concern, especially the situation of overlapping tract boundaries. The GIS component of the system would need to eliminate this possibility going forward.

Exhibit 9.5 GIS Tract Search

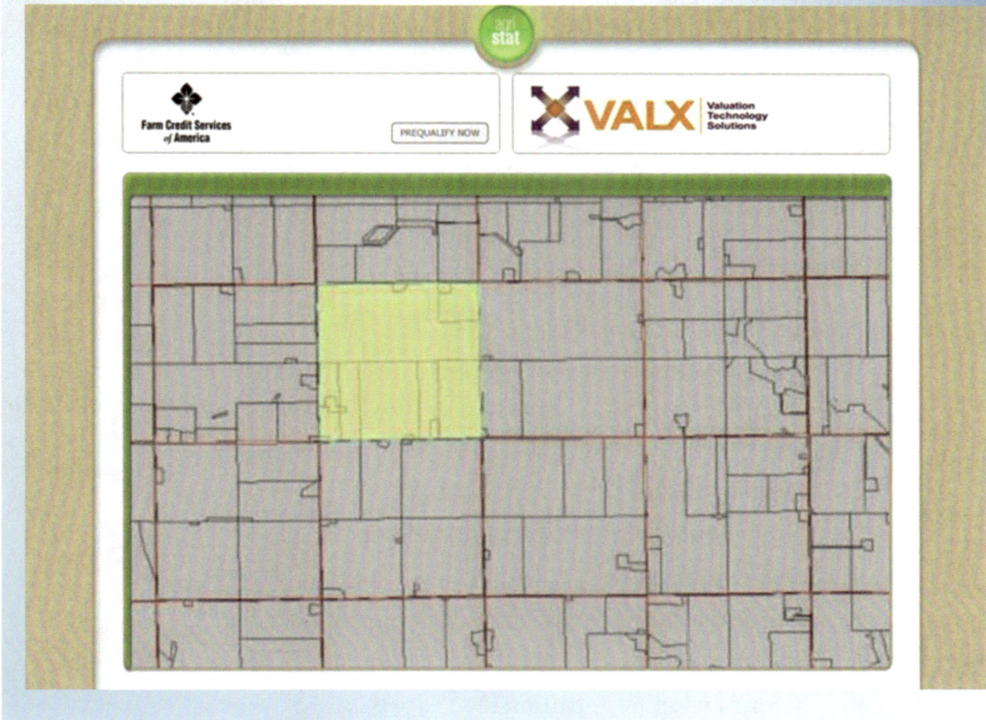

When a tract is located and its preexistence within ValueSource is ascertained, one of two events would logically occur:

1. If the tract is new to ValueSource, some amount of property information would be entered by the user.
2. Otherwise, displayed data fields would be automatically populated.

Rules concerning required data fields to be initially entered for a new tract would be identical to those required by ValueSource.

Since ValueSource already provides a rich and robust valuation environment, the design of AgriStat would rely on ValueSource routines and procedures to perform those functions that ValueSource was designed to perform, such as data entry and verification. The AgriStat design was intended to build upon and enhance–not replicate–ValueSource functionality.

From a purely AgriStat viewpoint, the appraiser would need to initially enter only those data elements that provide AgriStat sufficient information to respond to user requests. For example, if the appraiser wants to find nearby comparable sales of similar size, then acreage needs to be entered (given that the system already has a tract location at this point). However, to locate properties based on a tract similarity index, information on improvements and land use would be required. AgriStat was designed to find similar tracts and comparable sales based on the information entered. As more detail is added, those searches become more precise. When an AgriStat value estimate is desired, required data entries can be projected. Since AgriStat would likely model grain farms separately from feed lots or hog farms, the user would need to, at a minimum, select a general property usage from a system prompt. Other data entries would depend on the property usage and could not be delineated until the actual models were developed through interaction with the FCSA appraisal group.

AgriStat Functionality

When a tract is located and properly described in ValueSource, an appraiser would have a wide range of functionality available relevant to the selected tract, including

1. Comparable sales
2. AVM results from all relevant models
3. Soils comparisons
4. Comparison of nearby tracts and appraisals
5. Market trending
6. FCSA market penetration
7. Reporting
8. Measurements

Some functions, such as locating comparable sales, rely more on spatial relationship and display better in the GIS. Other functions, such as comparing attributes among subject and other tracts, may display better textually.

Comparable Sales

A primary focus of the appraisal process is finding comparable sales, i.e., locating the available comparables and selecting the most appropriate (Exhibit 9.6). AgriStat would provide the appraiser with multiple methods for comparable sales comparison, including

- Map display of actual comparables (Exhibit 9.7).
- Map/textual display of comparables based on price per acre.
- Display of comparables with similarity index ranking. AgriStat would be able to generate a default similarity index or allow the appraiser to add and remove index parameters or adjust weights to obtain results that the appraiser feels are more suitable to a specific tract.

Exhibit 9.6 AgriStat Comparable Selection Screen

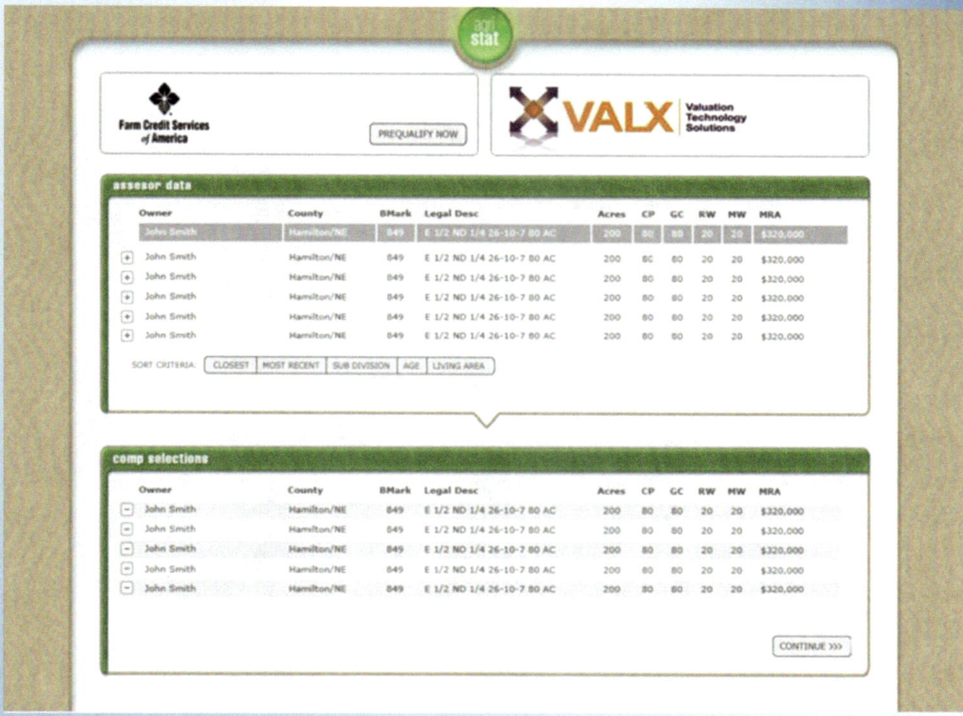

- Display of income estimates for comparable properties based on the state land productivity measure. Iowa uses Corn Suitability Rating (CSR), South Dakota uses Soils Productivity Index (SPI), Nebraska uses Soils Rating for Plant Growth (SRPG), and no measure is used for Wyoming.

The AgriStat interactive valuation approach allows the model to self-adjust initial AVM values based on the selected set of comparables to obtain a more precise estimate of value. This revised value estimate can appear instantly as the appraiser selects comparables and assigns them relative weights.

AVM Results

The user could obtain a value estimate for the subject or compare key parameters of proximate model groups. AgriStat would provide summary statistical analysis of the accuracy and reliability of the generated estimate(s), including R^2, coefficient of dispersion (COD), coefficient of variation (COV), price-related differential (PRD), residuals, and similar measures. All of these are key statistics that measure the overall predictive nature and accuracy of a model. Information

Exhibit 9.7 **Display of Selected Comparables**

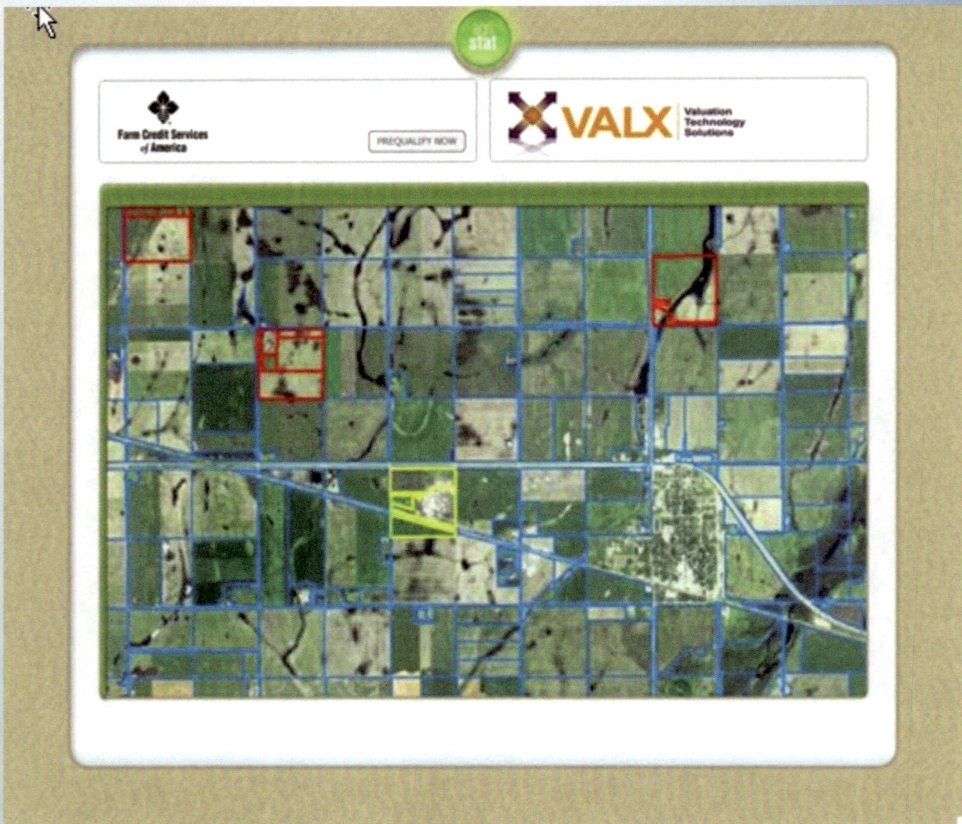

on projected rents, capitalization rates, and depreciation are available from the income and cost models. The modeling procedure also highlights outliers and cookie-cutter tracts. Outliers are tracts that are in some way dissimilar from other tracts in the model grouping and may call for special, hands-on attention. Cookie-cutter tracts are just the opposite, i.e., tracts so standard within the model group that no special attention is required. AgriStat can also examine previously appraised sales within a model grouping. The use of completed appraisal data has a number of implications, including the ability to evaluate the consistency and accuracy of individual appraisers. The ability for individual appraisers to independently specify and run models was not being considered in the proposal presented to FCSA. Model formulation requires a sophisticated skill set in both appraising and statistical methodology and should be confined to a few specialists.

Soils Comparisons

AgriStat could apply land productivity measures (SRPG, CSR, or SPI) to determine field productivity and derive tract income estimates.

Nearby Tracts

The user could query ValueSource and display/compare other nearby tracts to the subject tract. There may be relevant comparative appraisal data to consider, such as rents or capitalization rates, even though the tracts are not contemplated as comparable sales.

Market Trends

Based on ValueSource historical data and income analysis based on soil comparisons, AgriStat would be able to provide near- to intermediate-term forecasting by extrapolating recent price/value trending forward in time.

Market Penetration

AgriStat would display or report local or regional FCSA market share (total or broken out by market sector). Temporal changes and trends would be shown as well.

Reporting

AgriStat would have the ability to create a PDF file from a selection tool incorporating maps, interactive AMV results, and ValueSource reports and forms. Recognizing that appraisers and FOs may perform a fair amount of map tweaking (layer addition and subtraction, rescaling, and centering), AgriStat would incorporate a bookmarked query scheme to allow the user to temporarily bookmark a map for later retrieval.

Measurements

The field user would be able to perform and save linear measurements and calculate area subject to the accuracy of the data layers. Measurements that could be calculated in acres, such as roads and wasteland in a tract, can be reasonably approximated. Measurements of square footage, such as a building footprint, could produce unacceptable error. For such measurements to be applied, a standard of accuracy needed to be established. For example, a GIS measurement of building dimensions should not usually be allowed to overwrite a measurement that was made directly in the field.

Appraisers and Financial Officers in Disconnected Mode

Operating in disconnected mode can offer significantly greater functionality when the appraiser can do a little advance preparation before going into the field. The appraiser can then preload a regional package of relevant GIS, ValueSource, and other AgriStat data onto a laptop. Again, if the user is working in his or her regular "subscription" territory, this data would automatically be preloaded. Each time the user was connected to the FCSA network, the subscription data would be automatically synchronized. Within the preloaded region, functionality is comparable to connected mode if all available data layers are loaded.

Management in Connected Mode

The operating mode assumed for management was connected mode. Connected mode could mean in the office or at home connected by virtual private network (VPN). Management would likely focus on issues with enterprise-wide consideration for FCSA. These issues would typically extend far beyond a single county or market zone geographically, thus requiring connected mode operation for full GIS functionality and broad data access. Typical functions would include

1. Descriptive reporting
2. Market analysis
3. Risk assessment
4. Forecasting
5. Personnel decisions

Descriptive Reporting

The user can display maps containing tract-related summary information of georeferenced tracts stored within ValueSource and AgriStat. Tabular reports are also available. Possible display/report scales from which to choose could include

- All of FCSA
- State level
- Market zone
- County

Interregional comparisons are also possible. Historical trends can also be presented. One example would be price-per-acre trends compared among states, market zones, or counties over a period of several years.

Market Analysis

A wide variety of market analyses can be performed. A key example is portfolio analysis. AgriStat allows the user to revalue the entire agricultural portfolio as of a specific date. The date can be some time in the future or past. The portfolio can be limited to a specific geographic area if desired. The effects of rapidly changing market conditions can thus be measured rapidly and their implications on company profitability quickly understood. As another example, rough market penetrations can be estimated by comparison of FCSA portfolio tracts to acreage contained within the common land unit (CLU) layer. More sophisticated analyses will require importing some external data. For example, to know the market percentage for feedlots in a specific market zone or state, data on the overall population of feedlots is required to make the comparison. Temporal studies in areas like growth in market share can be performed on historical ValueSource data. The primary limitations to analysis are data and imagination.

Risk Assessment

Users can examine loan concentration by agricultural sector and region. Some risk measures may require the introduction of external data, such as foreclosure rates by farm sector and geographic region. Another example would be analysis of drought and flood risks.

Forecasting

To a limited extent, historical ValueSource data on economic variables such as land sales prices, crop prices, feedlot income, and the like can be extrapolated forward in time. Mid- to long-range estimates should really draw on regional or national econometric forecasting models.

Personnel Management

One simple tool would be to display overlays of appraisal work with FCSA offices to better optimize job assignment. Another would be to use AgriStat's flagging of outlier or unique tracts that require special attention to assign them to more experienced or better-performing appraisers. Similarly, cookie-cutter tracts, which are more common and require less attention, can also be indicated.

Modeling Development, Process, Procedures, Analysis, and Validation

The modeling process implemented in the pilot study was a simplified version of the process used in the full project. First–and based in part on the fact that the pilot focused on a small, regional grouping of relatively homogeneous Nebraska counties–an additive model was used. In the later, full project, a more robust multiplicative model was used. For the pilot, an additive model was desirable because of the ease of explanation. The valuation mechanism was simply this: adding positive elements added to property value, while negative elements detracted from property value.

Soil productivity (SRPG in Nebraska) was excluded because some minor issues in the way SRPG was implemented on a county basis in Nebraska were not worked out until the final project, when SRPG was used with a constraint. Site improvements were also excluded in the pilot model because they required a significant amount of investigation, although they too were included in the final project. Exclusion of improvements had little effect because most tracts had no improvements. Indeed, in the final project the total value of improvements came to less than 3% of average value.

The major outcome statistics used in the pilot were the coefficient of dispersion and coefficient of variation. Both are based on the average error of the model expressed as a percentage. The R^2 statistic was also examined, but typically this general goodness of fit statistic is not as important. Each model's residual patterns were examined by plotting the sale ratios across the sale period, sale price range, size of land or improvements, and other variables.

Initial Proof of Concept Analysis

FCSA and the development team determined that a smaller-scale pilot would be conducted to permit an evaluation of the potential accuracy of a modeling system prior to implementation of a system-wide modeling project. To demonstrate how an agricultural AVM system would work with a robust GIS component, modeling an eight-county market area in Nebraska was undertaken. The purpose of the analysis was to determine the market value of unimproved agricultural land ranging between 100 and 400 acres in this market. There were 190 sales spanning the period 2001 to 2004. Land use included irrigated land, dry cropland, CRP (Conservation Reserve Program) land, pasture, meadows, road/waste, and recreational use, which covered all significant land use categories in this market area. For simplicity, an additive model was used.

A sample subject parcel was selected, consisting of a total of 120 acres, with 50 acres of gravity irrigated land (i.e., land that is irrigated by water flow over the soil surface) and 70 acres of potentially irrigated (but currently not irrigated) land. The subject parcel was part of a larger parcel that totaled 220 acres and included significant improvements. The subject was completely productive with no highest and best use issues.

The scatter plot shown in Exhibit 9.8 indicates the distribution of properties by sale amount and land area for all 190 sales. The SPSS software-generated scatter plot indicates clustering of properties at 160 acres, which is not surprising given the predominance of 160-acre quarter sections under the Public Land (section-township-range) surveying method used in the western states.

The process for building the valuation model was a multiple-step function, which each step indicated as shown in Exhibit 9.9. The

Exhibit 9.8 Sale Amount and Land Area Scatter Plot

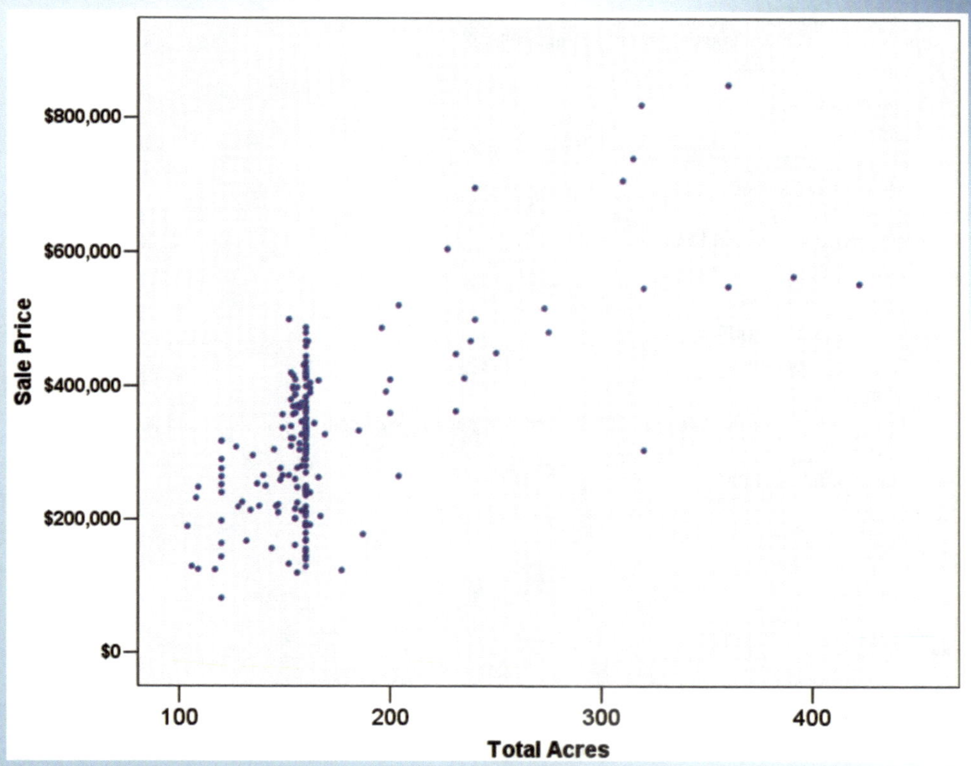

initial output from a statistical analysis (Exhibit 9.10) indicated the overall fit of the model (represented by the adjusted R^2 statistic of 0.904 and standard error of $39,481). Comparing the standard error to the mean predicted value of $322,953 indicates a good approximation of the coefficient of variation, or 12%.

To determine the accuracy of the model, the residual analysis from the valuation model uses sale ratios, calculated by dividing the estimate of value by the sale price of each sold property. Based on this construct, ratios at or near 1.0 indicate good accuracy. All of these ratios can first be plotted without GIS to determine the overall model accuracy, as shown in Exhibit 9.11.

Next, a spatial element can be added to the process, using GIS enhancement for further analysis, this time adding a geographic perspective to the model results. This outcome is demonstrated in Exhibit 9.12, plotting the most proximate sale ratios to the process.

This GIS representation (Exhibit 9.12) indicates the subject property (the star symbol) and the nearest comparable sale properties. By also including the sale ratios of these closest sale properties, the model

Exhibit 9.9 **Overview of the Analysis and Process**

Data Modeling Example

- Appraisal question: value unimproved agricultural land
- 8-county area
- Internal data only
- Subject
- County Parcel #0410021245
  240 acres (105 acres gravity irr., 130 potential irr.)
  Market Zone 362
  Unimproved Agricultural Land

Sales Data

- 8 counties in Nebraska
- Sales from 2001 to 2004
- Unimproved properties
- 190 total sales
- Land types: Pivot, Gravity, Potential, CRP, Upland Dry, Bottom Dry, Recreational, Road and Waste, Pasture, Meadow

Step 1: Look at Initial Data

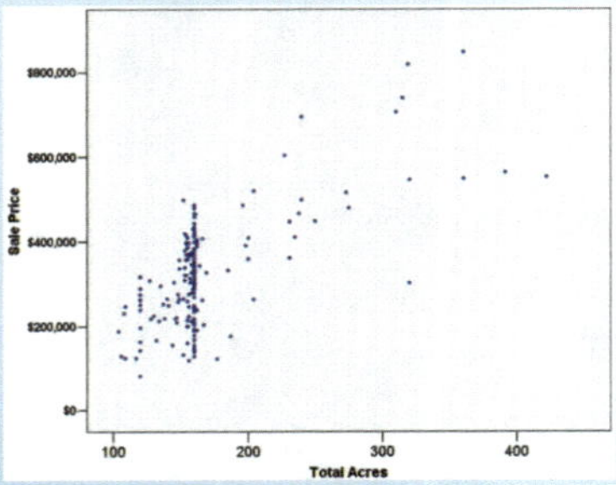

Step 2: Set Up Model

- Edit data (appraiser)
- Model specification–decide which variables might contribute to value (appraiser)
- Run regression model (statistician)
- Interpret output from model (appraiser)

Step 3: Property Elements Contained in Model

- Land area types: Pivot Irrigation, Gravity Irrigation, Potential Irrigation, Upland Dry, Bottom Dry, CRP, Pasture, Non-Ag., Recreational
- Sale year (2001 to 2003; 2004 is base)
- Area
  - County (Butler to York)
  - Market Zones (302, 364)
  - Market Zone and Counties overlap

Step 4: Property Elements Not Considered in Pilot Study Model

- Productivity units
- Soil type
- Rainfall
- Improved properties
- Sales not collected

Exhibit 9.10 Statistical Analysis

Model Summary[b]

Model	R	R Square	Adjusted R Square	Std. Error of the Estimate
1	.958[a]	.917	.904	39481.193

a. Predictors: (Constant), syear2001, Recreational, UplandDry13, GravityIrr21, IrrPotentialCrop53, NonAgDev, CenterPivotIrr33, Pasture, UplandDry11, york362, CLAY, GravityIrr23, CRP, seward306, GravityIrr, IrrPotentialCrop, syear2003, marketzone302, marketzone362, BottomDry, BUTLER, syear2002, UplandDry, CenterPivotIrr, BottomDry1516

b. Dependent Variable: S_PRICE

Coefficients[a]

Model		Unstandardized Coefficients		Standardized Coefficients	t	Sig.
		B	Std. Error	Beta		
1	(Constant)	45925.373	14915.385		3.079	.002
	CenterPivotIrr	2399.817	89.839	1.339	26.712	.000
	CenterPivotIrr33	-639.385	96.862	-.164	-6.601	.000
	GravityIrr	2076.424	86.433	.891	24.024	.000
	GravityIrr21	537.000	210.755	.070	2.548	.012
	GravityIrr23	-855.740	371.676	-.057	-2.302	.023
	IrrPotentialCrop	2106.743	110.598	.700	19.049	.000
	IrrPotentialCrop53	-604.720	211.265	-.070	-2.862	.005
	UplandDry	1183.684	101.718	.522	11.637	.000
	UplandDry11	608.140	187.089	.115	3.251	.001
	UplandDry13	-884.146	265.934	-.092	-3.325	.001
	BottomDry	921.557	538.971	.108	1.710	.089
	BottomDry1516	-231.528	678.119	-.025	-.341	.733
	CRP	749.558	247.197	.078	3.032	.003
	Pasture	289.640	150.686	.048	1.922	.056
	NonAgDev	1004.975	271.933	.089	3.696	.000
	Recreational	510.895	315.489	.040	1.519	.107
	BUTLER	49315.427	10465.769	.145	4.712	.000
	CLAY	-20689.6	11679.978	-.046	-1.771	.078
	marketzone302	-33240.3	19433.861	-.049	-1.710	.089
	seward306	17044.385	11162.053	.044	1.527	.129
	york362	45654.133	11581.526	.102	3.942	.000
	marketzone362	21590.960	8170.816	.082	2.642	.009
	syear2003	-24331.8	10349.696	-.086	-2.351	.020
	syear2002	-32739.7	10530.673	-.115	-3.109	.002
	syear2001	-42813.8	10309.884	-.156	-4.153	.000

a. Dependent Variable: S_PRICE

Residuals Statistics[a]

	Minimum	Maximum	Mean	Std. Deviation	N
Predicted Value	68664.50	809895.25	322953.16	122184.069	190
Residual	-100795	107224.8	.000	36777.422	190
Std. Predicted Value	-2.081	3.985	.000	1.000	190
Std. Residual	-2.553	2.716	.000	.932	190

a. Dependent Variable: S_PRICE

Exhibit 9.11 Sale Ratios

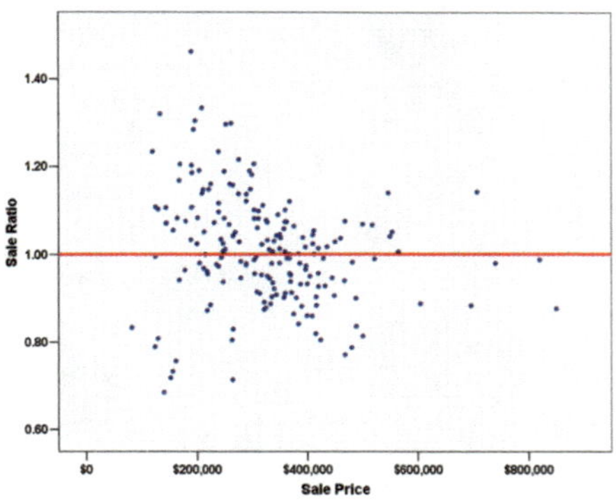

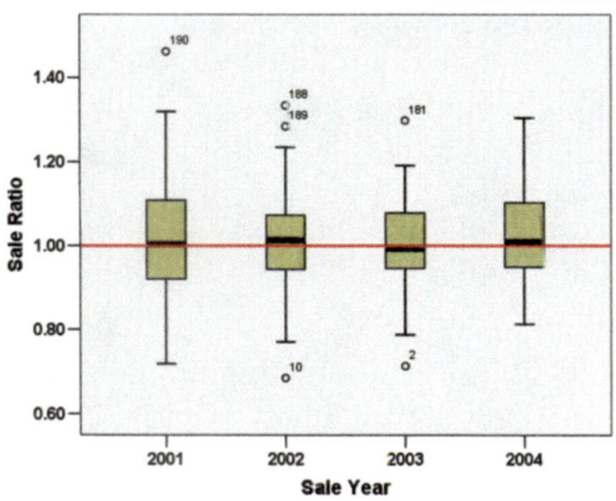

provides an indication of the local accuracy of the agricultural model. If all significant factors have been accounted for in model or GIS presentation, a user can apply an additional adjustment to the value estimate for the subject property based on these local sale ratios. For example, the model estimate for the subject prior to this GIS step was \$560,000. Using the median ratio from the above sales would indicate an adjustment of 3.5%, to an adjusted value of \$540,000.

Using GIS as part of agricultural valuation must make appraisal "sense" and be consistent with the appraisal process. There are a number of different opportunities for GIS to assist the appraiser in the valuation process. Spatial analysis can be used

- At the start of the valuation process, when defining modeling areas
- During the modeling process, when creating location-based variables in the model
- Finally, as a refinement step (Exhibit 9.13) in which the spatial element helps to identify the most relevant and appropriate sales for analysis

The next step in running an additive regression model using this sales data involves identifying the significant valuation elements that could potentially affect the market value of the sales described earlier. Elements included the land type described in Step 3 of Exhibit 9.9, the year of sale, and several location variables, including the sale

Exhibit 9.12 Plotting Sale Ratios

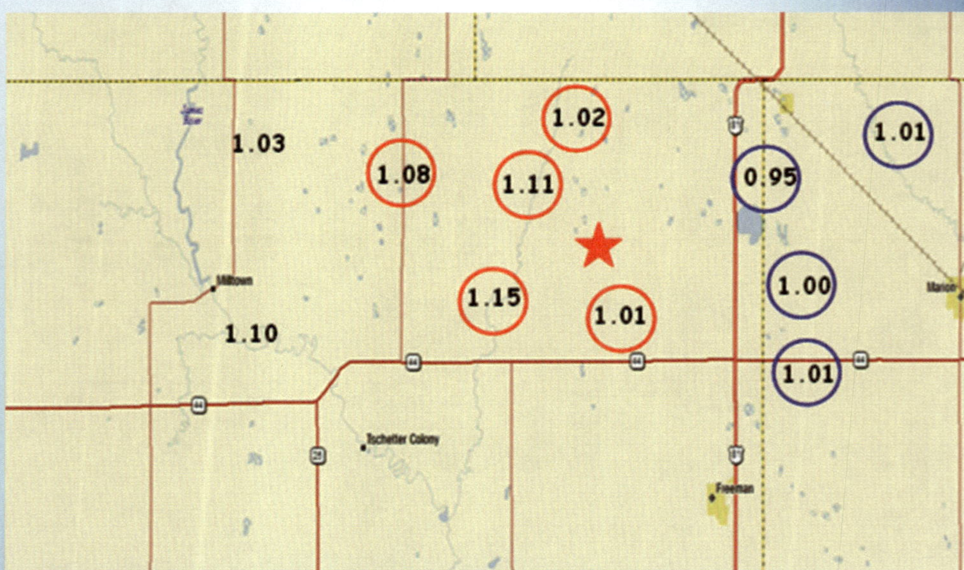

county and appraiser-derived submarkets reflecting common economic conditions in a market zone. The model generated the adjustments shown in Exhibit 9.14.

Exhibits 9.15 to 9.21 illustrate the model outcome and evaluation.

Exhibit 9.13 Modeling Sale Ratios

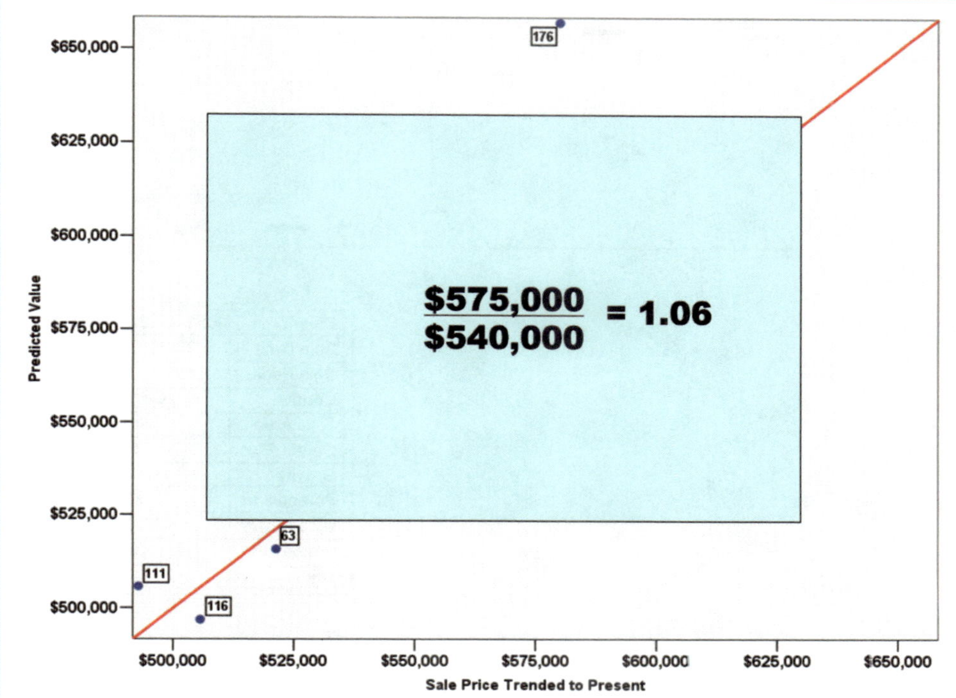

Exhibit 9.14 Model-Generated Adjustments

	Adjustment		Adjustment
(Constant)	45925	CRP	750
CenterPivotIrr	2400	Pasture	290
CenterPivotIrr33	-639	NonAgDev	1005
GravityIrr	2076	Recreational	511
GravityI4423	-856	BUTLER	49315
GravityIrr21	537	CLAY	-20690
IrrPotentialCrop	2107	marketzone302	-33240
IrrPotentialCrop23	-605	seward306	17044
UplandDry	1184	york362	45654
UplandDry11	608	marketzone362	21591
UplandDry13	-884	syear2001	-42814
BottomDry	922	syear2002	-32740
BottomDry1516	-232	syear2003	-24332

Exhibit 9.15 First Set of Comparable Sales: The Regression Model Itself

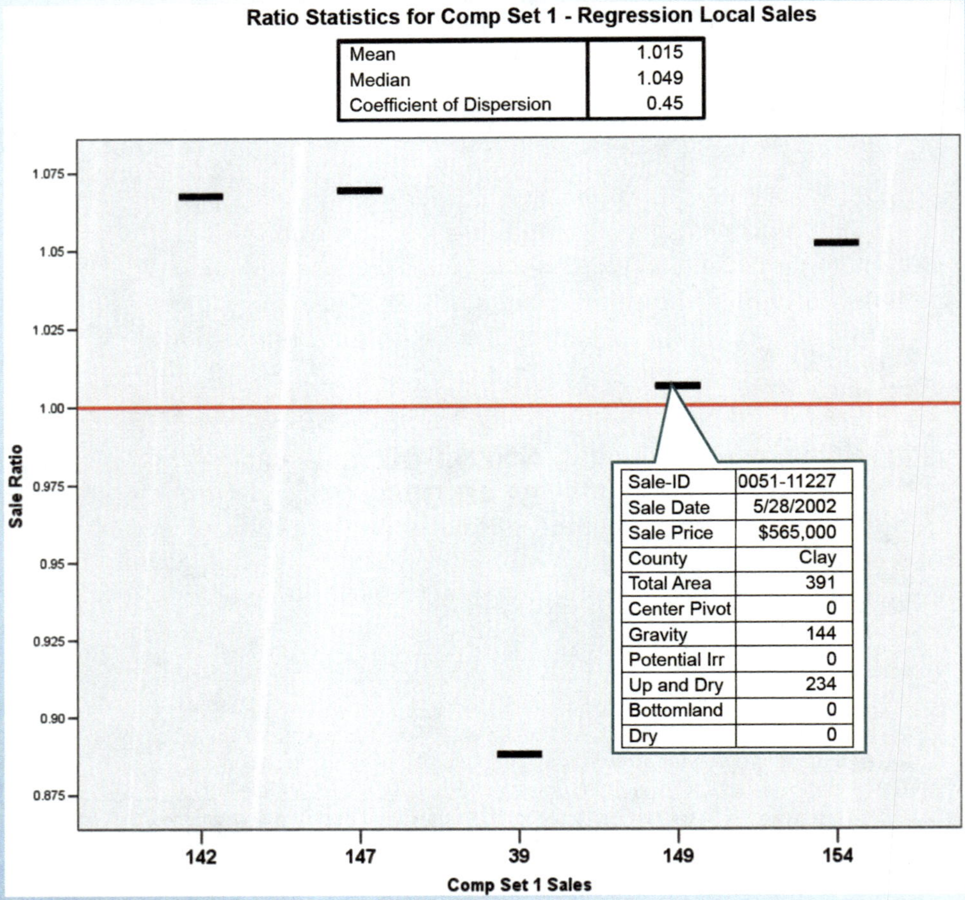

Exhibit 9.16 **Second Set of Comparable Sales: Total Land Area**

Ratio Statistics for Comp Set 2 - Land Area

Mean	1.019
Median	1.066
Coefficient of Dispersion	.047

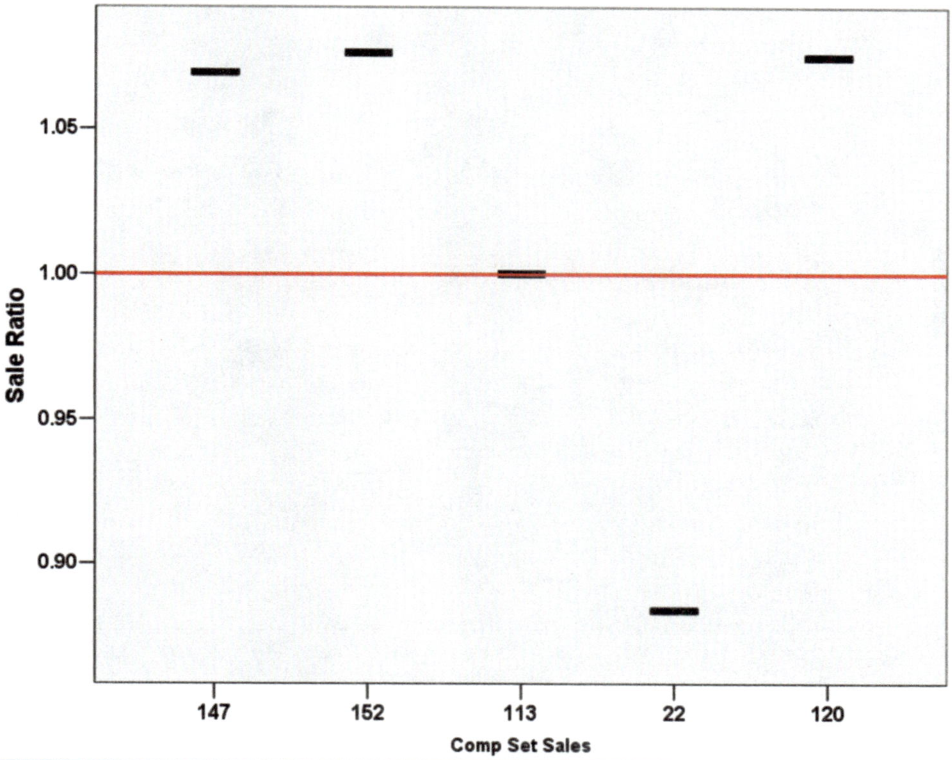

Exhibit 9.17 Third Set of Comparable Sales: Most Recent Sales

Ratio Statistics for Comp Set 3 - Most Recent Sales

Mean	1.023
Median	.995
Coefficient of Dispersion	.088

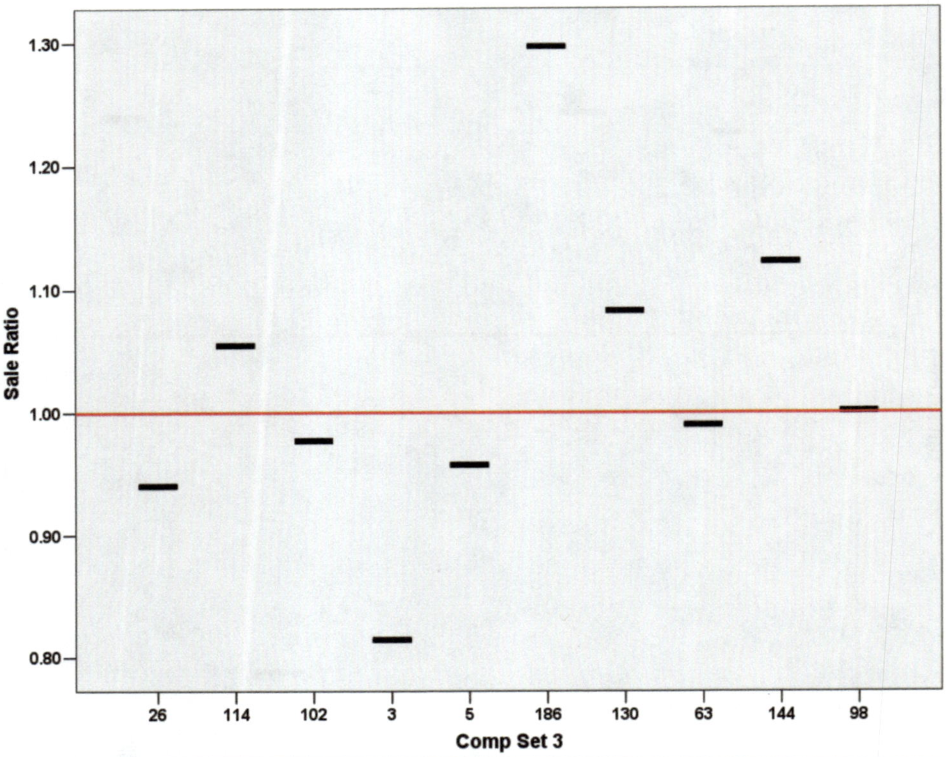

Exhibit 9.18 **Fourth Set of Comparable Sales: Land Type Mix**

Ratio Statistics for Comp Set 4 - Land Type

Mean	1.054
Median	1.036
Coefficient of Dispersion	.096

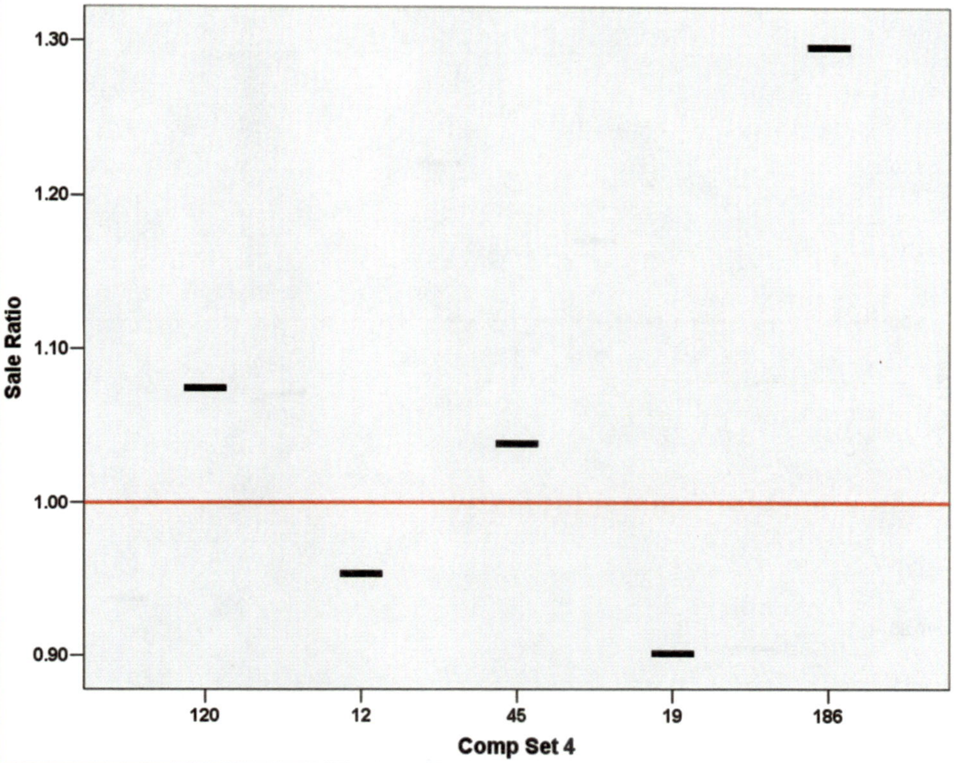

Exhibit 9.19 Fifth Set of Comparable Sales: Closest Sales

Ratio Statistics for Comp Set 5 - Closest Sales

Mean	.946
Median	.958
Coefficient of Dispersion	.105

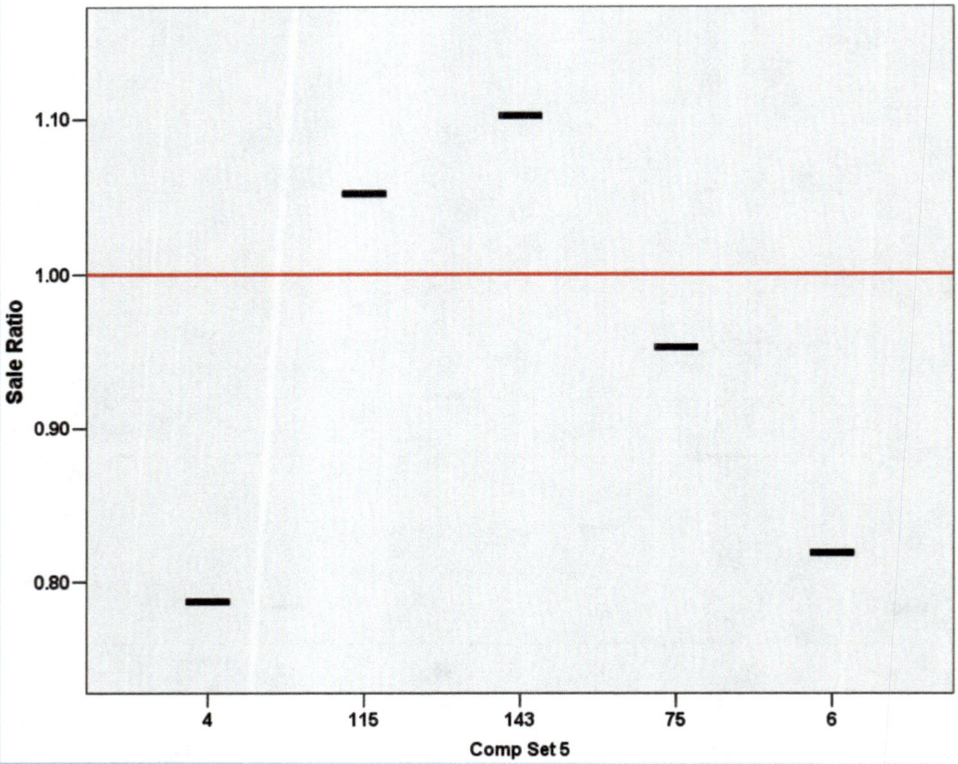

Exhibit 9.20 Sixth Set of Comparable Sales: Productivity Units

Ratio Statistics for Comp Set 6 - Productivity Units

Mean	.984
Median	1.011
Coefficient of Dispersion	.050

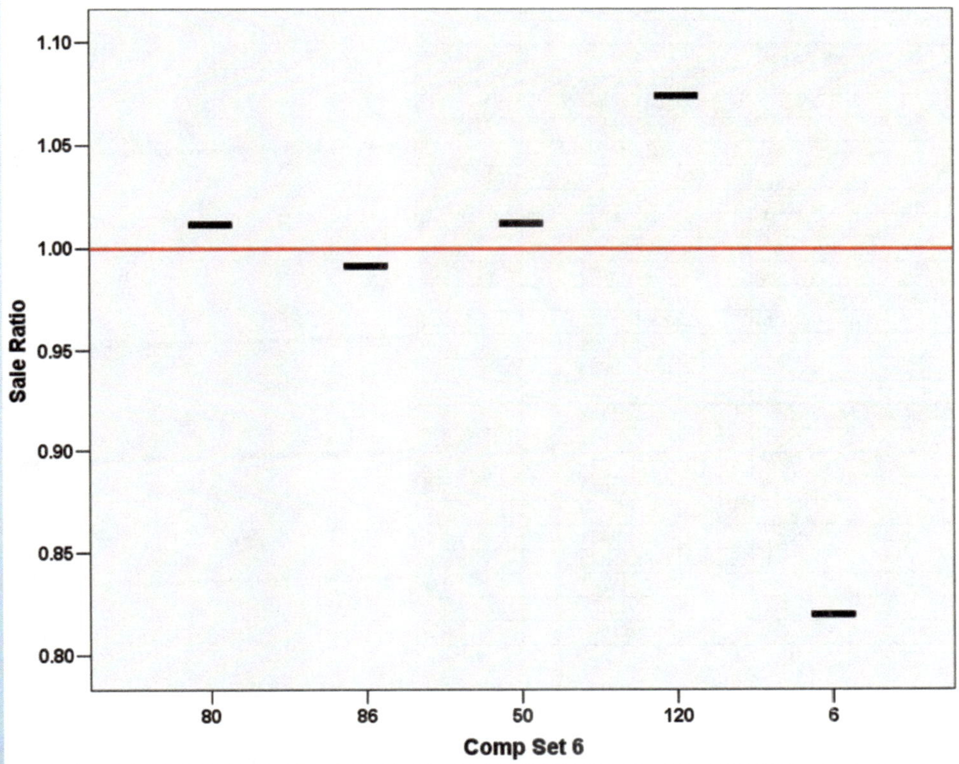

Exhibit 9.21 Conclusions

Comps Set	Sales	Sale ID	Sale Ratio	Med Ratio
1 - Regression	1	0051-11398	1.067	
	2	0051-11465	1.069	
	3	0001-10745	0.888	1.052
	4	0051-11277	1.006	
	5	0051-11382	1.052	
2 - Land Area	1	0051-11465	1.069	
	2	0051-11151	1.076	
	3	1171-12486	1.000	1.069
	4	0051-11265	0.884	
	5	0051-11237	1.075	
3 - Most Recent Sales	1	0051-11558	0.940	
	2	0019-10275	1.055	
	3	0051-11561	0.977	
	4	0051-11523	0.814	
	5	0019-10262	0.957	0.996
	6	0001-10726	1.297	
	7	0051-11509	1.083	
	8	0051-11531	0.990	
	9	0051-11529	1.122	
	10	0051-11575	1.001	
4 - Land Type Mix	1	0051-11237	1.075	
	2	0051-11147	0.954	
	3	0051-11381	1.039	1.039
	4	0001-10553	0.902	
	5	0001-10726	1.297	
5 - Closest Proximity	1	0051-11434	0.787	
	2	0051-11207	1.052	
	3	0051-11369	1.102	0.953
	4	0051-11053	0.953	
	5	0051-11100	0.820	
6 - Productivity Units	1	0051-11421	1.011	
	2	1171-12642	0.992	
	3	0051-11247	1.012	1.011
	4	0051-11237	1.075	
	5	0051-11100	0.820	

General Conclusions of Pilot Study

The FCSA pilot study provided an indication that valuation technology is at a crossroads between numeral-based analysis and a more visual-based solution, even beyond what traditional GIS offers. While this stretches beyond the scope of this chapter, it is clear that the appraisal profession is on a path that will move from older methods of property valuation to new technology. The addition of GIS to a valuation modeling process can clearly enhance the valuation process. Integrating these techniques will bring great promise and challenges for valuation practitioners.

Based on the system design parameters and the results of the initial pilot study, sufficient evidence on the feasibility of the system and its usefulness at the enterprise level for management decisions were determined to make the case for funding. Coefficients were reasonable and had the expected signs. After a 12-month period of analysis and discussion, FCSA made the decision to move forward and implement an AVM/GIS solution to empower its appraisers, reduce costs, realign internal appraisal team structure, and focus on increasing business capacity without increasing staff.

The next chapter details the actual implementation of the system and how its results were evaluated and realized.

10

Innovating the Approaches to Value

Incorporating Valuation Modeling Techniques into Traditional Appraisal Practice

Mark R. Linné, MAI, SRA, James Gresham, and James Russell Thimgan

One of the difficulties with innovation is dealing with the cost of changing the status quo. The case needs to be made at the outset that best practices alone are not a sufficient motivator to implement change, and that a true cost/benefit case needs to be made to justify the expenditures involved in implementing pervasive change.

Several key elements came together for the Farm Credit Services of America (FCSA) project described in the previous chapter that moved the process from concept to implementation. The first was the reality of an enlightened management that recognized that true innovation includes both enhanced business processes and the opportunity to extend best practices/best-in-class motivation as an added benefit. The availability of GIS specialists and valuation modeling experts with a strong background in both assessment and private sector appraisal helped to provide the empirical support and a "real-world" functionality to the process. Both the GIS and modeling experts had a background in mass appraisal at the assessment level, and understood that there were techniques that could be transferred from the assessment side to solve private sector challenges. Finally, the costs of data and implementation were constrained within the project to a point at which cost recovery/return on investment dynamics made practical sense. Primarily this was achieved through the use of readily available data–especially GIS–and, secondarily, the elimination of application functionality that ultimately proved unnecessary or redundant.

Enhancing the entire process was a truly interactive and supportive relationship between FCSA and the development team, which ended a four-year development and implementation process with an application that could do far more than had been initially conceptualized and ultimately achieved enterprise-wide benefits.

Reimagining the Approaches to Value

It is important to note that FCSA built its system by incorporating the best of traditional appraisal expertise and advanced valuation modeling techniques. The organization brought in experts in data modeling, integrated a cross-functional team of appraisers, GIS experts, and IT staff, and in the end developed a system that had buy-in from staff appraisers and senior corporate leadership. Rather than replacing appraisers with the advanced functionality and cost and time savings that the system provides, FCSA reinvested their savings into more effectively gathering and analyzing the data that powers perhaps the most advanced valuation system in the world.

As FCSA brought advanced spatial analytics into the valuation process, valuation professionals within the organization began to evaluate their best practices. In some cases, they saw a validation of existing processes. In others, they gained an understanding of the valuation elements that needed to be changed to harmonize with the realities of the market.

Practitioners can often succeed where others outside the field, or lacking the requisite skills, cannot. In the case of Farm Credit, some prominent academics at a regional state university flatly stated that FCSA's goal of building agricultural property models was impossible. Appraisers built the system and, in so doing, have now brought advanced valuation analytics into the agricultural marketplace. The moral of this story is that practitioners who work in the trenches and have to deliver solutions can drive innovation and effectively overcome obstacles.

An important inference can be drawn from the process of developing the FCSA valuation system. Rather than attempting to find new approaches to valuation, it simply makes more sense to reinvent and reinterpret the traditional approaches to value.

One example that can be made is the relevance and importance of the cost approach, which has declined in usage by the appraisal community in the last decade, in favor of the sales comparison and income capitalization approaches. The biggest contributor to the decline of the cost approach has been the diminished reliance by lending institutions and others involved in the financial marketplace. The cost approach has relevance in today's market and it provides an objective measure of both overvaluation risk and uniformity that has applicability in a number of situations, including portfolio analysis and market analysis. The cost approach can also be used as a relevant metric to gauge the status of markets. One strength of the approach is its ability to act as a red flag when value indications from the cost approach are significantly different than value indications from the other approaches to value, which is often seen in suddenly declining residential markets, for example.

In the previously mentioned FCSA example, the company's advanced valuation system incorporated sales, income, and cost data to develop the final value estimate. It is clear that the cost approach, by drawing data derived from the market, provided a critical validation to the true market value of properties that this lender will consider in its collateral risk analysis. By this same logic and an examination of the other approaches to value, the incorporation of advanced analytics was clearly a direction with promise. In the development of AgriStat, pushing advanced analytics into the income capitalization approach, for example, allowed the predictive power of the market to more precisely estimate the income-generating capabilities of agricultural property.

Looking at the Data

Land and Improvements

One of the first steps in the development of AgriStat was to determine what type of property would be valued. For this project only agricultural property was targeted. The criterion for selection was based on the primary highest and best use as assigned by the appraiser. Primary land uses included

- Cropland–dry
- Cropland–bottom
- Cropland–gravity-irrigated
- Cropland–sprinkler-irrigated
- Cropland–meadow/hay
- Pasture–all types
- Conservation Reserve Program (CRP) land

Once the basic land types were identified, all additional land types and improvements had to be accounted for. These items included such additional land types as

- Recreational land
- Timber land
- Roads and waste
- Land in transition
- Nonagricultural land

Improvement types included

- Dwellings
- Garages, shops, machine sheds, and general storage
- Grain storage and grain complexes
- Barns, arenas, and corrals
- Feedlots
- Hay/livestock sheds
- Sprinkler systems (e.g., center pivots)
- Fencing (quality only)

Agricultural Property Data versus Residential Property Data

Residential AVMs tend to rely heavily, if not exclusively, on public (county assessor) data. The parcel ID (counties may also use the term *schedule number* or *account number*) tends to be a primary property identifier. Attributes measured by the county serve as key inputs to the model. Occasionally, data from a local Realtors multiple listing service (MLS) may supplement county records, but that can entail additional expense and licensing issues. A well-funded modeling effort may include GIS, but obtaining the parcel layer and other files from the county is often prohibitively expensive.

The agricultural world, the development team learned very early, is different. In the residential valuation world, county assessor data reigns supreme, but the assessor takes a back seat in the agricultural world for several reasons:

- **Collateral is treated differently.** A residential owner may have two or more loans on a single property, but the loans are typically based on exactly the same collateral and assume positions of priority. Agricultural loans are commonly based on separate collateral components in a single parcel. For example, a farm owner may take out a loan on 40 acres of cropland that is part of a 160-acre tract, with the remaining 120 acres unencumbered.

- **Land is not just land.** When valuing 120 acres of cropland, knowing only that a parcel is located in Hamilton County, Nebraska, is not sufficient. Nor is it sufficient to know that the tract lies in Township 9N, Range 8W, and Section 35. The appraiser needs to know specifically which acres in Section 35 are being valued. Land productivity can vary significantly across a section (640 acres) of land, significantly affecting income potential. Not surprisingly then, agricultural appraising is beginning to rely heavily on GIS.
- **The availability of U.S. Government data.** Two important sets of data are provided by the USDA: 1) Soil Survey Geographic DataBase (SSURGO) soil quality maps from the Natural Resources Conservation Service (NRCS) and 2) aerial photography from the Farm Service Agency (FSA) that provides field boundaries. The former supports accurate estimates of land productivity, while the latter shows the acreage planted. Where center-pivot irrigation is used, the aerial maps can be used to segregate dry corners from actual irrigated land to increase accuracy in estimating productive potential.

Use of aerial imagery from the FSA is another reason that reliance on county data is diminished in model building for agricultural property valuation. The FSA imagery shows individual field boundaries–commonly known as a *CLU* for *common land unit.* CLUs show land use rather than ownership. It is very possible that more than one CLU exists within a given parcel. A typical example occurs when a tract is split between planted acreage and grazing land. As far as the county is concerned, both cropland and pasture could fall within the same parcel. The FSA imagery, however, will show two separate CLUs, or fields. The reverse is also quite possible. That is, the farmer owns or leases two separate but adjacent parcels planted as a single field. Aerial imagery would show the single CLU.

All of the data used in the AgriStat models was researched and collected by FCSA appraisers and sales specialists. Their research helped create an unusually rich and robust sales database, going far beyond the information contained in public records or even local MLS databases. Data collection covered a number of aspects pertaining to the sale as well as attributes of the tracts. Elements that were found to be useful in the modeling effort included

- **Buyer type,** such as 1031 exchange, local investor, tenant, and out-of-area investor. Only a few buyer types were used in the model.
- **Sale method,** such as real estate broker, public auction, private sale, and sealed-bid auction. The first three occur with about the same

frequency (approximately 30% of the time). Public auction was the most frequent by a tiny margin and was used for the base case.

- **Farm ease.** This element is essentially an estimate of the tract physical layout and topography, and how suitable it is for farming. Measured as above average, average, or below average.
- **Land quality.** Measured as above average, average, or below average.
- **Location.** Measured as above average, average, or below average.
- **Water quality.** Measured as above average, average, or below average.
- **Water quantity.** Measured as above average, average, or below average.
- **Road type.** This element describes the road surface, e.g., paved, gravel, or dirt. Only dirt roads had an impact.
- **Fence quality.** This element related more to grazing land, and only had an impact where the quality was below average (e.g., breaks in the fence lead to wandering livestock).
- **Hazards.** This element was used to state the presence of a particular hazard. These could include rocky soils, sandy soils, high water table, erosion, and other hazards.
- **Improvement inventory.** This element covers improvement inventories of all types, from corrals to grain bins to grain complexes. Each improvement had a number of descriptive elements, including condition, utility, remaining economic life, effective age, replacement cost, contributory value, and depreciation.

County data was also collected by FCSA staff and referenced for improvement information, especially in cases where a house existed. Year of construction, living area, baths, and other improvement data would then be used from the county. Even though counties maintain data on soil quality and land productivity, those estimates are not captured by the application. Indeed, an early attempt at modeling farms based on the county land productivity calculations showed them to be less reliable than productivities calculated by FCSA appraisers. The development team attributed this mainly to the particular care and attention that each FCSA sales specialist gave to each sale collected. A secondary reason is the fact that counties tend to provide an average rating covering the entire parcel, including acreage taken up by roads, waste, high productivity crop and land with marginal pasture.

In terms of the number of agricultural properties in the system, the AgriStat project may not seem impressive by current modeling standards based on size alone. The assessment division of any large urban county in the United States annually models a greater number of residential properties than there are farms within all four of the FCSA states. According to USDA count, there were almost 140,000 farms of at least 50 acres in the four-state territory in 2007. The FCSA appraisal database has five years of sales data for just over 19,000, or about 13.6%, of the total farms and ranches (Exhibit 10.1). These sales, in turn, support about 27,000 valuations (again, over a five-year period).

What is unique within the system, despite the smaller aggregate number of properties, is the fact that modeling and GIS had never been integrated, implemented, and modeled on this scale anywhere in the world.

Implementing Innovation

The fundamental goal of the FCSA project was to create models for each of the basic appraisal approaches (sales comparison, cost, and income capitalization). These models would mimic appraisal theory and approaches to allow appraisers to identify value and adjustments for the whole or for individual characteristics. The first two models concentrated on the sales comparison approach and were based on market sales and the data collected by FCSA staff. The first market model was a complete model that looked at both land and improvement influences and their interaction with market sales. The second model was an abstracted land sale model where improved values (from cost) were subtracted from market sales and then modeled for determination of the land portion of value. The second set of models was based on the income capitalization approach to value. Each aspect of the income capitalization approach was analyzed and models were created for each part:

- Potential gross income (*PGI*)
  - Cropland
  - Pasture land
  - Improvements
- Expense ratios
- Overall capitalization rates

Exhibit 10.1 Percentage of Total Farms Modeled

State	Total Farms	FCSA Sales	Percentage	Valuations	Percentage
IA	66,323	7,953	11.99%	11,084	16.71%
NE	38,861	5,825	14.99%	8,671	22.31%
SD	26,351	4,735	17.97%	6,791	25.77%
WY	8,413	553	6.57%	558	6.63%
Total	139,948	19,066	13.62%	27,104	19.37%

The final set of models was based on developing each of the basic items of the cost approach (replacement cost, obsolescence, and depreciation). FCSA staff gathered cost data and created tables for the replacement cost. Obsolescence and depreciation were determined through market sales using curve fit analysis within the regression analysis process.

Application of Regression Analysis

Because of the diversity of agricultural property, multiplicative regression models were created for all models except the expense model within the income analysis. This model type consistently produced better results than additive regression analysis. This model format further assumes that adjustments to value are multiplicative in nature rather than additive. The typical statistics used in additive modeling require all variables be converted to natural logarithmic numbers, which enables the modeler to use the standard additive model structure to calibrate the model. By taking logs of the variables, exponents become multipliers and multipliers become additive. After the calibration has taken place, the model can then be converted back into its multiplicative format. In an effort to further assist in understanding the model, it can be further converted into a starting base rate with appropriate multipliers for the independent variables, including size, location, and other factors.

Specification and Calibration of the Models

Multiple linear regression analysis was the method used to calibrate all but one model. During the modeling process, the overall accuracy of each model can be tested along with the accuracy of coefficients for individual property components. In all models each variable used can be analyzed to determine whether it is significant and whether the coefficient is logical. Insignificant variables or illogical coefficients can either be dropped or be constrained and modified so that they do not have an adverse influence on the value. Constrained variable adjustments should be based on comparisons to similar properties or relationships. The inherent goal of the model is to replicate actual sales (or income) data in order to accurately predict the value or component of value.

One of the characteristics of a multiple regression analysis is that all relationships are linear. However, in most cases, value influences do not follow a linear pattern. A certain amount of diminishing returns or discounting takes place. As an example, as more land is purchased, the price per acre tends to decrease. Because of this nonlinear pattern, data must be converted into mathematical relationships that fit into the regression model as a linear pattern. One of the easiest ways to accomplish this conversion is to convert the data into a log linear model structure. (The logarithmic conversion process was discussed earlier in the chapter.)

This conversion turns the specification of the model from an additive linear model approach into a multiplicative model. (The advantages of a multiplicative model were discussed earlier in the chapter.) One benefit is that the exact relationship for discounting can be identified, which is more difficult to calibrate using an additive model. A trial and error approach is the only way to identify the pattern. Secondly, through the use of logarithms, outliers have a smaller impact on the analysis. Also, for diverse property types, this type of model consistently produces better results than additive regression analysis. With the exception of the expense models, all models were calibrated using the log linear model structure to allow for a multiplicative model. The expense model was based on the percentage of expenses as an additive model.

The Sales Comparison Approach

In the development of the AgriStat system, the sales comparison approach to value was a complete model that looked at both land and improvement influences and their interaction with market sales. Unlike the income models, the sale model included stratification of additional land types and looked at the relationships among characteristics in a single model. The overall market model was created using 9,722 sales (overall model) covering the time period January 2006 through April 2008.

The market model performed well in terms of level and equity. The target median ratio is 1.00. The coefficient of dispersion was used to indicate equity in setting values. The COD for the market model was 14.87. It indicates, in this case, that predicted market values, on average, will fall within 14.87% of the median. The land-only market model also performed very well in terms of level and equity. Exhibit 10.2 shows the overall statistics for each model.

Exhibit 10.2 **Comparison of Overall Market Model and Land-Only Market Model**

	N	Mean	Median	WTD Mean	PRD	COD
Overall market	9,722	1.02	1.00	1.00	1.02	14.87
Land-only market	8,386	1.02	1.00	0.99	1.03	14.24

The Income Capitalization Approach

The income capitalization approach component of AgriStat integrated several models into one final outcome. The start of the process was to analyze the income stream to predict potential gross income for the property. The second step was to analyze expenses to the property. These expenses were levied against the income to produce a net income to the property. Once the net income was determined, market sales were used in conjunction with the net income to predict direct capitalization rates.

The income stream models were segregated into three analyses (cropland, pasture land, and improvements). The cropland model used 13,987 income streams covering the time period 2005 through 2007. The pasture land model used 3,813 income streams also covering the time period of 2005 through 2007. The improvement model used 1,305 income streams covering the time period of January 2005 through March 2008.

Only one expense model was used for the project. It encompassed 10,338 properties covering the time period 2005 through 2007. One capitalization model was also used for the project. That model encompassed 13,846 properties covering the time period 2005 through 2007.

Exhibit 10.3 Comparison of Income Models

	N	Mean	Median	WTD Mean	PRD	COD
Cropland PGI	13,987	1.01	0.99	1.00	1.01	10.44
Pasture land PGI	3,813	1.02	0.99	1.04	0.98	15.51
Improvement PGI	1,305	1.05	1.01	0.98	1.07	25.29
Expense ratio	10,338	1.00	1.00	0.95	1.06	16.04
CAP rate	13,846	1.02	0.99	0.98	1.04	15.81
Overall income approach	13,411	0.95	0.94	0.95	1.01	15.87

In most cases, the income models also performed very well in terms of level and equity. Exhibit 10.3. shows the overall statistics for each model.

Statistical Review

The income models performed very well in terms of level and equity. The International Association of Assessing Officers (IAAO) standards and guidelines for mass valuation methodologies, including the Standard on Ratio Studies and the Standard on Mass Appraisal of Real Property, established the standards to determine model quality. The target median ratio for the level of valuation was 1.0. The coefficient of dispersion was used to indicate equity in setting the final values. For example, the COD for cropland is 10.4. The COD output indicates that income streams, on average, fell within 10.4% of the median for the data set. For all models, the level (median) and uniformity (COD) statistics fell well within the standards set by IAAO, which indicate an ideal range for COD of 0.95 to 1.05. The one exception was the improvement model, which is within the standards set by IAAO except for the COD (25.3) falling above the standard of 20.

Trend Analysis

Because the data collected covers a continuum of time, trend analysis is needed to more accurately identify values. Several methods can be used to analyze trends. For this project, the income streams, expenses, capitalization rates, and sales data were analyzed to determine if any trends should be applied. All trends were created as variables within the regression model. The advantage of this approach is that it isolates the effect of time within the model without being biased by other influences. In both land income models, the time period studied is January 2005 through December 2007. The improvement model includes income from January 2005 through March 2008.

Cost Approach

The application of the cost approach in AgriStat involves a market-derived land model, cost surveys of local building costs, and depreciation and obsolescence factors. Three separate analyses were performed to calibrate the cost approach. The land model was listed above in the market approach (i.e., the land-only model). The improved cost values were developed by FCSA staff, who gathered cost data and created tables for the replacement cost. Obsolescence and depreciation were determined through market sales using curve fit analysis.

Curve fit analysis was the technique used to determine the best curve to fit to the depreciation. In this approach all major curves were analyzed to determine the best fit to the data. The curve was then extracted in order to build the depreciation tables. The table and graphs in Exhibits 10.4, 10.5, and 10.6 show an example of the curve fit model and the resulting table creation. This curve fit analysis determines the best fit for center-pivot sprinklers.

Compliance Testing

For each of the models built, comprehensive analysis was performed to determine the degree of accuracy in terms of both the overall model and the characteristics of the property (whether used or not in the

Exhibit 10.4 Model Summary and Parameter Estimates

Dependent Variable: ACCRUED_DEP_PER

	Model Summary					Parameter Estimates			
Equation	**R Square**	**F**	**df1**	**df2**	**Sig.**	**Constant**	**b1**	**b2**	**b3**
Linear	.724	392.512	1	150	.000	.216	.020		
Logarithmic	.707	361.791	1	150	.000	-.012	.215		
Inverse	.422	109.649	1	150	.000	.620	-.800		
Quadratic	.758	232.795	2	149	.000	.125	.036	-.001	
Cubic	.763	157.577	3	148	.000	.165	.024	.000	.000
Compound	.578	205.527	1	150	.000	.226	1.050		
Power	.723	390.932	1	150	.000	.110	.584		
S	.639	265.378	1	150	.000	-.429	-2.643		
Growth	.578	205.527	1	150	.000	-1.489	.049		
Exponential	.578	205.527	1	150	.000	.226	.049		

The independent variable is EFFAGE2.

Exhibit 10.5 **Accrued Depreciation (percentage)**

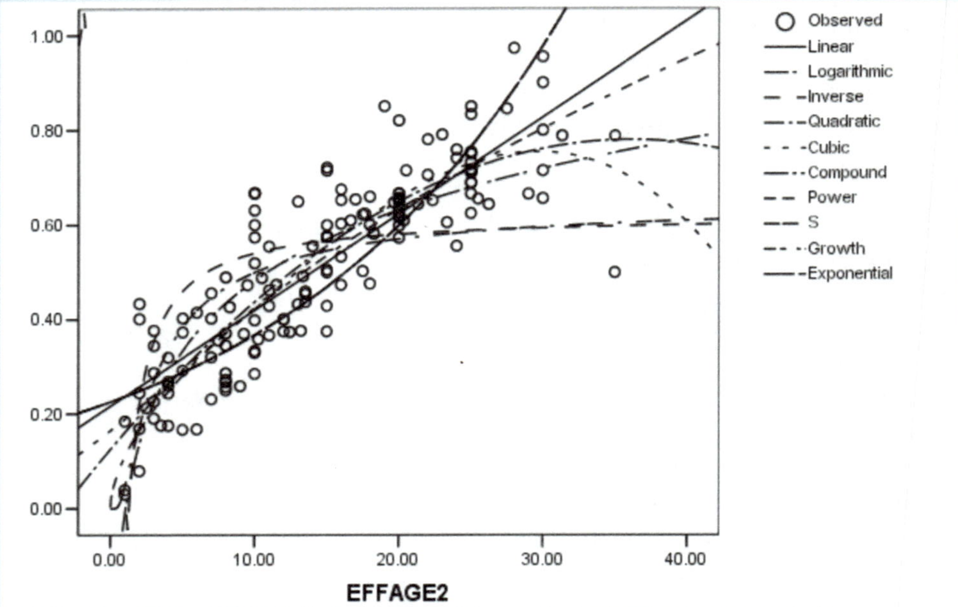

Exhibit 10.6 **Depreciation of Center-Pivot Sprinkler Properties**

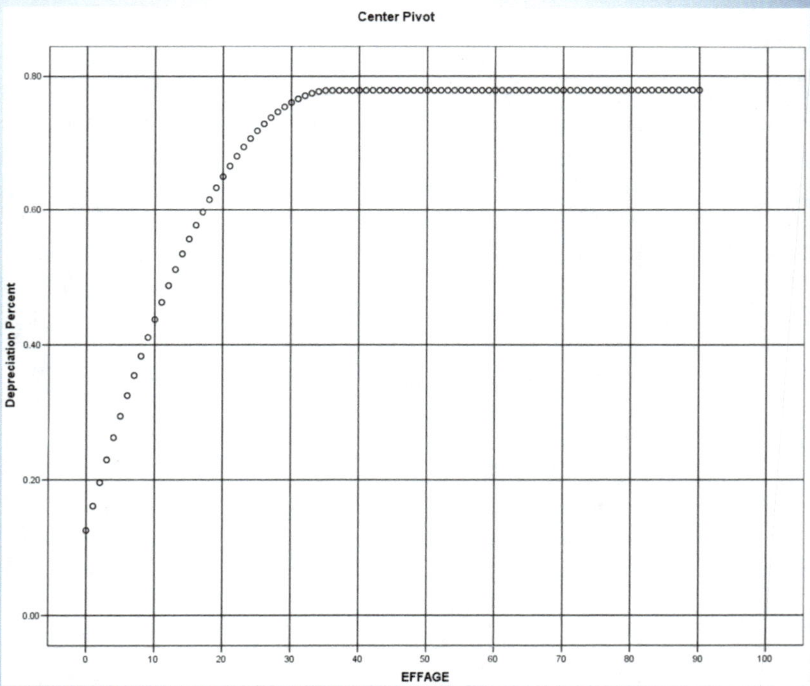

Exhibit 10.7 Cropland *PGI* Model Equity Analysis

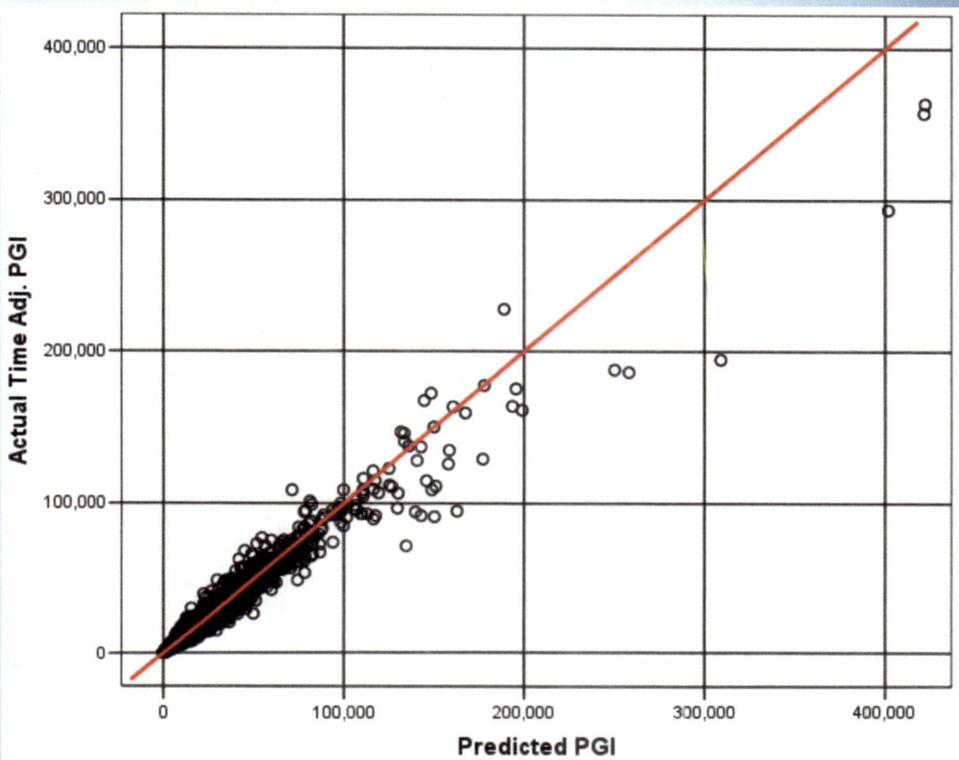

model). An analysis of the ratio of the predicted value to its actual sale, income, expense, or capitalization rate was developed while building each model. These ratio studies provide comprehensive analysis to help indicate both strengths and weaknesses of the model. Exhibit 10.7. shows an example of the performance of the cropland PGI model.

In addition to ratio studies, GIS analysis of the results provided better feedback to the modeling process for location factors. Exhibit 10.8 shows a small area portion of analysis of the market model relative to location (from southeast South Dakota). The analysis looks at the ratios of the predicted values relative to actual sales to determine if there is any location bias.

Developing Meaningful Metrics and Valuation Process Enhancements

Automated Comparable Sales Selection

One of the project objectives was to assist FCSA's staff appraisers in their work by providing an automated comparable sales selection

Exhibit 10.8 **Market Analysis**

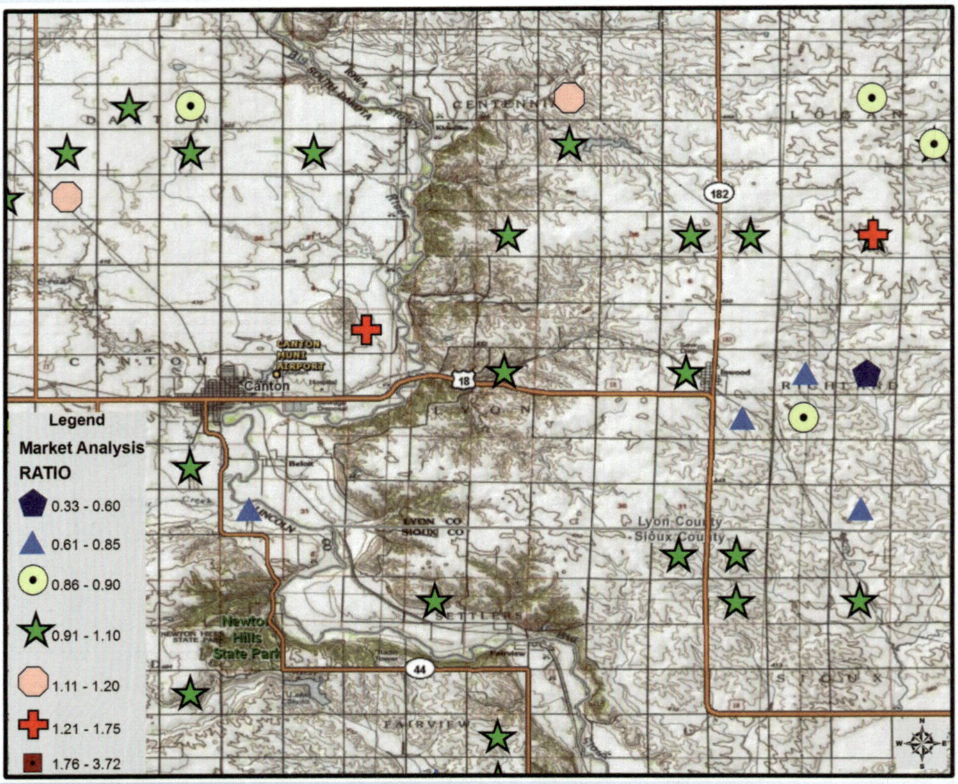

tool. The sales selection tool could be used whether the appraisal took advantage of AgriStat or relied on more traditional methods. This feature was designed to present the top 25 comparable sales in a suggested rank ordering, from a preliminary set of up to several hundred. The appraiser, however, was free to replace one or more higher-ranked comparables with a lower-ranked comparable, or even a previously unranked comparable with justification.

Automated comparable selection is a common feature of many residential real estate applications, even where no valuation modeling exists. A simple system may select comparable sales based on date of sale (most recent is better), proximity to subject, and size of living area. More sophisticated systems will avoid selecting a condominium sale as a comparable for the sale of a detached home, and vice versa. More sophisticated systems may also consider age differential, residential style, and whether the comparable sale is in the same subdivision as the subject, among other possibilities. In general, a similar selection process can work for agricultural land as well, and the AgriStat development team started from a similar point. In its first incarnation, the comparable sale selection process of AgriStat was based on five attributes:

1. *Time*, with the most recent sale ranked highest.
2. *Distance from subject*, with the nearest sale ranking highest.
3. *Similarity in acreage*, using the absolute value of the difference, so, for example, a comparable sale with five acres more than the subject ranked exactly the same on this scale as a comparable sale with five fewer acres.
4. *Similarity in land productivity*, as in acreage, using absolute value of the difference.
5. *Similarity in tillable percentage*, again using absolute value again.

The comparable sales were required to have the same highest and best use as the subject. The ValueSource system includes

- Agriculture (farms and ranches)
- Recreational
- Residential
- Agricultural facilities (commercial)

Agriculture–by far the predominant type–was the only highest and best use modeled.

Distance was capped at 25 miles, although this variable was changeable at the administrator level. *Land productivity* had a different measure in each state:

- Iowa uses Corn Suitability Rating (CSR).
- Nebraska uses Soils Rating for Plant Growth (SRPG).
- South Dakota uses Soils Productivity Index (SPI).
- Wyoming has no measure and therefore had only four comparable attributes.

The comparable sales selection process did not allow selection across state lines, so the difference in *land productivity* measures among states did not matter.

For each attribute, the highest-rated comparable sales received a score of 100, the next received 99, and so on. If two comparable sales tied for a position, the scores were averaged. So if two comparable sales tied for best score, each was assigned 99.5. Therefore, a comparable sale that ranked first by each attribute could have a theoretical maximum of 500 (without weighting). Because the attributes were unlikely to be of equal importance, the system developers also allowed the attributes to be weighted (at the administrative level) on a scale from 1 to 5. This method is simple, yet provides great flexibility. For example, if *proximity* is determined to be much more important than the other attribute measures, it could be given a weight of 3, holding the others at one, allowing proximity to carry three times the impact of any other attribute . However, if proximity were determined to be only slightly more important, with the others about equal, proximity could be assigned a weight of 5 and the rest a weight of 4, giving the proximity attribute a weight of 25% more than the other attributes.

During the modeling process and especially during the preliminary data analysis stage, several instances were uncovered in which the ValueSource data inputs needed improved standardization. Determination of *land type* was such a case. When sales data was entered, land type selection was menu-driven, so only predefined types could be entered. However, when income data was entered, free-form input of *land type* was allowed. As is typical with free-form input, the possibilities grew exponentially, and about 22 possible land types suddenly became over 2,000, with well over 100 variations of "dry crop" alone.

When developing AgriStat support tools and methods, the development team had an opportunity to rethink the manner in which data would be entered in to the system to permit greater standardization. One good example occurred in an early presentation of the comparable selection tool to the collected appraisal staff. To provide a realistic demonstration, a comparable subject tract in the northeast corner of Calhoun County, Iowa, was used. The "best" comparable sales selected by AgriStat included two in Pocahontas County to the north and one in Webster County to the east (see Exhibit 10.9). The subject is indicated in red and the three comparables in black.

Traditionally, FCSA has grouped counties that are both economically similar to and nearby each other in *market zones.* The appraisers typically work in a small set of market zones, perhaps only one. They often work in the field, with laptops loaded with a small portion of the Appraisal Database relevant to their assigned territory. Based on their territory, the appraisers "subscribe" to specific market zones.

After the comparable selection presentation, the appraiser who actually valued the Calhoun tract noted that he would never see the Webster County comparable because it was in a Market Zone (501, in green) to which he did not subscribe. The development team was about to reconsider the selection filtering when the appraiser noted that, had it been available to him, he would have definitely preferred to use the Webster comp. Rather than change the AgriStat application, a change to the subscription process was indicated.

Tract Similarity Index

During the modeling process, it became apparent that *tillable percentage* was not statistically significant, which suggested that using that variable to select comparable properties was probably not overly meaningful either. Furthermore, in discussions with FCSA, the development team came to a consensus that an attribute measuring tract similarity was needed.

Constructing a good similarity index is relatively intuitive. For example, consider Exhibit 10.10. Comparable 1 is twice as big as the subject but has exactly the same relative (percentage) land mix. Comparable 2 is the same size as the subject but has a significantly different relative mix. Comparable 1 is more similar to the subject, even without considering that total acreage difference is a separate selection criterion. Having acreage differential as a separate attribute is important to situations like this because, while comparing a 250-acre tract to a 500-acre tract is not a great stretch, comparing a 40-acre tract to a 1,000-acre tract might be. Clearly, then, a comparable property with a comparable relative land mix should have a high score in a well-designed index.

The measure developed was the result of some very fruitful collaboration between the FCSA and development team members. In the process, two similar but subtly different indexes were developed: one focused on differences between tracts, the other on similarities. Testing by FCSA suggested that the latter performed better. Retrospectively the development team realized that the similarity index performed better when examining

Exhibit 10.9 Iowa Comp Selection

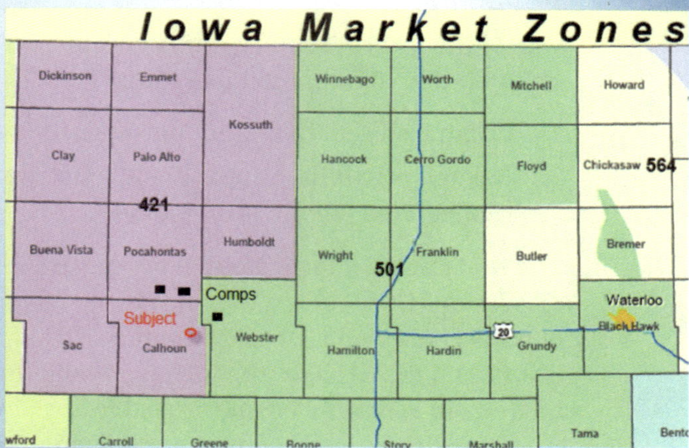

Exhibit 10.10 Tract Similarity

	Subject Tract		Comparable 1		Comparable 2	
Inventory	Acres	Percentage	Acres	Percentage	Acres	Percentage
Dry crop	50	20%	100	20%	125	50%
Pasture	80	32%	160	32%	75	30%
Meadow	120	48%	240	48%	50	20%
	250		500		250	

tract similarities than when examining tract differences. While the retrospective explanation may be obvious, it was not so beforehand.

The index construction was conceptually fairly straightforward. The main idea was to focus on the primary productive land types. To eliminate minor influences and simplify the calculation, any land type that had acreage less than 10% of the tract total was ignored. This tended to eliminate non-productive and lower productive acreages like roads and waste, building sites, and dry corners. For the remaining acres, the relative (percentage) portion of each land type between the subject and comparable property was compared, taking the lower of the two values. For example (see Exhibit 10.11), the subject property has 20% dry cropland while Comparable 2 has 50%. The lower of the two is 20%, which is assigned to the similarity column. It did not matter whether the subject or comparable property has the higher percentage for a given land type–the lower number is the one selected. If Comparable 2 had no dry cropland, the similarity score for that land type would be 0%. Thus, Comparable 1, which has exactly the same relative percentages as the subject, has a perfect total similarity score of 100%.

Reliability Index

One of the project requirements was to devise and construct a reliability index (RI) that did two things:

1. Compared reliability among sales, income, and cost models
2. Related reliability to specific inventories (dry land, pasture, irrigated land, etc.) within a model

The first criterion turned out to be less useful because the coefficient of dispersion (COD) among models tended to drive the ultimate selection of model based on COD alone: if the COD of the sales model for Iowa is 0.10 and the COD for the income model (for Iowa) is 0.13, the sales model will generally be more reliable.

After casting about a bit for a reasonable basis for a metric, the development team made the decision to rely on sensitivity analysis, which had previously been used to test whether empirical results were sensitive to built-in assumptions but not in developing a reliability score. Research uncovered an extended

Exhibit 10.11 Tract Similarity Computation

	Subject Tract		Comparable 1		Comparable 2	
Inventory	Acres	Percentage	Percentage	Similarity	Percentage	Similarity
Dry crop	50	**20%**	20%	20%	50%	20%
Pasture	80	32%	32%	32%	**30%**	30%
Meadow	120	48%	48%	48%	**20%**	20%
	250			100%		70%

abstract of a Slovenian PhD dissertation that covered the comparison of resistivity analysis to reliability score development, "Estimation of Individual Prediction Reliability Using Sensitivity Analysis of Regression Models."[1] Fortunately the abstract was in English. The dissertation, coupled with the practical nature of the knowledge that the team gathered from an examination of the market and the factors that inherently influence the reliability of the analysis, provided the basis for the reliability estimate.

A well-designed reliability index (RI) should return a higher score for tracts that the sales model values well and a lower score where the model struggles. Construction of the RI should consider significant attributes of a tract–land, improvements, and location. These variables should in fact be the primary components of the model itself.

When tracts sell, those with high RIs should, on average, have sale prices closer to the value estimated by the AVM than those with lower scores. There should be a correlation between the RI and sales ratio. As the ratio approaches 1.0, the score should generally increase. For tracts estimated within 10% of the (time-adjusted) sale price, scores should be higher than tracts falling in the 15%-20% range.

One significant caveat on the reliability score is that although there is an expectation of a (negative) correlation between COD and reliability, a high correlation would not occur if the model is properly specified and calibrated. If there was a high correlation between, say, the variable dry land percentage and COD, the first question would be why the variable had not been included in the model, or why the model was not properly accounting for dry cropland. Minor trends can be uncovered with a properly constructed model.

The approach used in developing a reliability score was sensitivity analysis. The model was stratified by various key parameters and examined through a consideration of how the COD changed given the presence or absence of a parameter (*center pivot*, for example) or as those parameters changed–*cropland/dry land ratio* or *land quality* (above average, average, or below average). Identifying a pattern did not by itself guarantee that the associated variable was used to calculate the RI. If the inclusion of the variable failed to increase the overall correlation between COD and reliability, there was no real reason to include it.

Reliability Index Calculation

An initial COD was considered based on location. If sales were sufficient, location was set at the county level; if not, analysis was expand-

1. Z. Bosni, "Estimation of Individual Prediction Reliability Using Sensitivity Analysis of Regression Models (in Slovene)" (PhD dissertation, University of Ljubljana, Faculty of Computer and Information Science, 2007), available online at http://lkm.fri.uni-lj.si/zoranb/dissertation.htm/.

ed to the market zone level. In a couple of western areas, state CODs were used due to the paucity of data. The location COD became the initial RI, to be adjusted as better information was obtained. For each variable, the relative difference was considered between the location COD and the COD for the factor to be added. If the location COD was 0.10 and the factor COD was 0.12, the ratio was 0.12/0.10 = 1.2. When all factors were examined, they were multiplied by the original:

$$\text{Score} = \text{Original COD} \times \text{Factor 1} \times \text{Factor 2} \ldots \times \text{Factor } n$$

The final result was subtracted from 1. Thus, a final score of 0.15 became an RI of 85%. Exhibit 10.12 shows the individual steps in more detail.

Reliability Scores

Exhibit 10.13 shows the development of the reliability index line by line, in the same order discussed in Exhibit 10.12. The numbers in the table do not reflect the actual RI or COD, but show the correlation between how close the sales ratio is to 1.0 and the combined score COD. In some cases, a state score does not change from one line to the next because that variable was not used for that state and thus had an adjustment of 1.

Validation of Reliability Index

The reliability index provides an indication of accuracy, but not a guarantee. The RI can only indicate that conditions suggest that the estimated value should be close to or far from the sale price. The scores calculated by the sales comparison approach suggest that the reliability index does exactly that. For example, as shown in Exhibit 10.14, consider cases where the estimated value misses the sale price by more than 20%. Where the score is greater than 80, the miss rate is 22.5%. Increase the score to a minimum of 85 and the miss rate drops to 17.3%, while at 90 or more the rate drops to 9.15%. At the same minimum score of 90, the odds of missing the sale by 25% drop to 4.1% and by 30% or more to 1.85%. The higher the reliability score, the better chance of estimating closer to the actual sale price.

Exhibit 10.15 lists land types and improvement types that were identified for potential inclusion as a means of developing the reliability index.

Project Outcome

When the AgriStat project was completed in 2009 and research was being completed for a presentation to the URISA Technologies Conference, an interesting discovery was made: the concept and idea for a regression-based agricultural valuation system had first been proposed in 1922 by George C. Haas, as a means for the valuation of

Exhibit 10.12 Sales Model Calculation

1. Start with *county* COD as a base. For the model, a basic rule was set that any COD-based measure must have at least 18 observations. If a county had fewer than 18 sales, the base was set to the default *market zone* COD and the same test of 18 was applied. If the market zone failed, the *state* COD was used.
2. Next was *land type*. Since the sales AVM was primarily a land model, including several common land types seemed to make sense. However, multiple examinations of each land type along with groupings of similar types were made: only *dry land* showed a beneficial effect (with *irrigated potential* grouped in). *Land type* was included by considering its relative proportion to tract size in 10% groupings: less than 10%, 10%-20%, and so on. Some groupings were combined because the pattern only became apparent at a more aggregated level. Including the dry land percentage improved the RI in every state but Wyoming:
   a. In Iowa, tracts with less than 10% dry land showed the highest COD. The next two groups (10%-20% and 20%-30%) were similar and thus grouped together, as were 30%-40% and 40%-50%, 50%-60% and 60%-70%, and 70%-80% and 80%-90%. Last was tracts with more than 90% dry land.
   b. In Nebraska, there were fewer groups: <10%, 10%-50%, 50%-90%, and >90% dry land.
   c. In South Dakota, groupings were exactly the same as Iowa.
   d. Wyoming defaulted to the *state COD* in every case because results were poor and numbers were too few to develop any reliable patterns.
3. Location desirability. *Location* improved the RI in every state except Iowa, where it made the score worse. Interestingly, the possible values—above average, average, and below average—produced inconsistent results among states, although they were consistent within states.
   a. Iowa: The *location* measure exhibited no useful pattern, so the adjustment was made 1.0 for all possible values (i.e., no change to score).
   b. Nebraska: Average *location* had a COD almost identical to the state COD. Above average *location* was actually associated with a higher COD, while below average had a lower COD. This only means that the model did a slightly better job of valuing below average locations in Nebraska.
   c. South Dakota: Average *location* modeled somewhat better than either above or below average and had a better COD (i.e., a multiplier of less than 1), while above and below had higher CODs.
   d. Wyoming: Both average and below average had lower CODs than the state average, while above average *location* was higher.
4. Land quality. Adding *land quality* improved every score except Nebraska. Unlike *location*, this measure worked very much the same in the other three states: above average *land quality* always had the lower COD while below average always had the higher. In Nebraska, average quality modeled much worse than the other two. However, including this variable made the score worse, and it was not included.
5. Although the development team examined the existence of *improvements* in several combinations, most degraded the RI because of weak correlations. The only case where an improvement's inclusion improved the score was dwellings in Nebraska. In this case the measure was: does at least one dwelling exist, where a dwelling could be house, mobile home, modular home, cabin, or bunkhouse? Interestingly, in Wyoming, this variable made the score much worse. In South Dakota and Iowa, including this factor only degraded the correlation slightly. So if there was a significant business reason to include whether or not a dwelling exists, the analysis would not be badly distorted.

Exhibit 10.13 Reliability Scoring by State

	IA	NE	SD	WY	ALL
state/county	0.229831374	0.203924381	0.226106703	0.021807245	0.2931
+ land	0.276989089	0.215245855	0.239627477	0.021807245	0.305994
+ Loc2	0.276989089	0.215603338	0.243839751	0.022183864	0.306826
+ LandQual2	0.277732889	0.215603338	0.253992696	0.05244932	0.299936
+ Dwelling2	0.277732889	0.220478461	0.253992696	0.05244932	0.301648

Exhibit 10.14 Reliability Validation

Chances of having a reliability of > 90, 85, 80
And missing sale by > 20%, > 25%, > 30%

Score	> 90	> 85	> 80
Count	2636	6338	8782
Miss rate > 20	240	1096	1977
	9.10%	17.29%	22.51%
Miss rate > 25	109	642	1274
	4.14%	10.13%	14.51%
Miss rate > 30	49	349	788
	1.86%	5.51%	8.97%

Exhibit 10.15 Potential Land Types and Improvement Types

Land Types	Improvement Types
Dry cropland	Dwellings
CRP (Conservation Reserve Program) land	Garage/shop/storage
Irrigated land: gravity	Grain storage
Irrigated land: sprinkler	Grain dryer
Meadow/hay	Corral
Building sites	Feedlot
Pasture: 7 types	Hay/livestock sheds
Recreational	Machine shed
Timber	Grain complex
Roads and waste	Center-pivot system

farmland in the Midwest in his doctoral thesis, published as "Sales Prices as a Basis for Farm Land Appraisal" by the University of Minnesota Agriculture Experiment Station.[2] Even though proposed more than 86 years ago, the use of regression analysis for private sector agricultural valuation had never been implemented.

AgriStat is the first enterprise-wide implementation of an "interactive valuation modeling" product in which the appraiser, the data, and a GIS/modeling infrastructure are tied together in a manner that enhances valuation and brings a greater transparency and consistency to the valuation process. It should be acknowledged, however, that the concept of an "appraiser-centric" product was not a significant component of the original development initiative. The original game plan was to merge the best of the GIS and modeling environments to provide better analysis.

There was an even greater change in existing structure and policy than was anticipated as part of the evaluation and development process for FCSA. Initially, the concept of developing an "appraiser-centric" product was discussed but was not a focal part of the project development. After going through the two-year development process, a more appraiser-sensitive management style has been adopted by FCSA, with all of the requisite benefits that such a system and philosophy pro-

2. G. C. Haas, "Sales Prices as a Basis for Farm Land Appraisal," Technical Bulletin 9 (St. Paul, Minn.: University of Minnesota Agriculture Experiment Station, 1922).

vides. FCSA concluded that empowering appraisers should be a core strategy of the organization and would result in the implementation of best practices in appraisal. Now, the loan officer is out of the collateral valuation process, which helps FCSA in its reaction to the current economic environment, with a push towards appraiser independence and separation between the loan origination and appraisal function.

Any outcome analysis would be incomplete without understanding the effect on the various stakeholders at FCSA. Specifically, appraisers, appraisal managers, and senior leadership all benefitted from the implementation of AgriStat. Appraisers gained a greater appreciation of standardized data, and a greater understanding of what constitutes "good" data and how they could improve their collection, verification, and analysis. After an initial distrust of the system and a fear of how their positions could be affected, the appraisers throughout the Farm Credit organization now see the potential of the system and how it can improve the valuation process and their job functions, according to John Connelley, Appraisal Systems Manager for FCSA. Kirk Manker, FCSA's Chief Appraiser, notes that there is also strong support for the move from generalization to specialization within the organization, in effect creating a "highest and best use" focus for each appraisal position in the department. Additional benefits came from automating mundane and routine tasks, allowing a focus on the valuation problem that required a solution and a better handle on understanding and analyzing the land inventory.

FCSA's Senior Leadership Team (SLT) found a number of best practices arising from the AgriStat project, including instantaneous portfolio updating, better use of resources, better use of data, a greater appreciation of the importance of the collateral valuation process, and an understanding of the need to restructure to create even greater efficiencies, according to Manker.

One of the goals of the Appraisal Leadership Team (ALT) at FCSA was to ensure that management, appraisers, and ultimately customers had trust in the system, the process, and the outcome. The leadership team saw a number of positive outcomes from the implementation of the AgriStat application, including greater efficiencies and a more effective use of resources. Manker notes that the Appraisal Leadership Team was able to leverage the existing staff and handle significantly more appraisal work during a large increase in new business in 2008. Manker also notes that FCSA sees the potential for increased productivity with a combination of structure, processes, data, and technology that includes the AVM/GIS solution represented by AgriStat.

Some key changes FCSA is experiencing include the application of data standards across the organization and a greater appreciation of data and its importance and accuracy. FCSA, already a strong proponent of clean and accurate data, is now acutely aware of the importance of this element, and a greater focus on the importance of data has been achieved in a realistic manner throughout the organization. Appraisers, sales specialists, and management all appreciate the need for data that has been appropriately scrubbed, checked for accuracy, and validated. Due to the increased scrutiny on sales and data scrubbing, the organization now has a larger number of usable sales than it previously had available with which to power the AgriStat system. Prior to the implementation of AgriStat, the total appraisal staffing was 49 appraisers. The number of sales processed within the system has increased from 4,000-5,000 annually to 6,500-7,000, and overall appraisal turnaround times have decreased.

At the time of this writing, the current staffing of FCSA included eight sales specialists, six real estate analysts, and 31 certified appraisers. One of the key changes made to the appraisal function within FCSA was the creation of a new position–the "appraiser paralegal"–that focuses on more straightforward valuations. Through use of the ValueSource/AgriStat system, these real estate analysts can handle as many as 1,200 to 1,300 simple analytical assignments per year, in situations where a full appraisal completed by a staff appraiser is not needed. The appraisers are then available to take on those properties that are more complex and represent higher risk.

Another new position is the sales specialist, who gathers and validates sales data by speaking with auctioneers, real estate brokers, assessors, and others involved with or knowledgeable about market activity, thus ensuring the most accurate information on the sales that power the AgriStat application. Some of the FCSA appraisal staff were initially unhappy with the changes because the sales specialists were diverted from an appraisal support role, but ultimately all concerned recognized that the outcome was superior to the status quo prior to the implementation of the application. The appraisers, who did not initially like the changes, have now worked in the new data-focused environment for two years and report that the new organization of the department provides them with significant enhancements over the earlier structure, with greater specialization and better quality data. For example, appraisal team members, including appraisers, sales specialists, and analysts, use tools such as the sales ratio analysis as they examine sales because the sales ratio functions as a quality control tool, alerting them to abnormalities in sales that would not normally be evident, such as the predictive nature of the regression model in a given area.

The Appraisal Leadership Team adapted to the influx of technologies by reorganizing the leadership functions of the department based on an implementation of role specialization, moving from generalists to specialists. The new core functional teams are

- Fee appraisals
- Agribusiness finance
- Modeling/technology
- Retail lending

One critical outcome that the project produced was that FCSA was able to gain a better appreciation of the role and importance of data. Manker and Connelley both noted that FCSA realized that what they had was not as good as they thought. After two years of working on the development of AgriStat, there is a greater consistency in the data collected and standardization of the property characteristics.

Conclusion

In his doctoral thesis, Haas noted the following:

> The method of appraisal here outlined can be of great practical value for many purposes, chief of which are the following: (1) as a basis for mortgage loans; (2) as a basis for assessment and taxation; (3) as a basis for buying and selling; (4) for inventorying a farm business or an estate; (5) as a basis for estimating rent on different farms and different classes of land where rents are needed for cost accounting; (6) as a basis for determining benefits and damages for drainage ditches, roads and other local improvements; (7) as a basis of settlement of court disputes involving questions of land values; (8) for making valuation of farm land used in railway right-of-ways and the like.
>
> The Federal Land Banks, the joint stock land banks, other banks, mortgage companies, insurance companies, and others who loan money on farm land are in need of a scientific measure of farm land value such as is here presented. Land credit can not be put on an equitable basis unless farm values are appraised accurately. The lack of this value information is the principal reason . . . the farmer is not able to get all of the credit he desires. This is not only of interest to farmers, but also to the investors. The investor wants to be certain that the values placed on the mortgaged land are their true values. If the investing public was convinced that such valuations represented true market values, it would no doubt accept a lower net yield on the investment in lieu of the decreased risk due to accurate appraisal.[3]

Though this vision took 86 years to see practical implementation, a system that leverages GIS, robust property data, and the judgment of a knowledgeable local appraiser will serve as a powerful and compelling model for the future.

3. Ibid., 24.

11 Practical Integrations of Valuation Modeling and Geographic Information Systems

First Steps toward a More Robust Market Analysis and Valuation Methodology

Patrick M. O'Connor and Michael Ireland

Many appraisers have used computers for over 30 years as word processors, for access to the Internet, for communication with appraisal clients, and to review information and calculate property values. Government mass appraisers who have used mainframe computers for valuation purposes are the precursors to computer-assisted valuations by private-sector appraisers. To remain or gain competitive advantages, appraisers have adapted many new technologies to help provide appraisals to clients quickly. However, when confronted with technology related to valuation issues, some appraisers are unwilling or lack the requisite education to upgrade their skills to accommodate the requests of their clients. The purpose of this chapter is to discuss and demonstrate the use of computers to assist and support appraisers in improving their valuation skills.

Appraisal Tools

The real estate market is dynamic and constantly changing. In the same way, appraisers' ability to provide market analysis in an efficient, methodical, cost-effective manner must evolve. Tools of the trade for appraisers have changed from pen and paper to automated value estimates based on market analysis. As the business has changed, many appraisers have refused to change their appraisal practices by mastering new technologies to meet the demands of a fast-paced market. Many appraisers did not anticipate the need to use a digital camera, laser measuring device, or financial calculator. Many more have been slow to make full use of local and national databases, which became available through subscription services or multiple listing services (MLS) in the mid-1990s.

Clients are asking questions relating to quantitative and qualitative analysis that many fee appraisers are unable to answer. For example, a client may ask for statistics on the confidence interval (range of probable value) around the appraiser's estimate of market value. Finding these answers, rather than simply developing an opinion of market value, requires the use of databases. The old tools are not going to work as they did in the past. This is the time for appraisers to say, "Goodbye, financial calculator; hello, personal computer; welcome, desktop software." The good news is the expanding skill sets developed by the use of financial calculators are transferable to the use of the computer.

Appraisers now have the computer hardware to support real estate market analyses. They only need to purchase statistical software, a geographic information system (GIS), and databases for the counties in which they perform the majority of their appraisal services. Once these databases are established, appraisers can summarize several property characteristics using statistical measures such as ranges, averages, and medians in chart formats. Appraisers can also provide reliable statistical summaries of the market area surrounding subject properties.

Appraisers who accepted these technological changes in the 1980s and 1990s are ahead of the learning curve and may skip the rest of this section or move on to the discussion of statistical and geographical software. For those who have resisted technology, it is your time to increase your knowledge base.

Software Choices for Appraisers

The following discussion suggests cost-effective methods for conducting market analysis using quantitative and qualitative methods. Software used for completing standard appraisal forms or narrative reports and for reporting the results of market analysis are different discussions. The analytical software discussed here includes spreadsheets, statistical packages, and geographic applications.

Spreadsheets

A spreadsheet is defined as a computer application that simulates a paper worksheet. A spreadsheet displays multiple cells that together make up a grid of cells containing either alphanumeric text or numeric values or formulae. Exhibit 11.1 shows a blank spreadsheet. The boxes in the grid are the cells referred to in the definition. A reference to a cell uses a column heading (alpha) and a row (numeric). For example, cell D5 is at the center of Exhibit 11.1.

Exhibit 11.1 **Blank Spreadsheet**

Spreadsheets offer many advantages to appraisers over calculators. They are relatively easy to use and have a moderate learning curve, and with experience users will be able to create sophisticated analyses. The elegance of a spreadsheet does not depend on the sophistication of the user, but rather is a function of the ease of use of the tool itself. Spreadsheets support the math and statistical functions that the appraiser uses to develop market values. This type of statistical analysis includes those performed on a financial calculator such as calculating the median value of sale set. Appraisers who understand these concepts from work on financial calculators will see a seamless transfer of skills.

Creating graphic displays for better clarification of numeric summaries is just one benefit of the spreadsheet over the use of the calculator. Computer-based tools, such as spreadsheets, allow data to be entered once and used many times. Many MLS providers allow their data to be saved directly to a spreadsheet or to a file that can then be imported into the electronic spreadsheet without the need to reenter the data. Sorting, filtering, and querying data are all part of most spreadsheet functions. For the appraiser, the primary limiting factor in the use of spreadsheets is rooted in the individual's mind, not in the software. A spreadsheet can summarize large amounts of data, perform basic statistical functions, obtain a data analysis routine with a low-budget impact, and share the results with others. Exhibits 11.2 and 11.3 are examples of simple graphic reports created with a spreadsheet from Microsoft Excel 2007.

Spreadsheet software is helpful to appraisers who must complete Fannie Mae Form 1004MC, a form designed for reporting market conditions in a specific area. The 1004MC Market Conditions form is now required as part of the appraisal report for Fannie Mae and Freddie Mac loans. Analyzing the direction of market conditions is

Exhibit 11.2 Transaction Count by Price

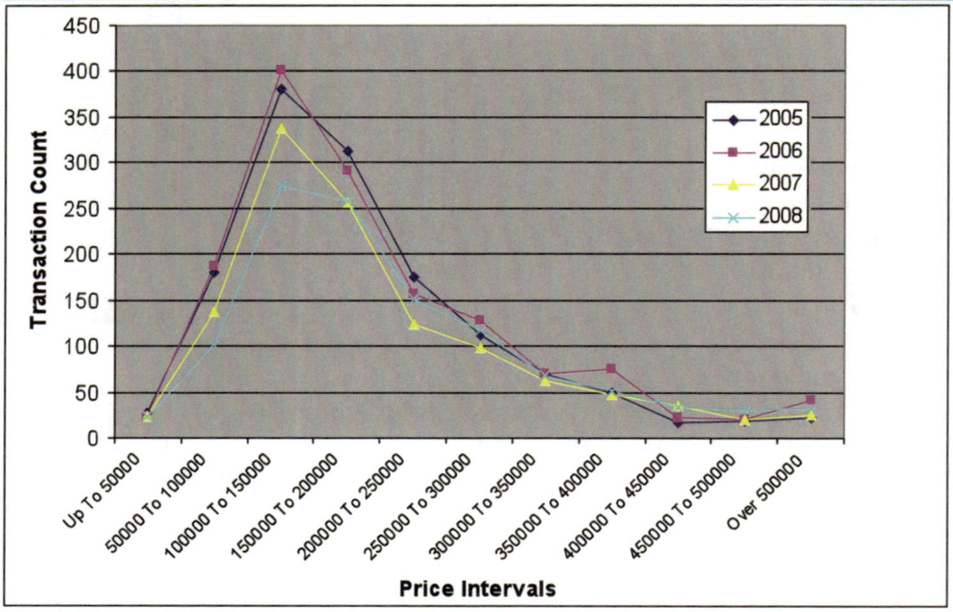

Exhibit 11.3 Median Sale Price per Square Foot of Living Area

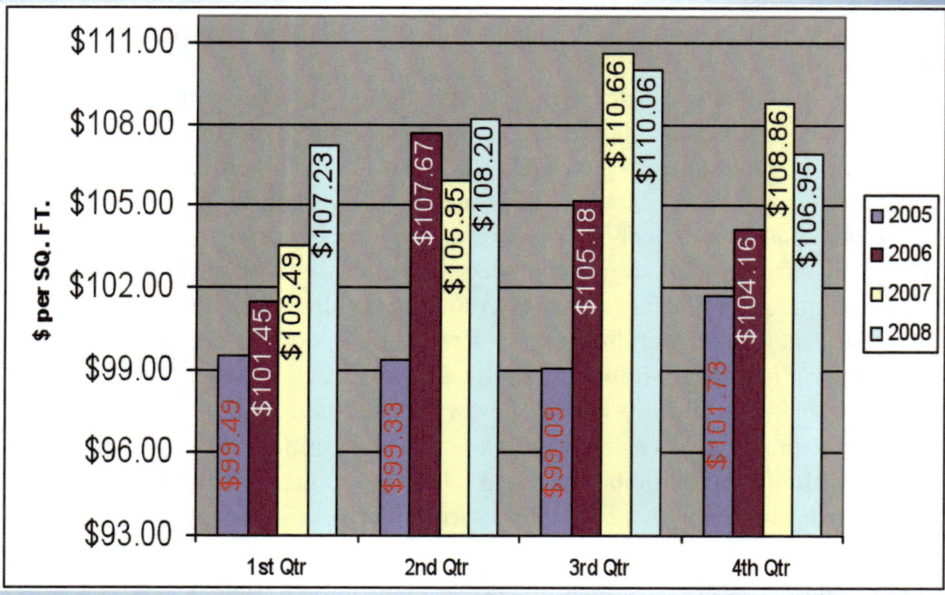

an important task that all appraisers must complete, regardless of the intended user of the appraisal.

Exhibits 11.4 through 11.7 show how an MLS was used to evaluate changes in market conditions in a neighborhood.

- Exhibit 11.4 gives a history of market changes in the neighborhood using three-month time increments from the date of appraisal.
- Exhibit 11.5 shows the change in price along the vertical axis from April 2007 to May 2009. The solid blue line accented with diamond shapes represents sale price, the solid red line accented with squares represents the list price, and the dotted blue line is a polynomial fit, or prediction curve, of the sale price. The sale price peaked in April 2007 and began a decline through the fourth quarter of 2008 before a recovery began.
- In Exhibit 11.6, sale price has been converted to a price per square foot by dividing the sale price by the gross building area. The sales are plotted using the month of sale (April 2007 = A-07) on the horizontal axis. The red line is the prediction curve.
- Exhibit 11.7 compares the number of sales (red bar) to the number of listings (blue bar). The dotted lines are the polynomial fit curves to the respective data.

Exhibit 11.4 Appraiser's MCA Worksheet

APPRAISER'S MCA WORKSHEET File No. ________

Property Address: ________

City: Bloomington **State:** IL **Zip:** 61704

	Prior 22-24 Months	Prior 19-21 Months	Prior 16-18 Months	Prior 13-15 Months	Prior 10-12 Months	Prior 7-9 Months	Prior 4-6 Months	Prior 0-3 Months
Total # of Comparable Sales (Settled)					(13-24 Mos.	Increasing)	(0-12 Mos.	Declining)
	1	2	3	6	15	7	8	9
Absorption Rate (Total Sales/Months)					(13-24 Mos.	Increasing)	(0-12 Mos.	Declining)
	0.3	0.7	1.0	2.0	5.0	2.3	2.7	3.0
Total # of Comparable Active Listings					(13-24 Mos.	Increasing)	(0-12 Mos.	Stable)
	0	6	11	13	18	19	21	21
Months of Housing Supply					(13-24 Mos.	Increasing)	(0-12 Mos.	Increasing)
	0.0	8.6	11.0	6.5	3.6	8.3	7.8	7.0
Median Comparable Sale Price					(13-24 Mos.	Declining)	(0-12 Mos.	Stable)
	$452,875	$327,446	$457,000	$274,400	$280,300	$334,000	$288,500	$319,000
Median Sales Days on Market					(13-24 Mos.	Increasing)	(0-12 Mos.	Increasing)
	0.0	31.0	78.0	55.0	1.0	9.0	51.0	152.0
Median Comparable List Price					(13-24 Mos.	Stable)	(0-12 Mos.	Stable)
	$304,950	$304,900	$320,000	$302,000	$339,900	$314,900	$320,000	$332,500
Median Listings Days on Market					(13-24 Mos.	Increasing)	(0-12 Mos.	Declining)
	69	127	197	176	272	272	155	155
Average Sales Concessions Paid per Sale (%)					(13-24 Mos.	Stable)	(0-12 Mos.	Stable)
	0.000%	0.000%	0.000%	0.000%	0.024%	0.000%	0.060%	0.063%
Median Sale-to-List Price Percentage					(13-24 Mos.	Stable)	(0-12 Mos.	Stable)
	100.0%	99.5%	99.4%	102.6%	100.0%	98.9%	100.0%	99.7%

Exhibit 11.5 Price Trends

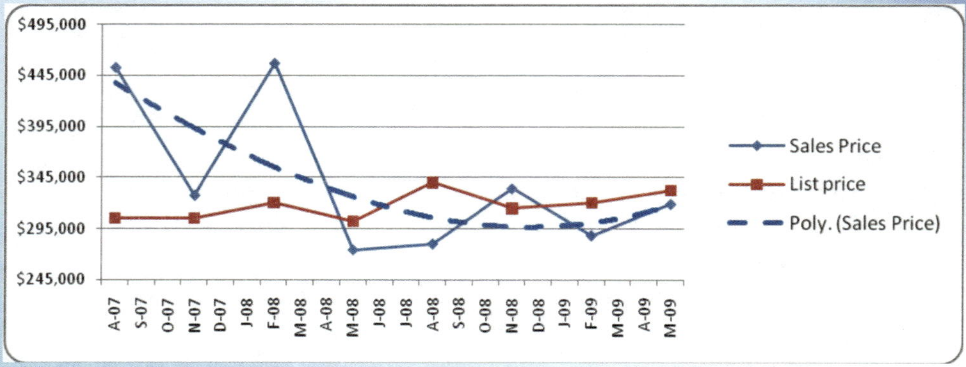

Exhibit 11.6 Sale Price per Square Foot Trend

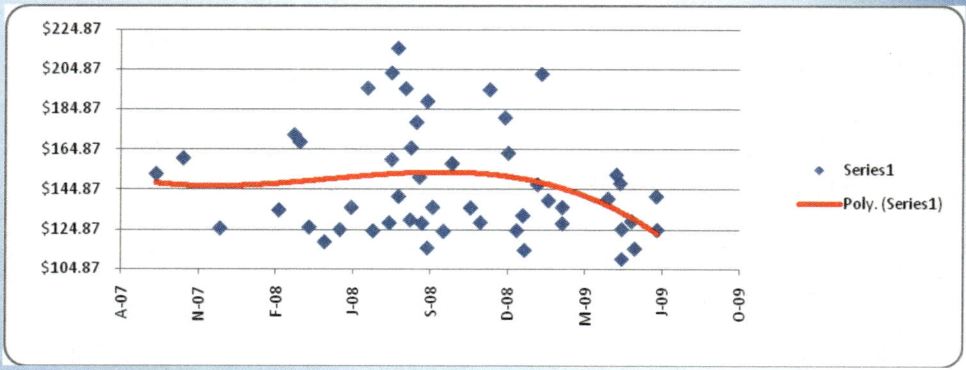

Exhibit 11.7 Volume Trends

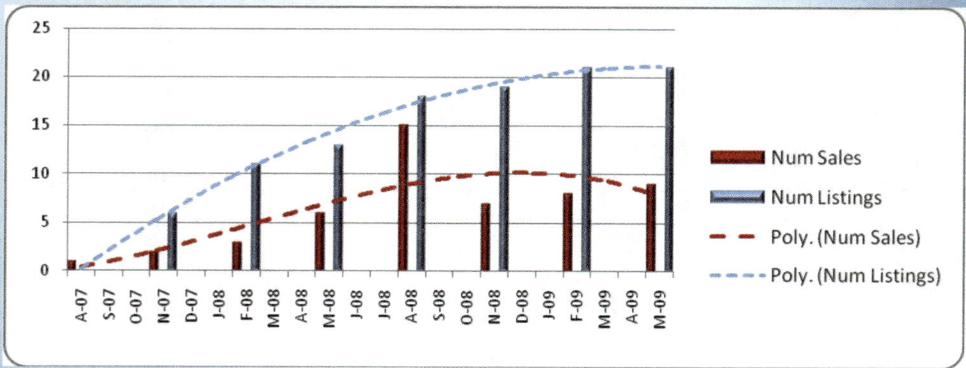

These exhibits demonstrate how a spreadsheet template uses multiple data sets to generate an analysis that will help the client and appraiser understand changes in market trends.

Statistical Software

Spreadsheets are an essential tool for appraisers and can accomplish much with minimal effort. For the appraiser who provides market projections, more analytical and processing power will be required. Chapter 1 discussed the definition and use of AVMs. The process of using computers to develop valuation models using statistics was around long before AVMs were introduced in the 1990s. Since the late 1970s, the International Association of Assessing Officers (IAAO) had been teaching practitioners these methods, known then as *computer-assisted mass appraisal (CAMA)*. Regardless of the name, the preferred tool used to develop valuation models from real estate data is a statistical software package.

A model produces mathematical formulas that can be changed and store components. Statistical software bundles statistical programs that perform multiple functions and provide a variety of analytical results, It enables appraisers to obtain the results of standard statistical procedures and statistical significance tests, without requiring low-level numerical programming. Most statistical packages can be used for data management.

A statistical package has many advantages over a spreadsheet for the development of a valuation model. Specifically, a statistical package has all the same spreadsheet functionality, but is built with more analytical capacity and more advanced modeling tools. Also, statistical software provides the user with a graphical user interface (GUI) and the option of using command-line or syntax-driven operations.

In addition to the basic statistics available in a spreadsheet package, these packages will include calibration modules, such as multiple regression analysis (MRA), and may include a nonlinear regression (NLR) module as well. Calibration modules like MRA and NLR are typically the statistical programs used to develop credible AVMs. The functions available with an NLR module may not be included in some standard statistical packages and may need to be added.

A statistical package includes all the basic analysis table and graphics formats found in spreadsheets. Data results can be illustrated using cross-tabulation reports, breakdown or contingency tables, cluster analysis, box plots, contour maps, or 3-D scatter plots, to name just a few. Data management in a statistical package will usually be an enhancement that is not typically available in a spreadsheet. When working with large or multiple data sets, a statistical package can be a timesaver. Most statistical packages allow for a variety of file formats, including direct importation from external databases.

Statistical software is available in a wide range of open-source, public domain, freeware, proprietary, and add-on packages for spreadsheets. While many appraisers associate spreadsheets with Microsoft products, a wide variety of other products are available. In many cases, add-on packages are developed for Microsoft Excel because this spreadsheet software has become an industry standard.

Not all statistical packages are created equal. Some are Web-based[1] and others require a mainframe computer, but most are PC-based packages such as SPSS and NCSS. As with any program, a learning curve exists. For an appraiser new to the world of using statistical software, consider a statistical package that naturally transitions from methods and skills learned first using an electronic spreadsheet. The appraiser who has recently taken a course in statistics or plans to take such a course in the near future should consider using the same statistical package in his or her appraisal practice. Using the same statistical package in practice and in class will give the appraiser access to built-in tutoring and process manuals as well as a network of fellow students.

Like so many choices of software, no single choice is universally the "best." What is best depend on the individual appraiser's needs as part of his or her business practice. Mastering how to use the software is important, but price is also an important consideration. Freeware, open-source software, and public domain software will be the most attractive from a price point, but users of these statistical packages should beware because the learning curve may be extreme and the software may not easily integrate data from multiple sources.

Two proprietary packages–Number Cruncher Statistical System (NCSS) and the Statistical Package for the Social Sciences (SPSS)–have some built-in real estate–specific modules to help appraisers get started. NCSS is the only statistical package providing an automated comparable sales approach module with adjustments, a significant time-saver not readily available in other statistical software. Exhibits 11.8 through 11.12 were generated using NCSS software (www.ncss.com).

Exhibit 11.8 shows an example of output from a statistical package. The 20 sales in this neighborhood are grouped by the sale price, in $5,000 increments. The frequency distribution counts the number of observations in each group and provides a cumulative count, a percentage for each group, and cumulative percentage. A simple graph shows the data distribution.

1. See the online statistical packages available at http://glass.ed.asu.edu/stats/analysis/; http://davidmlane.com/hyperstat/analysisf.html/; and http://www.statcrunch.com/.

Cross tabulation reports illustrate a two-way frequency distribution. First the data in Exhibit 11.9 is filtered by the line of parameters in the heading. Then the remaining data presents the counts by categories of exterior wall type, compared to the exterior condition of the property The data in the table can be changed from counts to percentages in a single process step using the statistical package.

Breakdown tables are a good tool for showing market activity using a specific property attribute. Exhibit 11.10 displays changes in sale price and sale price per square foot, based on changes in condition. The table can also be depicted graphically using box plots The pattern shown graphically in Exhibit 11.11 is the data from the breakdown table in Exhibit 11.10. Exhibit 11.12 shows a 3-D scatter plot with interactive view rotation, which allows the user to dynamically view the relationship of size to bath counts and sale price.

Exhibit 11.8 Frequency Distribution of Sale Price

<table>
<tr><th>Sale_Price</th><th>Count</th><th>Cumulative Count</th><th>Percentage</th><th>Cumulative Percentage</th><th>Graph of Percentage</th></tr>
<tr><td>Up to 89,999</td><td>1</td><td>1</td><td>5.00</td><td>5.00</td><td>| |</td></tr>
<tr><td>90,000 to 94,999</td><td>2</td><td>3</td><td>10.00</td><td>15.00</td><td>| | | |</td></tr>
<tr><td>95,000 to 99,999</td><td>4</td><td>7</td><td>20.00</td><td>35.00</td><td>| | | | | | | |</td></tr>
<tr><td>100,000 to 104,999</td><td>6</td><td>13</td><td>30.00</td><td>65.00</td><td>| | | | | | | | | | | |</td></tr>
<tr><td>105,000 to 109,999</td><td>4</td><td>17</td><td>20.00</td><td>85.00</td><td>| | | | | | | |</td></tr>
<tr><td>110,000 to 114,999</td><td>2</td><td>19</td><td>10.00</td><td>95.00</td><td>| | | |</td></tr>
<tr><td>115,000 to 119,999</td><td>1</td><td>20</td><td>5.00</td><td>100.00</td><td>| |</td></tr>
</table>

Exhibit 11.9 Cross Tabulation Report

Filter GoodSale = 1;flag = 1,2,3;Saleyear > 2004;SaleYear < 2008;lot_sf < 25000

Counts Section

	EXTR_WALLS									
	1	**2**	**3**	**4**	**5**	**6**	**7**	**8**	**9**	
EXTR_COND	**BRK**	**WOOD**	**MASON**	**SID**	**STUCCO**	**BRK& FRM**	**BRK FR**	**ASPHLT**	**OTHR**	**Total**
1 Poor	0	9	0	13	0	1	0	6	0	29
2 Fair	2	46	1	75	3	0	1	8	2	138
3 Average	24	99	3	254	10	0	5	6	1	402
4 Good	33	46	5	148	1	6	12	0	0	251
5 Very Good	98	228	8	954	2	15	148	0	1	1455
6 Excellent	10	66	3	320	5	11	71	0	0	486
Total	167	494	20	1764	21	33	237	20	4	2761

The number of rows with at least one missing value is 0.

Exhibit 11.10 Breakdown Table of Counts, Means, and Medians

	Extr_Cond						
Variables	**Poor**	**Fair**	**Average**	**Good**	**Very Good**	**Excellent**	**Total**
Sale price/counts	29	138	402	251	1455	486	2761
Mean	$62,146.55	$74,597.15	$101,494.57	$137,139.13	$175,223.07	$199,546.46	$159,090.38
Median	$56,500.00	$74,150.00	$93,900.00	$123,500.00	$166,715.00	$195,000.00	$149,500.00
Per sq. ft./counts	29	138	402	251	1455	486	2761
Mean	$53.40	$68.09	$79.94	$95.62	$96.56	$95.88	$92.06
Median	$52.77	$70.66	$79.15	$100.26	$96.80	$95.92	$93.87

Exhibit 11.11 Box Plot

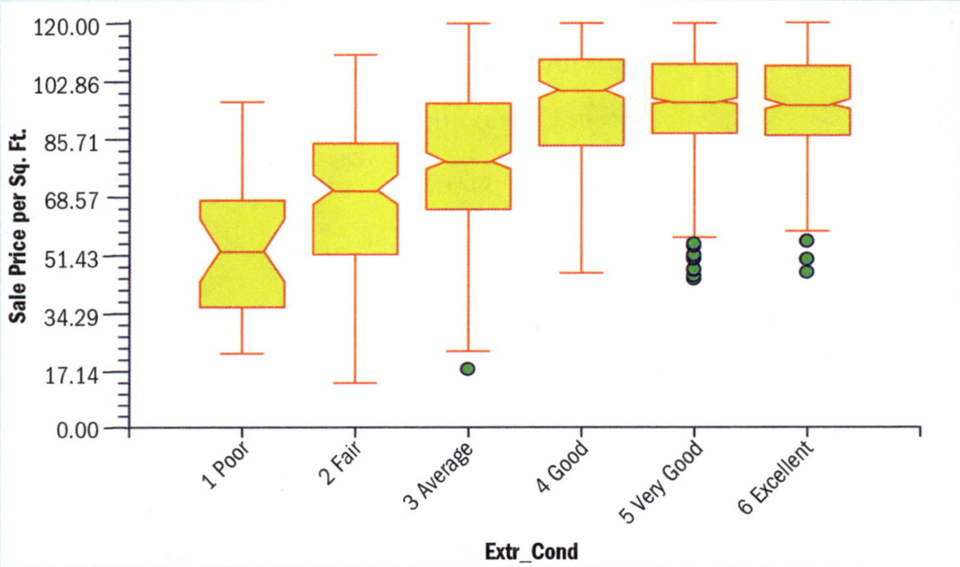

Statistical software packages are capable of the same functionality as spreadsheets, but they add another dimension to the data, literally and figuratively. Many spreadsheet skills can be transferred for use in a statistical package. Furthermore, using a statistical package will inherently communicate to the end user that the user understands how and when statistical analysis is best used. Statistical vocabulary and comprehension will also improve as the appraiser becomes more familiar with the statistical tools and analysis needed.

Geographical Information Software

An introduction to the use and history of geographical information systems (GIS) was provided in Chapters 5, 6, and 7. GIS refers to programs that allow the use of digitized maps to develop and report

Exhibit 11.12 Scatter Plot

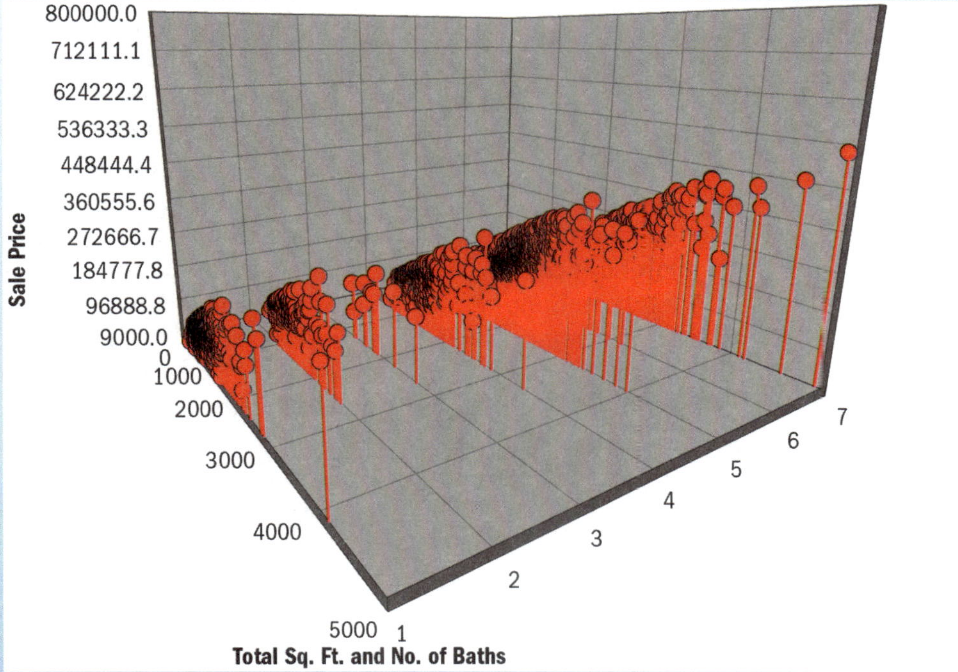

the same results found through the tabular analysis of spreadsheets and stats packages. In brief, the general term refers to software used to create thematic mapping. Even though thematic mapping is only a small segment of what can be accomplished using geographic information systems, this section of the chapter will focus on the use of thematic mapping to help explain and display real estate data and market analysis.

A geographic information system captures, stores, analyzes, manages, and presents data that is linked to location. Most, if not all, of the data used in the appraisal process is linked to a specific location that can be georeferenced. GIS, therefore, is another power tool to consider using for market analysis when locations are known or can be identified. As with spreadsheets and statistical packages, thematic mapping has evolved. Early GIS applications have migrated from mainframe computers to the PC platform, opening up their use to the masses. Internet mapping services such as Google Maps, Yahoo Maps, Live Earth, and others have exposed the general population to the many advantages of thematic mapping.

Internet mapping services can be deployed as a "poor man's GIS" using a technique known as a map *mashup*. This type of "neogeography" allows users of Web-enabled map services to integrate data and create GIS with limited experience and expense. An example of a mashup is the use of cartographic data from Google Maps to add location information to real estate data, thereby creating a new and distinct Web service that was not originally provided by either source. Mashups avoid the cost of purchasing GIS software but not the cost of application development. The alternate choice is to obtain GIS-specific software through freeware, public domain, or proprietary sources.

In GIS, all data must be georeferenced to a map. Numerous methods exist to georeference data, with address matching being the most universal. The International Association of Assessing Officers[2] and the Urban and Regional Information Systems Association[3] can provide detailed educational and application information on geographic information systems and integration with appraisal tools and techniques.

GIS thematic mapping is a tool that helps the appraisal analyst and clients have a better understanding of real estate market dynamics. Users of real estate services are always interested in where a property is located and what surrounds it. Exhibit 11.13 shows an example of a map mashup used to display the subject property (marked by the red X) and surrounding neighborhood amenities. Adding and removing an amenity is as simple as a mouse click on the bottom of the screen.

With a simple change to the search criteria, a new map is produced showing listings, sales, and pending sales in the area. The full power of this GIS application can only be appreciated from a live demonstration. (See the output shown in Exhibit 11.14.)

A simple summary view is shown in the left panel of the screen, but the real power is that the data is also available in a spreadsheet (Exhibit 11.15) ready for more analysis as required, including change to the map view. Thematic mapping of this type helps the appraiser and client see different market data using a geospatial view. The locations of active, sold, pending, and expired properties are shown in Exhibit 11.14. Sold and pending properties outnumber the active listings, indicating that demand is strong in the area. Similar maps can be included in an appraisal report to help the client understand the appraiser's reasoning.

This integration of real estate data and geographic software provides detail about market trends not previously visible to the analyst or their clients. These examples required only access to data from an MLS and intermediate computing skills. Fundamental concepts and skills are enhanced using GIS software and the appraiser's primary data.

---

2. http://www.iaao.org/.
3. http://www.urisa.org/.

Exhibit 11.13 Map Mashup

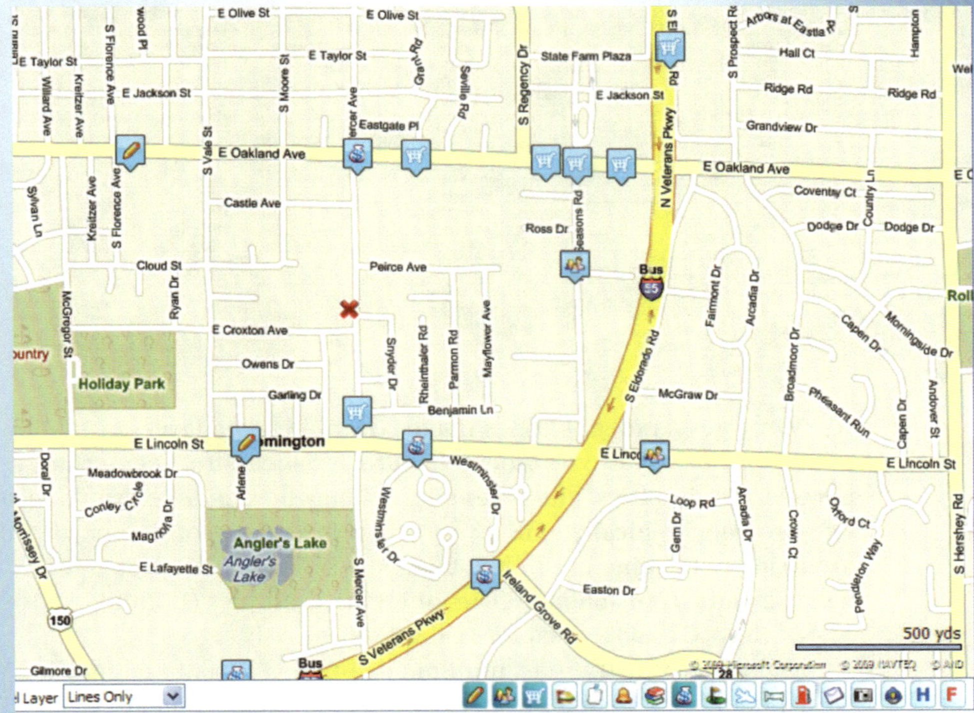

Exhibit 11.14 Map Showing Listings, Sales, and Pending Sales

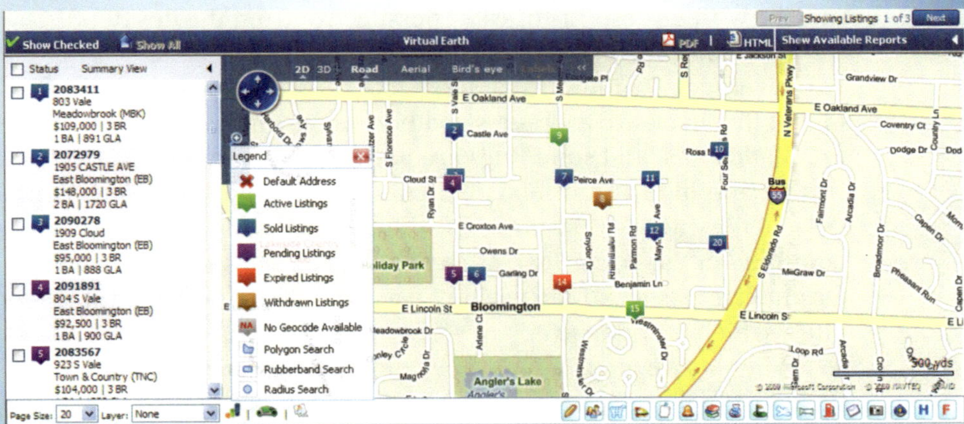

Exhibit 11.15 **Database of Pending and Closed Sales**

Show Checked | Show All | Pending vs Closed | PDF | HTML | Show Available Repo

	MLS #	Class	Status	Address	City	Area	List Price	Selling Price	Listing Date	Pending Date	Closed Date
15	2091891	RESIDENTIAL	Pending	804 S Vale	Bloomington	East Bloomington	$92,500		6/9/2009	7/17/2009	9/18/2009
16	2090278	RESIDENTIAL	Closed	1909 Cloud	Bloomington	East Bloomington	$102,500	$95,000	2/2/2009	2/4/2009	2/27/2009
17	2091343	RESIDENTIAL	ACTIVE	2204 E Lincoln	Bloomington	East Bloomington	$97,750		5/4/2009		
18	2073584	RESIDENTIAL	Closed	1901 Castle	Bloomington	East Bloomington	$115,900	$102,000	9/18/2007		3/31/2008
19	2083567	RESIDENTIAL	Pending	923 S Vale	Bloomington	Town & Country	$104,000		10/27/2008	6/24/2009	
20	2083695	RESIDENTIAL	Closed	810 Parmon	Bloomington	Town & Country	$109,900	$107,000	11/11/2008	4/3/2009	5/15/2009
21	2083021	RESIDENTIAL	Closed	816 Mayflower	Bloomington	Town & Country	$114,900	$108,000	9/4/2008	11/10/2008	12/4/2008
22	2074556	RESIDENTIAL	Closed	904 Parmon	Bloomington	Town & Country	$106,900	$108,900	12/21/2007	4/4/2008	5/13/2008
23	2083411	RESIDENTIAL	Closed	803 Vale	Bloomington	Meadowbrook	$110,000	$109,000	10/7/2008	10/20/2008	11/25/2008
24	2092067	RESIDENTIAL	ACTIVE	2208 E Lincoln	Bloomington	East Bloomington	$110,000		6/22/2009		
25	2074330	RESIDENTIAL	Closed	2117 Skyline Court	Bloomington	East Bloomington	$115,900	$110,000	11/26/2007	1/18/2008	2/29/2008
26	2080988	RESIDENTIAL	Closed	1911 E Cloud	Bloomington	East Bloomington	$115,500	$110,500	3/31/2008	6/5/2008	7/21/2008
27	2082321	RESIDENTIAL	Closed	2106 Pierce	Bloomington	Town & Country	$114,900	$112,000	6/30/2008	7/7/2008	7/31/2008
28	2081954	RESIDENTIAL	Closed	903 Arlene	Bloomington	Town & Country	$116,900	$112,000	6/10/2008	3/6/2009	3/30/2009
29	2083890	RESIDENTIAL	Closed	909 Arlene	Bloomington	Town & Country	$124,900	$113,000	12/5/2008	5/2/2009	6/5/2009
30	2081194	RESIDENTIAL	Closed	3 Ross Dr #8	Bloomington	East Bloomington	$119,800	$113,500	4/16/2008	1/4/2009	3/19/2009
31	2083672	RESIDENTIAL	Closed	2201 Peirce	Bloomington	Town & Country	$124,900	$118,000	11/9/2008	3/22/2009	4/22/2009
32	2073310	RESIDENTIAL	Closed	922 S Vale	Bloomington	Town & Country	$119,900	$119,300	8/26/2007	1/19/2008	3/4/2008

Many government entities make GIS-ready data layers available at little or no cost. Environmental characteristics, street centerlines, parcels, planimetrics, and zoning layers may be available from a local GIS provider. Typically, GIS layers will already be georeferenced and are ready to be joined or linked to any specific real estate data that needs be analyzed spatially. The following examples (Exhibits 11.16-11.19) show GIS data used for real estate analysis.[4]

In the four illustrations, multiple listing service data is "mashed" with a map from Virtual Earth, providing appraisers and analysts spatial views of tabular data sets. Market activity is viewable with direct relationships to distance. Planimetric data such as building footprints developed from digital aerial photography provide the appraiser with an "eye-in-the-sky" perspective from the desktop.

In Exhibit 11.16, the block demographic is immediately available, with average-sized homes on large lots being the norm. On the west end of the block are average-sized homes on small lots. Dashed lines are platted lots, and solid lines are parcels. Before zoning codes, large lots were split to accommodate higher building density. A one-car detached garage is common on the small lots, with two-car garages common on the larger lots. A few of the properties have no garage. Original access to garage parking was intended to be from the 20-foot alley, which is still active.

When using maps of this type, appraisers are cautioned to note the date the planimetrics are referenced. Changes may have occurred since they were developed. For example, the building footprints shown in Exhibit 11.16 are from a 2006 photo. A history of planimetric layers can reveal development patterns.

4. Maps and data were generated using Paragon 4 MLS online (August 2009).

Exhibit 11.16 Planimetric Layer Showing Building Footprints, Lots, and Parcels

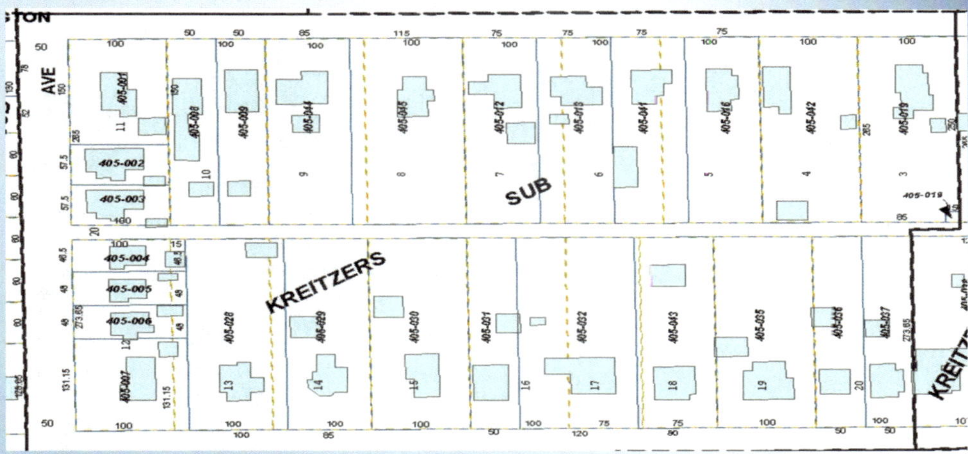

Source: City of Bloomington Township Assessor, Bloomington, IL, 2007-ESRI, ArcMap Ver. 9.2.

Exhibit 11.17 Neighborhood Sales by Story Type and Year of Sale

Exhibit 11.18 Quarter-Mile Concentric Distance from Subject Property Location

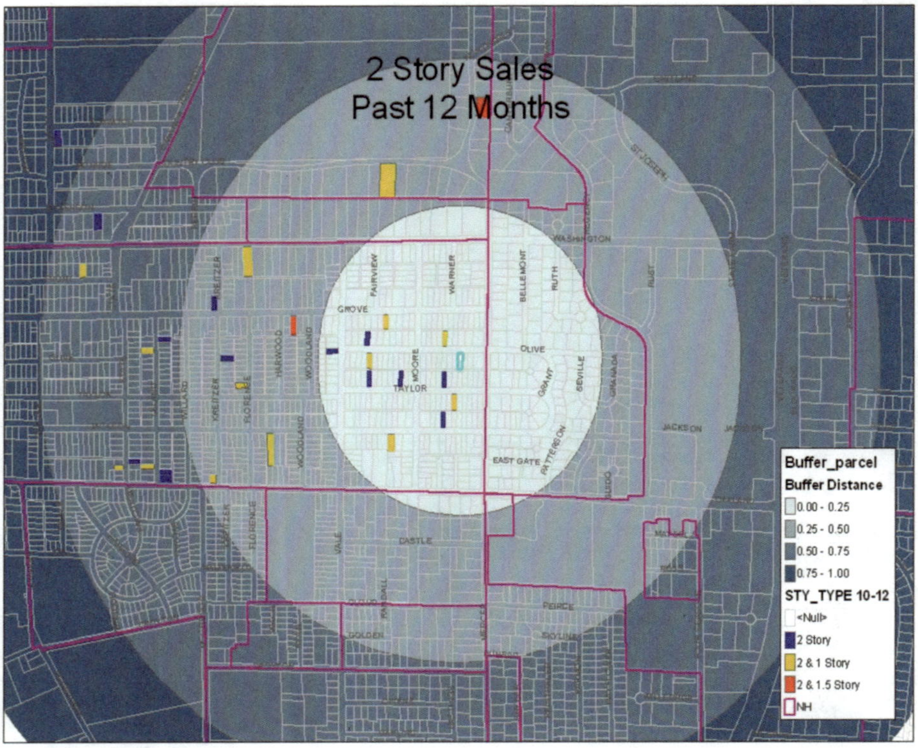

In Exhibit 11.17, GIS is used to show neighborhood sales by story type and year of sale. Exhibit 11.18 shows the distance between the subject property and other residential properties by type. Exhibit 11.19 shows the nonconforming uses by parcel. The light-colored parcels fail zoning requirements and are nonconforming uses based on the current zoning code. Properties with nonconforming uses have limited future economic life because local restrictions may prohibit the property from being modified or rebuilt.

Market Analysis

Residential real estate is an economic good in which ownership rights are transferred from seller to buyer. Appraisers attempt to describe these transactions based on the physical characteristics of the real estate. Residential appraisal forms provide for ancillary information about three properties similar to the subject property but do not provide for any market analysis of the subject property's market area, which is the foundation of the valuation. In the absence of

Exhibit 11.19 Nonconforming Uses by Parcel

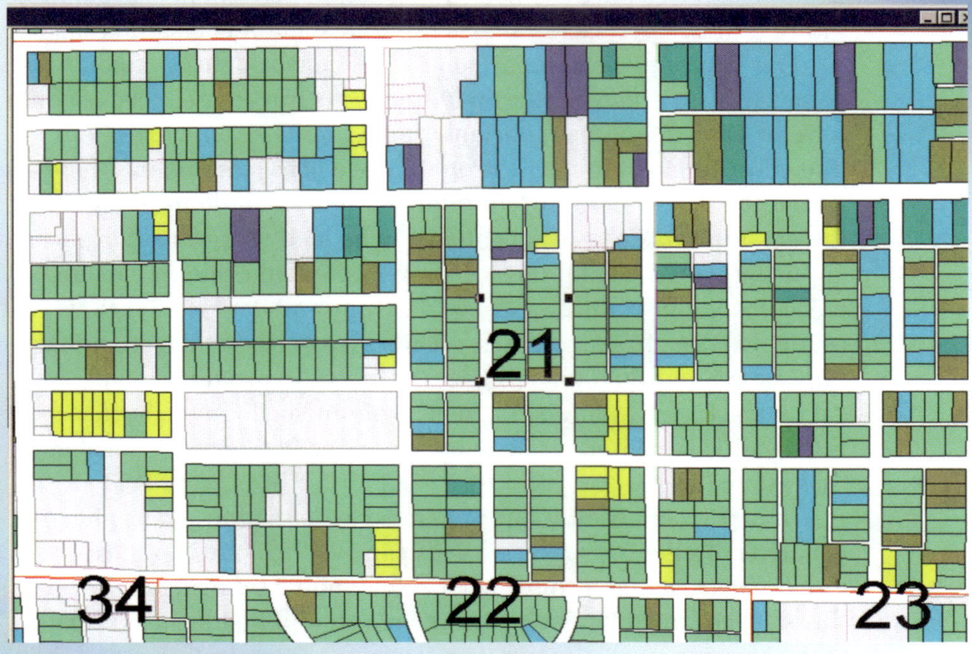

information about the market area, the subject property's value can only be developed in isolation from the surrounding marketplace, a disconnect that can fuel the rapid inflation and deflation of prices of property. In appraisal reports, the appraiser's opinions of value may not necessarily be related to the adjacent property markets or reflect existing trends, which would be a serious deficit.

Professional appraisal organizations state that market analysis requires the use of data that reflects the values of a particular area. This information can be taken from the individual appraiser's files or integrated with local government assessment offices, third-party data warehouses, or multiple listing services. Unfortunately, market analysis is often assumed and ignored. Neighborhood analysis is typically based on the experience and local knowledge of the appraiser rather than statistics derived from property record databases. Although third-party information derived from data warehouses is sometimes used, appraisers should have access to the whole universe of property information to provide accurate information about market areas and neighborhoods.

Without access to electronic real estate information, appraisers cannot be accurately informed. Therefore, their appraisal reports will not fully inform their clients about potential future changes in market conditions. The secondary mortgage markets (Freddie Mac and Fannie

Mae), AVM providers, appraisal management companies (AMCs), and data warehouses have extensive information that can be used to project future trends. Appraisers can use this information in a similar fashion.

For example, the comparison of U.S. census data on median incomes to rising house prices within census tracts or zip codes could have predicted the unsupportable rapid increases in market values that contributed to the 2007 drop in home prices. In the late 1980s, a similar comparison in the commercial market could have indicated the potential risk of market collapse. The lack of appraisal expertise applied to various computerized real estate and economic databases has contributed to the "roller coaster" ups and downs of various real estate markets. In many cases, clients turn to national or regional forecasts rather than to appraisers, who could produce local area estimates. Unlike appraisers, most economists do not have access to the detailed local real estate databases needed to provide local area real estate forecasts.

In market analysis, appraisers need to define the geographic market areas of subject properties. This can best be done in a narrative that relates to a map printed from a geographic information system. The data collected for each market area should include a summary of the housing stock, the number of sales over the last year, and the number of current real estate listings in the market area. Information on housing stock (based on population estimates) is available in electronic files from the municipal government assessor's office or third-party real estate data warehouses. In areas covered by multiple listing services, most sales transaction information can be downloaded in digital format.

A market area report could provide the actual ranges and median year built, size, and prices of the housing stocks. With this information, the client should be able to understand the local market area and the factors that relate to price. For example, when a smaller, older house sells in the upper range of prices for its market area, then additional review, explanation, and support is necessary.

Market areas need to be identifiable to the general public. In many cases, market areas are defined by political or school district boundaries. Communities are also identified by planning districts, census boundaries, or neighborhood-derived boundaries that are not compatible with the political subdivisions and school districts. Market areas may include, but are not limited to, towns, villages, small cities, school districts, or individual school boundaries. In urban areas, several market areas subdivide large cities into more homogeneous areas.

Appraisers can purchase local market information from various third parties, who develop and maintain real estate market reports.

These third-party reports may be too general, but the appraiser should be able to compile several reports into a basic market analysis. The agreement of clients (e.g., lenders) on standardized information will provide the incentive for third parties to create special or customized real estate market area reports. Whether appraisers develop their own reports or purchase third-party reports, they must understand that market areas change over time. As a result, research must be ongoing. Market areas should be reviewed using statistical software and GIS.

Submarkets and Neighborhoods

Just as individual properties cannot be valued in isolation from the market area, neighborhoods must be considered in relation to the market area. Neighborhoods are clusters of like properties within market areas.

Market areas may have variations in the range of prices that can be identified as location-based (or location-related) effects. These variations are a function of neighborhoods, which delineate geographical areas within the market area that conform to the norm or vary toward the higher and lower end of the price range within the market area. Location-related value variances can be due to variations in property characteristics, clusters around external influences (such as lakes, hills, access to transportation, and religious or social activities), or nonharmonious areas (such as those next to commercial properties, highways, and other undesirable neighbors).

The delineation of neighborhood boundaries within market areas involves more art than science. Neighborhoods need to be small enough to encompass the local characteristics but large enough to have the minimum number of sales to support its variance from the median location adjustments for the market area. Without a sufficient number of sales occuring within one year or possibly two years of the sale of the subject property to support the neighborhood adjustment, the assignment of a neighborhood adjustment is based on the professional opinions of each appraiser. Professional opinions do not provide the clarity needed by lenders to make decisions on mortgage amounts and to perform risk analysis.

While identification of market patterns can be done without the assistance of GIS, supporting neighborhood geographic areas can be very difficult, and clients can have even more difficulty visualizing the geographic area and price ranges within neighborhoods. Several geographical areas are already defined and available to both the appraisers and clients, including the following:

- Census tracts/blocks
- Zip codes

- Local government-assigned neighborhoods
- Assessment jurisdiction-assigned neighborhoods

Most of these designations are not designed to relate to clusters of homogeneous property values. For example, zip codes are part of most addresses and are familiar to appraisers and clients alike. However, zip code delineations can represent large geographic areas. "Zip+4" codes represent smaller areas, but the area covered by a zip+4 code may be too small for neighborhood identification. The subject property's census tracts also represent geographic areas, thus census tracts/blocks could reduce the geographic area to a reasonable size. Conversely, several census tracts could be clustered to provide sufficient economic information (sale transactions) to support market areas and neighborhoods. The main advantage of zip codes and census tract/block information is that national clients will be able to identify additional economic information using these geographic areas.

GIS provides the best evaluation tool to identify the general location that identifies the geographic references (i.e., location) with a reasonable number of sales within a specified radius from the subject property. Using standard distances from the subject property may help the appraiser avoid having to redefine the geographic neighborhood boundaries, which may shift from year to year. The buffer (e.g., 1 to 3 miles) should provide the closest properties that are comparable sales to support the estimate of value for the subject property. A GIS database for desktop and laptop computers can be distributed between networked computers and be accessible over the Internet. Various GIS maps identify property boundaries, streets, and aerial photography maps that can be added to the appraisal reports.

In order to analyze trends, GIS provide one way to identify specific trends by different levels of geographic areas (e.g., neighborhood, city, and region). State or national real estate trends may also become part of market analysis for individual property values. For example, mortgage rates and the availability of mortgage funds are external to the market area but will have an influence on property value. National economic trends such as inflation will affect house prices. When a large employer closes a facility within a large city, it may affect property values in several market areas within or nearby the subject property city.

Model Specification

One of the many benefits of appraisal organizations is that they provide standardized methods to explain the valuation process performed by appraisers. The valuation process has been standardized into specific structures and formulas known as *model specification*, which is the development of a model structure in a statement or equation based on analysis and appraisal theory.

Since the 1930s, North American appraisal societies have defined and refined three valuation approaches: cost, sales comparison, and income capitalization. These precomputer age processes were designed to illustrate the appraiser's thought process and support the value estimate. These approaches created a big step towards transparency of the appraisal process, as opposed to simple declarative statements of value. For many years, appraisal organizations have trained appraisers with practical formulas to explain appraisers' valuation opinions. When introducing new technology, such as statistical models, the ability to relate to existing appraisal model structures is very important for acceptance by appraisers.

A basic real estate model specification is

$$MV = [(\text{land size} \times \text{unit price}) + (\text{building size} \times \text{unit price})] \times \text{location percentage adjustment}$$

where

MV = market value

In this case, the location adjustment applies to both land and building values. Depending on market conditions, the location adjustment could be applied only to the land, or distinct values could be applied separately to the land and to the building.

Additional land and building characteristics, such as the number of bathrooms or condition (age), can be added to the basic land and building size–related values. Additional buildings can be added as a separate component to the model specification. If the appraiser is appraising a property with an extremely unique feature that is not present in sufficient cases to be significant in the market model, these unique features can be added based on the appraiser's experience or cost valuation.

In the 1970s two university professors, Robert Carbone and Richard L. Longini, proposed a new dynamic valuation model specification that is identified in IAAO's textbooks as a *hybrid model specification*. The basics of the hybrid model are

$$MV = \pi\, GQ \times [(\pi\, BQ \times \Sigma\, BA) + (\pi\, LQ \times \Sigma\, LA) + \Sigma\, OA]$$

where

MV = estimated market value
π GQ = the product of general qualitative variables
π BQ = the product of building qualitative variables
Σ BA = the sum of building additive variables

π LQ = the product of land qualitative variables
Σ LA = the sum of land additive variables
Σ OA = the sum of other additive variables[5]

This general model specification provides for separate land and improvement components, additive (dollar) and multiplicative (percentage) adjustments, and the consideration of other improvements beyond the main building structure. Interestingly, this general model specification with some modification can replicate the cost and income capitalization approaches and develop coefficients for the sales comparison approach. This formula provides the basis for the review of small or larger amounts of data within a specific model specification.

When appraisers understand when and how to use the formula, this formula's valuation estimates may become more dynamic through the uses of data that reflect current market conditions. With the advancement in statistical models from the academic community and their application by professional appraisers, there is a deliberate movement of appraisers toward using statistical analysis to develop more precise valuations.

Many scholars and instructors of advanced appraisal courses continue to stress the use of linear model specification as the primary method of advancing the appraisal profession into the statistical age. Linear and log-linear calculations have been performed on slide rules since the seventeenth century. Linear modeling is a well-proven model specification with many applications. However, developing comprehensive model formulas requires extensive training and experience. Unfortunately, linear model specification is counterintuitive to appraisal theory and the reasoning process taught to appraisers by professional appraisal societies. Linear model specification is a "nonstarter" for most appraisers, as they are concerned with how they will explain the model specification to clients or in court.

Appraisers prefer to use model specifications, formulae, or structures that demonstrate the relationship of economic information such as sale price, price per square foot, and income to property characteristics. Model specification is part of most appraisal reports that show the process to arrive at the valuation (e.g., point estimates or value ranges).

To appraisers, model specifications can seem to be intuitive. The appraiser's ability to support adjustments directly from market data is limited by the availability of data as well as the appraiser's ability to manipulate databases and demonstrate significant correlation between the economic variable (i.e., price) and the property charac-

---

5. International Association of Assessing Officers, *Property Appraisal and Assessment Administration* (Kansas City: IAAO, 1990), 319.

teristics. Most buyers of residential properties only consider a limited number of property characteristics when purchasing a home. For example, a buyer may select a property and agree to a price based on location alone. Depending on the scope of work and intended use of an appraisal, a complex cost approach formula may be required. The cost approach model specification may need to be simplified to match the type and availability of data, while the appraiser is still required to explain the variance in the prices of comparable properties.

A model specification should be:

- Accurate–the statistical quality results should demonstrate the quality of the model specification.
- Rational–the model specification and calibration coefficients should be understandable.
- Explainable–the appraiser or model provider should be able to present (and explain) the model and its results.
- Frugal–once the model calibration has reach its best quality statistics, the appraiser should remove the least significant property characteristics one at a time until the quality statistics begin to noticeably decline.[6]

The desire for frugality in the model specification is at odds with the typical appraisal process in which the appraiser may be attempting to allocate value to as many property characteristics as possible. Frugality is important as property characteristics that have low significance (*t*-test values of less than 2.00) contribute very little to value. But the chance of human error in recording the data increases exponentially with each additional property characteristic included in the model specification. So increasing the number of model variables may cause errors in the application to one or more subject properties. If the variables are chosen well, fewer model variables may be able to provide the same level of accuracy.

Unlike the traditional "three comparable property" sales comparison approach, market models need a reasonable number of cases for the property characteristics that are to be studied. As a general rule, for each variable studied in a model, 15 sales should be used in the market model. Appraisers do not need to develop individual models for every property assignment. Where the appraiser is geographically competent, market models–especially for one- to four-unit "uniform standard" residential appraisal reports–can be developed, used for

6. Michael Ireland and Patrick M. O'Connor, "Introduction to Valuation and Spatial Analysis" seminar, presented in conjunction with GIS/CAMA Technologies Conference sponsored by URISA and IAAO (March 8, 2010).

a period of time, and updated as market conditions change. Market models and calibration provide the statistical support for the model coefficients, which can be time-adjusted for up to six months or a year based on the rate of change over time within the market area.

As local appraisers learn more about developing valuation models, the market areas served by these appraisers can be expanded to other/larger geographies–town, city, county, or state. In fact, when reliable secondary data is available, experienced model builders may be able to construct national models.

Model Calibration

Model specification development includes defining the structure of the property characteristics that will be analyzed. Model calibration is the process that tests the specification against known economic indicators such as sale prices. Model specification and calibration are an iterative process. Model specification is tested by calibration, which leads to respecification and recalibration. This process continues until the best statistical quality available from the data set is achieved.

Model calibration is the development of market adjustments (coefficients) based on market analysis. For years, appraisers have been reducing the variance in data sets by limiting comparable sales to those in close proximity to the subject property or the subject's neighborhood, by selecting comparables that are about the same age and size as the subject property, and by using similar techniques. However, to provide support for adjustment grid conclusions, appraisers must explain the variance in property characteristics. For example, to determine the contributory value of a swimming pool to a house, the appraiser must have current sales transactions of homes with pools and other comparable sales of properties that do not have pools.

Statistical model specification must also consider the variance in data. This statistical process assumes variance in the data set. Typically, appraisers attempt to use data sets that are filtered down to a few properties that are very much like the subject property. Regression needs the variance in property characteristics to determine the relationship of that variance to the variance in sale prices.

The most used statistical calibration technique is multiple regression analysis (MRA). Regression calibration finds the "best fit line" through the center of the data. As such, the regression analysis tends to provide coefficients that represent the central tendency values of each contributing variable. For example, if every house in a modeling file has brick exterior walls, the market model will not assign any significance (i.e., no added value) to the brick walls of housing in the model. In climates where central heating units are required for year-round occupancy, almost every house will have central heating. Thus, the market model will assume that heating is part of the value per square foot of living area and will not apply a separate coefficient value to heating, as there in little or no variance. The lack of variance does not mean that whole house heating does not contribute value to the property. Rather, the heating variable will not significantly contribute to explaining the variance in market value in the market model. Thus, variance in the model is important to explain how attributes such as brick walls or central heating are valued in markets. Models that encompass larger geographic areas usually have more diversity of data and tend to have more sales transactions. In order to more accurately reflect the market using a regression calibration, small geographic areas with limited variance are not preferred. In this case, the appraiser would prefer to expand the market to include larger areas (market areas) with more variance in the data.

When appraisers use statistical programs to model data, the most common calibration techniques are based on multiple regression analysis models. While there are many regression programs, linear, multiplicative, and non-linear regression will be presented in this chapter. Regression models are available within Microsoft Excel, but this is not the preferred program. The regression model in Excel is a generic regression model and does not reflect the type of specification and model adjustment required to analyze real estate data. Some appraisal software systems include regression programs. Most statistical software packages include regression programs. At least two statistical software companies, SPSS and NCSS, have modified their software to accommodate appraisers with appraisal models and sales ratio routines.

The selection of the regression calibration program is based on model specification, just as the development of the model specification is based on the available or selected regression calibration program. Additive models require linear regression calibration programs. Additive models may also be called *linear regression models* because the process develops the best straight-line fit through the data set's sale price point compared to tested variables. Using the same basic characteristics as used earlier in this chapter, the additive model specification is

MV = constant + (lot size × unit price) + (building × unit price) + (location × percentage)

Linear regression is similar to links in a chain. The value of each cluster (characteristic multiplied by rate) is added to the next cluster (component). Appraisers are concerned with the use of linear regression models because the technique suggests that the constant does not relate to any particular property characteristic. The constant requires some explanation outside of typical appraisal theory. Qualitative property characteristics like location are not related to major components of the subject property, such as land, building, or both. In this model, location value is added without any additional specification within the chain of calculations. The land and building rates are single numbers (that are applied to all sizes) within the range of the data set. This could mean that the building value per square foot is applied without unit value adjustments. In order to correct for potential or perceived problems with additive models, modifications such as stratification or curvilinear variables are used. For an appraisal modeler, these perceived problems can be overcome with experience in model development.

The second type of regression model is multiplicative:

Log_MV = log_constant + (log_lot size × unit value) + (SQRT_building × unit value)
+ (log_location × coefficient)

Log represents the natural logarithm function and SQRT is the square root function. Multiplicative value estimates (Log_MV) are converted back to dollars by using the exponential function. Generally, multiplicative models are calibrated using linear regression calibration programs. The models become multiplicative when the dependent variable and independent variables are converted to curvilinear functions. For example, the building size is converted to the natural log or the square root of building size. The most typical conversion is a log-linear model where the dependent variable, sale price, is converted to the natural log of sale price. With the curvilinear functions already built into the property characteristic variables, linear regression then tests the model structure against the log of sale price.

Along with the development of estimated curvilinear functions, the log linear model structure reduces the variance within the continuous variables, such as square foot of living area. The relationship of price to property characteristics (size and count) tends to be curvilinear. Thus, multiplicative regression models tend to provide the best quality statistics. This model structure has many of the same problems as the additive model from the appraiser's perspective. In addition, thinking in scientific notations can be difficult for appraisers and their clients. One problem with a multiplicative model is the requirement to predetermine the scientific functions necessary to create the curvilinear variables. These perceived problems can be overcome with experience in model development.

The IAAO's hybrid model specification requires an additional regression calibration program that is available in several statistical software programs. This model specification requires calibration programs that have additive components, such as price per square foot, and nonlinear components, such as the building size adjustment for price per square foot.

A basic hybrid model specification is

MV = ((land size × unit price) + (building size × unit price))
× (location adjustment percentage^power)

This model specification indicates that the calibration will provide the statistically appropriate curvilinear function (coefficient) by adjusting the location percentage (relative value differences between neighborhoods) by a power function. This process is the result of the model calibration rather than a preconceived curvilinear function, as in the multiplicative regression model. The use of the power curves can function as the size adjustment for land and building areas or any other property characteristics that are numeric continuums. The hybrid model structure permits the application of appraisal theory and training to specify how the model reflects the values related to real estate trends and property characteristics:

MV = ([(land size × unit price) × (land size adjustment^power)] + [(building size × unit price)
× (building size adjustment^power)]) × (location adjustment percentage^power)

For land and building size, the adjustment is the individual property's size divided by each variable's median size. The power identifies the nonlinear relationship of size to sale price differences.

Similar to real estate appraisal, modeling specification and calibration is "part art and part science." Specifying models is not a perfect process. It depends on the available data to be examined, the quality of the data set, and the experience of the appraiser. Depending on the scope of work, appraisers may develop model specifications and run calibrations for every appraisal assignment. When selecting comparable sales for each appraisal assignment, an appraiser may have a tendency to minimize the variance between the subject and comparable properties. When appraisers reduce the variance within their data sets too much, the remaining data may not have sufficient variance in property characteristics to accurately identify differences in sale prices.

When the scope of the market analysis (modeling) is to use market formulas, market analysis may geographically include one or many market areas. When modeling by market area, monthly time adjustments can be assigned for the months between the model calibration runs. As a practical matter, market areas have larger data sets than neighborhoods. Adding a few new sales per day or month will not significantly change the market coefficients. Thus, time analysis of one variable (time) is more practical than frequently rerunning the multiple regression analysis on many variables.

The scope of work for the appraisal assignment will determine the required frequency of model specification and calibration (market analysis). Appraisals for mortgage lending on many properties will have a different scope of work than appraisal reports for expert witness testimony.

Automated Sales Comparison Method

Automated valuation models (AVMs) rely on comparable sales to calibrate a model to local real estate trends. In market areas where properties have a high degree of homogeneity, the AVM will produce value estimates with a high degree of precision. The sales comparison model format is as follows:

$$MV = SP_c + ADJ_c$$

where

MV = market value

SP_c = sale price of the comparable

ADJ_c = adjustment to the comparable ±

In the application of automated comparison sale programs, as the variability of the data used to calibrate an AVM increases, the precision or accuracy of the value estimate often decreases, becoming less reliable. This may seem contradictory to regression modeling. Regression is performing market analysis to find the property characteristics coefficients. Automated comparison sales programs apply the property characteristics coefficients to individual properties. Running automated comparison sales programs is not market analysis. Lenders like the immediacy of AVM products, which meet the requirement of instant loan approval from the desktop, following the example of credit approval. AVMs are often used as value checks to detect appraisal fraud and poor appraisal practices. Lenders also use AVM estimates to reduce the time between the loan application date and the approval date. Lenders that keep statistics on loan application and closing dates often see that a long interval between those dates reduces the likelihood of a loan actually closing, and each loan application that does not reach the approval and closing stage is lost revenue.

Gregory J. Landretti discusses the three criteria needed to produce credible values by the sales comparison approach:

1. An arm's-length sale of the subject property is the best evidence of market value.
2. In the absence of an arm's-length sale of the subject property, sales of reasonably comparable properties are the best evidence of the market value.

3. In the absence of an arm's-length sale of the subject and of reasonable comparable properties, the assessor should consider all other data collectively.[7]

Landretti even developed the procedure in Lotus 1-2-3 in 1986. Convincing the private sector of the advantages of an automated sales comparison approach rather than appraiser-selected and -chosen adjustments that are not consistent from appraisal report to appraisal report has been a challenge for two reasons. First, acquiring the data in a format ready for use in a spreadsheet or similar software program is not always possible. Second, too much of the effort to move appraisers towards use of the automated sales comparison approach was spent on convincing private appraisers to store their data in a spreadsheet or database rather than developing or improving upon the traditional methods of creating AVMs. When properly executed, automating the sales comparison approach does more to support the appraiser in meeting the USPAP requirement of appraisal analysis that is "independent, impartial, and objective."

The same sales used to develop an AVM are used again as each method relies on sales verified and available for electronic processing. Executing the automation of sales comparison is a three-step process:

Step 1. Selecting the most comparable sales to the subject property available.

Step 2. Adjusting the selected comparable sales to the subject property.

Step 3. Reconciling the adjusted sales into an opinion of value.

This all should sound familiar, and it is. Each step and how it is processed are explained below.

Step 1. Selecting the Best Comparable Sales

To select comparables sales, numerous techniques are used in practice, with the exception technique being the most popular. The most efficient and objective technique is the use of a distance metric that assigns a dissimilarity index to each sale in the file. The more similar a property is to the subject, the lower its metric value. Two indexes frequently used are the Euclidean distance metric and the Minkowski metric. These distance metrics rely on a selection model or algorithm specified by the appraiser to find the most comparable sales for adjustment.

The Euclidean distance metric is expressed as follows:[8]

$$\Sigma[W_j(X_{sj} - X_{ij})\sigma_j]^2$$

7. Gregory J. Landretti, "Commercial/Industrial Property and Sales Comparison Approach," *Property Tax Journal*, vol. 5, no. 4 (December 1986): 235-250.
8. International Association of Assessing Officers, *Mass Appraisal of Real Property* (Kansas City: IAAO, 1999), 201.

Some analysts may recognize this as having its roots in the Pythagorean Theorem. Where X_{sj} is the value of the subject attribute, X_{ij} is the value of the same attribute for the comparable property being analyzed. The symbol σ is the standard deviation for the specific attribute j and W_j is a weight assigned by the appraiser for each specific attribute used in the selection. Weighting of selection attributes aids the selection model in overcoming the lack of a distinguishing scale when discrete variables are used for selecting the comparable sales.

Exhibits 11.20 and 11.21 demonstrate how the metric calculation works. For Sale 1, the sum of the standardized weighted squared differences is 4.19657, based on the three selection attributes chosen. For Sale 2, the sum of the standardized weighted squared differences is 8.31715, based on the three selection attributes chosen. Based on the metric, Sale 1 is considered more comparable, and while it would not necessarily require fewer adjustments, the total adjustment amount would be less.

Standardizing the attributes reduces the magnitude of difference between attribute values, placing them all on a similar scale, analogous to a Z-score. Using $(X_i - \overline{X}/\sigma)$ where X_i is the value of the attribute and $\overline{X}$ is the mean, the sigma σ represents the standard deviation of the variable, which produces the Z-score.

Assigning the attribute weight is the art of using this selection method. For example, if the appraiser found that building age was more important than size, increasing the weight on age to 3 and giving

Exhibit 11.20 Example of Euclidean Distance Metric—Sale 1

Attribute	Subject	Sale	Difference	Standard Deviation	Standardized Difference	Appraiser-assigned Weight	Weighted Difference	Squared Weighted Difference
Living Area	1500	1660	-160	225.7	-0.70891	2	-1.41781	2.01019
Age	8	4	4	4.1	0.97561	1	0.97561	0.95181
Quality	4	5	-1	0.9	-1.11111	1	-1.11111	1.23457
								4.19657

Exhibit 11.21 Example of Euclidean Distance Metric—Sale 2

Attribute	Subject	Sale	Difference	Standard Deviation	Standardized Difference	Appraiser-assigned Weight	Weighted Difference	Squared Weighted Difference
Living Area	1500	1700	-200	225.7	-0.88613	2	-1.77226	3.14092
Age	8	10	-2	4.1	-0.48780	1	-0.48780	0.23795
Quality	4	6	-2	0.9	-2.22222	1	-2.22222	4.93827
								8.31715

living area a weight of 1 would make Sale 2 (Exhibit 11.21) the more comparable property. Using metric selection for assessing the comparability of the property attributes is a more impartial and objective process. The selection model grants every true comparable an equal chance of selection based on the degree of comparability. Exhibit 11.22 provides additional context in setting up an automated metric selection with selection weights assigned to specific property attributes.

Exhibit 11.22 **Distant Weight Example from NCSS Statistical Software**

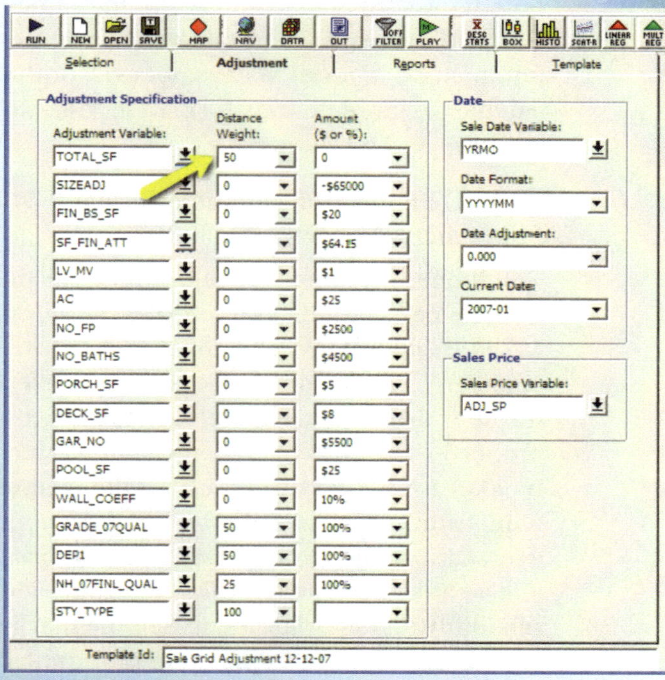

Source: NCSS 2007 Ver. 07.1.13 Comparable–Sales Price, released Nov. 5, 2008, J. Hintze.

Defining the metric selection model is a skill and an art. Using distance metric selection will improve the appraiser's impartiality and replace judgment with an objective-based method.

Step 2. Adjusting the Selected Comparable Sales to the Subject Property

As part of an AVM or CAMA system, adjustments to the selected comparable are ideally based on statistical model coefficients derived from the model calibration process previously discussed in this chapter. The same sales that have been validated and used for model development are also used in the automated comparable sales application. The market coefficients for the AVM are derived from these sales and validated through the modeling process. The use (and adjustment) of sales provides a higher level of detail and precision than the use of AVM values alone. Since the coefficients are derived from the sales and have been tested with various statistical methods within the modeling process, the appraiser gains yet another level of independence, impartiality, and objectivity.

In the example of automated comparable sales selection, the third column holds the rate of adjustment for the specific property attribute. The rates of adjustment are the same coefficients used in the AVM and were derived from a nonlinear multiple regression calibration method. If an AVM does not exist or the coefficients seem illogical due to, say, data transformation, "market intuitive" adjustments are used as a substitute.

Step 3. Developing an Opinion of Value from the Selected Sales after Adjustment

Traditional valuation education suggests that the appraiser arrive at a single point value opinion using the sales comparison approach. *Reconciliation* is the term for the process of intuitively weighing how applicable each comparable is to the subject. One textbook instructs the appraiser to place the greatest reliance on the "most comparable property," which is usually the comparable with the fewest adjustments.[9] A different text describes the method for reaching a single point estimate by the sales comparison approach and includes the explanation, "It is important that the appraiser consider the strengths and weaknesses of each value indication derived, examining the reliability and appropriateness of the market data compiled and the analytical techniques applied in the comparative analysis."[10] Little guidance is provided in either discussion to help the appraiser understand what "most comparable property" actually means. Words like "reliability" and "appropriateness" are subjective terms and add nothing to the appraiser's independence, impartiality, or objectivity. Only as AVMs have become more commonplace has the idea of using statistical measures to evaluate comparability become an acceptable alternative to the previous subjective method.

The distance metric selects the sales most comparable to the subject based on attribute difference. The top three to five comparable sales are identified and, based on user preference, listed in an adjustment grid in the order of comparability. The first sale on the grid is typically considered the most comparable. The remaining comparables are scaled based on a measure of similarity to the best comparable, or an algorithm using each comparable's total metric value is used to define overall comparability. Each comparable sale is assigned a comparability index to give the appraiser an objective measure for each sale on the grid prior to adjustment. The single point estimate can then be based on a weighted comparability index.

9. International Association of Assessing Officers, *Property Assessment Valuation*, 2nd ed. (Kansas City: IAAO, 1996), 124.
10. *The Appraisal of Real Estate*, 13th ed. (Chicago: Appraisal Institute, 2009), 312.

In this way the derived value is influenced by the best of the comparable sales. In fairness to the traditional methods, judgment can still be the preferred choice in selecting and ranking the comparables. See Exhibit 11.23 for a full example of the output from an automated comparables program.[11]

Special conditions can also limit comparability when appraisers are required to use comparables with location or time limits. The report in Exhibit 11.23 used the selection variables from sales that are within one mile of the subject. Special conditions required by clients often limit the available comparable pool, limiting better comparable indicators in some cases. The comparables on the grid are mapped for reference (Exhibit 11.24).

Once the three-step process is complete, use of an automated comparable sales selection method will save valuable time and improve consistency in a comparable's adjustments. The sale comparison technique provides traditional explanations over the more complex development of a statistically modeled AVM. Many clients require comparable sales on an adjustment grid. Appraisers have been entering comparable sales data into forms software since the mid-1980s, yet few of the forms software developers have added value to their products by including an objective, impartial technique for reusing and selecting comparables. Comparable sales do not require a database. In 1986, Landretti demonstrated that a database is not needed by completing the entire exercise from comparable sales selection to display and adjustment on a grid using Lotus 1-2-3.

The time has come for automated sales comparison programs. It is just a matter of who will be the first to start using automated sales comparison to meet valuation standards and USPAP requirements.

Quality Assurance

A large majority of appraisal assignments are requested from lender clients. Many lenders hold a general perception that appraisals do not "add value" to the lending process because many opinions of value reflect the sale price transaction amount. Some lenders have indicated that appraisals are seen as no more than a regulatory requirement, rather than a professional analysis necessary for mortgage commitments on risk. The appraisal profession will not survive simply as a government requirement.

The traditional residential appraisal report is most often a value opinion with three examples of recent sales transactions. It is impos-

11. Generated using NCSS 2007 Ver. 07.1.13 Comparable—Sales Price, released Nov. 5, 2008, J. Hintze.

Exhibit 11.23 Comparables Reports—Area 21

PARCEL_ID	21-03-431-021	21-03-462-007	NO. 1	21-03-429-007	NO. 2	21-03-452-029	NO. 3	21-03-381-024	NO. 4
	Value	Value	$Adj	Value	$Adj	Value	$Adj	Value	$Adj
ADDRESS	1420 E OLIVE	408 FLORENCE AVE		104 S MOORE		200 KREITZER		215 WILLARD	
YRMO SALE	200505	200705		200604		200705		200605	
STORY	1 Story	1 Story		1 Story		1 Story		1 Story	
GRADE	C+10	C+10		C+10		C+5		C+10	
WALL	Alum/Vinyl	Brick		Brick		Wood Siding		Brick	
SF CRAWL	0	0		0		0		0	
Yr Built	1925	1940		1925		1925		1925	
LOT SF	8,057	7,324		6,538		4,126		5,790	
NH	21	21		21		21		21	
Comparability		100%		100%		99%		97%	
Sale Price		$137,900	($1,600)	$150,000	$3,915	$118,000	($1,369)	$149,000	$3,457
TOTAL SF	1064	1019	$0	1143	$0	768	$0	980	$0
SIZEADJ	1.1024	1.1157	$875	1.08	($1,428)	1.208	$6,873	1.128	$1,675
Fin. Basmt	0	0	$0	0	$0	260	($5,200)	550	($11,000)
Fin. Attic	0	0	$0	0	$0	0	$0	0	$0
Land Value	29355	30654	($1,299)	24846	$4,509	21564	$7,791	24879	$4,476
% AC	1	1	$0	1	$0	1	$0	1	$0
Fire Place	1	1	$0	0	$2,500	0	$2,500	1	$0
# Baths	1	1	$0	1	$0	1	$0	2	($4,500)
Porch SF	200	130	$350	206	($30)	144	$280	160	$200
DECK SF	0	0	$0	0	$0	0	$0	0	$0
Garage Stalls	2	1	$5,500	2	$0	1	$5,500	2	$0
Pool SF	0	0	$0	0	$0	0	$0	0	$0
Wall Adj	1	1.1105	($8,613)	1.1105	($9,692)	1	0	1.1105	($8,710)
Quality Adj	1	1	$0	1.00	$0	0.9774	$1,130	1	$0
% Good	0.8	0.8	$0	0.8	$0	0.78	$1,363	0.75	$3,365
Location/NH	1.0929	1.0929	$0	1.0929	$0	1.0929	$0	1.0929	$0
Story Level	1 1	$0 1	$0	1 $0	1	$0			
Net $Adj.			($4,787)		($225)		$18,868		($11,037)
Sum $Adj.			**$16,638**		**$18,158**		**$30,683**		**$33,925**
Adj Sales Price	**$139,646**		$133,113		$149,775		$136,686		$137,963

Exhibit 11.24 Comparables Sales Map, Area 21

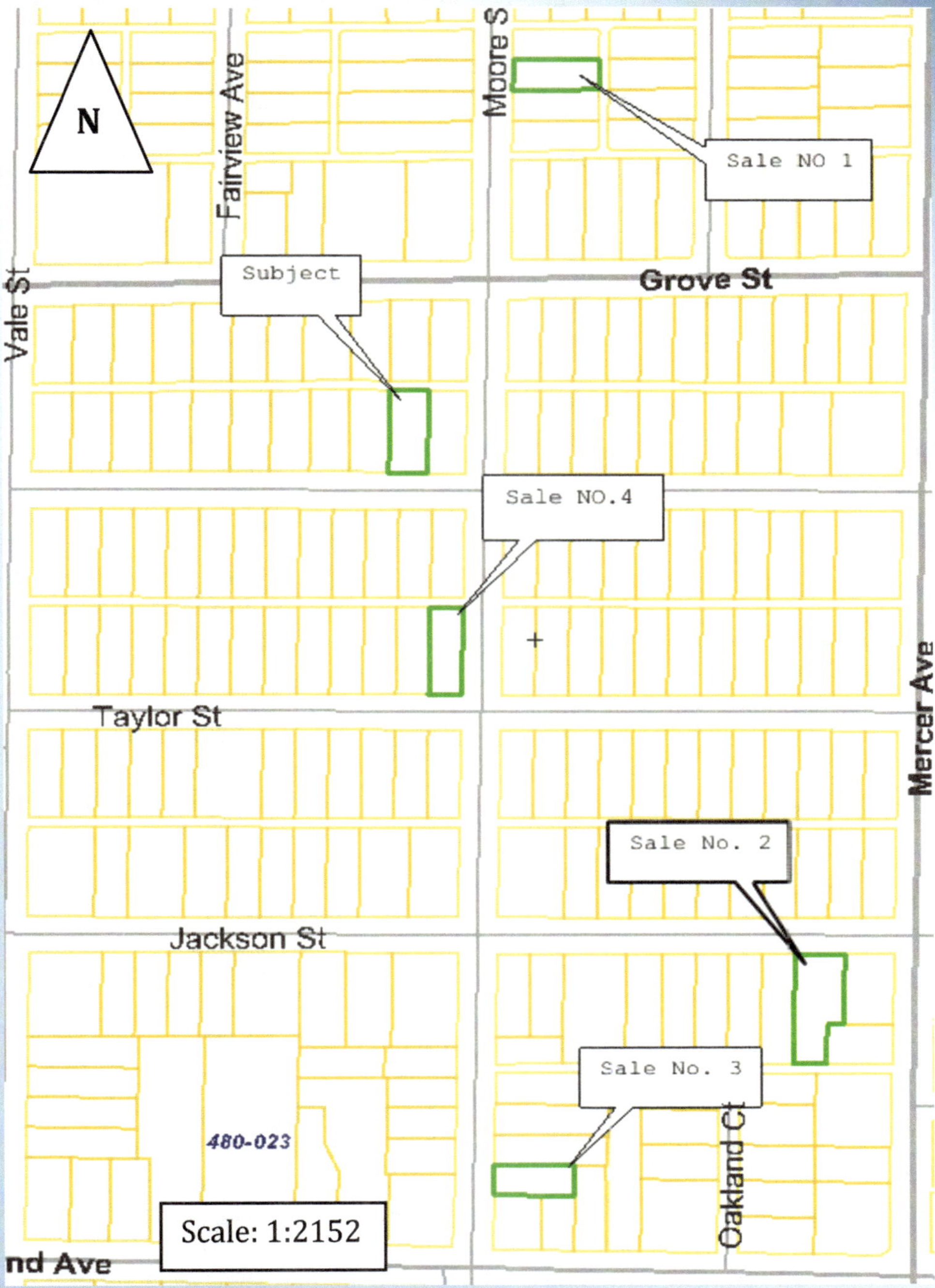

Source: www.wevaluebloomington.org, M. Ireland, 2009, and www.mcgis.org

sible to review the market value in one appraisal report in isolation, but once an AVM valuation model has been calibrated to the market, the formula can be applied one property at a time or to groups of properties. Groups of appraisal reports from a loan portfolio can be reviewed statistically, providing lenders with supported statistics on the loan-to-value ratios at the time of mortgage origination with updates to current dates. These statistical reviews provide better indications of the quality of each appraiser's valuation opinions than visual review of individual property appraisals. Statistical reports are useful for identifying loan officers or mortgage brokers who have been accepting risky and questionable loan-to-value applications at inflated transaction prices. The statistical review of loan portfolios will provide "the missing link" in a loan quality rating to improve the position of lenders wishing to sell the portfolios to financial markets.

Statistical software provides several measures of central tendency, such as mean, median, and weighted mean. These are based on point estimates. The statistical review of recent (within one year) sale prices by market area or neighborhood can indicate price trends either below or above the central tendency. Appraisals and sale transactions (loans) significantly above the central tendency of sold properties within the neighborhood may indicate greater risk for the loans or potential fraud. Certainly these loans should be more closely reviewed by the lenders.

The IAAO maintains standards for statistical review in the organization's Standard on Ratio Studies. For example, the coefficient of dispersion (COD) is designed for abnormal data distributions. The COD is similar to the coefficient of variation that requires normalization of data. Unfortunately, actual data sets of sale transactions are not evenly distributed, so nonparametric statistics must be used. The IAAO Standard on Ratio Studies indicates the acceptable range of CODs for the comparison of market value estimates to sale prices. A COD that is too large indicates poor relationships between appraisers' opinions of market value and actual sales prices. However, if the COD is too small, the appraisers may be "chasing sale prices in their opinions of value."[12] Both situations indicate potential problems for lenders that should be reviewed. Unfortunately, lenders prefer to shift the risks related to property valuation to the appraisers, so that appraisers are under pressure to hit the transaction prices. A comparison of the appraiser's estimate of market value to standardized market values (AVM values) can be used to find potential problems in loan applications.

12. International Association of Assessing Officers, *Standard on Ratio Studies* (Kansas City: IAAO, 2009), 49, Table 2-3.

Benefits of Computer-Assisted Market Analysis

Appraisers must learn to marshal more facts by improved methods of market analysis. In most geographic areas, electronic databases of residential properties are available at low prices from government assessors or data warehouses. Multiple listing services provide sales transfer information to their members. Appraisers can provide more statistical and geographic support to better explain their value opinions. Consistent and frequent study by appraisers of their service areas will increase their knowledge of those markets. By using this knowledge along with fact-based studies, statistical trends may highlight changes in market conditions that are not yet common knowledge.

Most of the data presented on valuation forms is based on the appraiser's observations and experience but must be supported with factual information. For example, estimating average neighborhood sale prices and effective age should require statistical review but is normally based on experience and opinions. By using statistical software and GIS, appraisers have support for their opinions. Appraisers performing appraisals for mortgage purposes ought to review their prior appraisal reports within a neighborhood before making adjustments on current assignments. However, expedited opinions of value may be emphasized more than quality in appraisal work. Unfortunately, some loan officers only want to approve the loan applications and do not evaluate the trends in appraisal estimates from the same region. This tendency may have changed recently because operational departments in lending institutions are forced to defend lending practices to government entities. In order for lenders to sell their loan portfolios, the appraisals may be used to indicate some level of quality within the portfolios.

As appraisers develop statistical skills, use databases, and learn how to maintain consistency in their valuation adjustments, appraisers may come to develop their own AVMs. Appraisers who understand their market areas will intuitively know what works and does not work in developing their models and estimating adjustments based on their own analyses. Local appraisers should not violate statistical principles, but they should be able to create local AVM valuation formulas that work in their market areas. Once AVM market value formulas are developed, estimates of market value will be readily produced on demand and can be provided to clients. The development of an appraiser's AVM formula is preferable to spending time on comparable checks or broker price opinions (BPOs).

Some lenders do not want to wait for proper validation of the transaction price but rather want to rely solely on credit reports to approve loan applications. Appraisers complain that lenders are pushing for quicker valuation confirmation at lower prices. Rapid confirmation of transaction prices can be generated by appraisers with command of the facts from databases of their market areas and simple AVM formulas. The ability of the appraiser to routinely expedite value opinions depends on the availability of data and improved e-commerce. Complete databases of large cities can be downloaded to laptop computers, and Internet access is available throughout most of the country. Once the market analysis and formula are derived, appraisers can more readily respond to lenders' requests. In some cases, an AVM-generated value may not be adequate for the scope of work of the assignment, and a property inspection or a full appraisal report may have to be put together as well. But the initial value estimate (supported by local data and AVMs) will permit lenders to move the loan application faster toward closing. In order to achieve this new level of efficiency, appraisers must add new skills and services to their professional offerings.

Generally, government assessors provide their updated information on property characteristics and values once a year. Multiple listing services permit appraisers to download their information as required. Third-party data warehouses send out monthly updates, but these are constrained by the annual updates of government assessors, even if new sales transactions are available monthly. Most assessment jurisdictions have a stable set of improved property characteristics on existing housing. Getting new construction information depends on MLS databases until the next assessor update, so the available electronic databases have limits. Freddie Mac and Fannie Mae have electronic databases that should be made available to the appraisal and lending communities to enhance their databases. Despite those limitations, appraisers can download sales information daily, but monthly downloading is probably more practical because a monthly update would be less time-consuming and sufficient new data would not be available daily. Once appraisers have established AVM models for market areas, time adjustments can be reviewed monthly by simple checks against the AVM's predicted values.

One way for appraisers to enhance their standing in the real estate community is to publish reports or white papers on local market trends. These can be published on the Web or mailed out in paper format or as e-mail attachments. Statistical software can quickly produce summary statistics that are not available to anyone without the local databases. Even with only the MLS files, summary statistics of sales transactions by neighborhood, zip code, or some other local market delineation will enhance the appraiser's standing within the local real estate and lending communities. Instead of waiting for lender-generated economic indicators, appraisers may be sought to provide value indicators based on their primary data and AVMs that are customized for their market.

Conclusions

Real estate appraisers who are familiar with technology and the use of statistical modeling have led the way toward the creation of "appraiser-centric AVMs." By using statistical models that were developed to model market behavior, appraisers have been able to use these tools to add statistical support for their opinions of value. Over the last 30 years, innovations in hardware and software have made statistical and GIS software systems more user friendly. The appraisal profession's early adopters have embraced these technological advancements and received significant financial benefits from the experience. The time has come for all appraisers to upgrade their skill levels to meet the current challenges in the appraisal profession. Partly because the appraisal profession has not modernized its educational products, appraisers are facing new regulations such as the Home Valuation Code of Conduct and AMCs that dramatically reduce the appraiser's portion of fees for single-property residential appraisals and require expedited appraisals. Today appraisers must meet increased regulation and perform additional analysis without a direct return on time spent. All professional appraisers need to provide more statistical and GIS support in their appraisals to meet the USPAP requirement of rendering opinions of value that are "independent, impartial, and objective." Appraisers must be less concerned with matching the transaction price and more concerned with providing statistically supportable opinions of value.

The Federal Reserve Board guidelines state that lending institutions must have programs that include "selection criteria and procedures to evaluate and monitor the performance of individuals who perform appraisals and evaluations."[13] Lending institutions, which have extensive computer systems for financial transactions, feel that the manual review of appraisal reports (i.e., a person looking at the physical objects) is all that is necessary to evaluate and monitor appraisals. Many lenders, like appraisers, have not been interested in applying computer-age techniques to appraisal reviews. Appraisal management companies (AMCs) and lenders must be more con-

13. Board of Governors of the Federal Reserve System, Interagency Appraisal and Evaluation Guidelines, SR 94-55 (FIS), October 28, 1994.

cerned with quality and accuracy, rather than approving the sales transaction without proper justification. Both AMCs and lenders need to use statistical software to review the work quality of appraisers. Appraisers' opinions of values that consistently match or are above transaction prices are suspect. Appraised values that are consistently below transaction prices are also of concern because the appraisals may eliminate too many potentially good transactions. Statistical software and GIS will improve the ability of AMCs and lenders to review appraisal reports and the appraisers that develop supported opinions of value.

As AMCs and lenders use statistical software and GIS in more effective ways, appraisers need to improve their use of these same appraisal tools. Appraisers will not be considered competent or professional if they use outdated valuation tools. Appraisers must learn to apply their years of experience and appraisal principles through the use of new tools such as statistical methods and GIS to advance their valuation processes and make these tools an integral part of their business practices.

The appraisal profession still requires both art and science to analyze real estate markets. Unfortunately, the science end of the equation has been missing from too many appraisers' reports. Incorporating statistical software and GIS does not require a doctoral degree, and appraisers can become competent in using these software tools. Appraisers without these skills only have to look to USPAP's Competency Rule for the solution to their lack of knowledge: "Competency can be acquired in various ways, including, but not limited to, personal study by the appraiser, association with an appraiser reasonably believed to have the necessary knowledge and/or experience, or retention of others who possess the necessary knowledge and/or experience." Moving an appraisal business into the computer age only requires the will and the effort to improve appraisal skills.

12

Innovation in Residential Valuation Tools

Integrating Data and Analytics

Mark R. Linné, MAI, SRA

In February 2004, a group of industry thought leaders representing practitioners, academics, and vendors met at the KW401 Conference to discuss the challenges facing the valuation profession and to investigate proposed solutions to ensure a successful course of action for the future of the profession. Several participants felt that no compelling factor would ensure that advanced tools would be introduced to the market without the involvement of key stakeholders and the emergence of champions to push the process further.

In 2006, several key participants from the KW401 conference first proposed and identified an analytics engine that empowered an appraiser to determine an independent valuation conclusion, rather than forcing the appraiser to accept the valuation of a vendor or AVM process. This engine–an interactive valuation model (IVM)–would be an entirely new manner of addressing the use of data and analytic and computing power by an appraiser. The key characteristics of an interactive valuation solution require that an experienced appraiser be in control of the data, analysis, and conclusions.

The work of these two groups and the key stakeholders who recognized the need and the opportunity would form the basis of an effort to build an entirely new software application to meet the needs of a data-rich environment.

In 2009, two conferences identified and addressed the key issues and challenges facing the industry:

- The Collateral Risk Network (CRN) conference
- The Collateral Matters Congress, sponsored by the Appraisal Institute

A white paper titled "Reengineering the Appraisal Process" summarized the findings of the CRN conference, and at the same time the Appraisal Institute produced the Collateral Matters Congress Final Report. Some of the nation's leading appraisers, thought leaders, stakeholders, and clients from the industry participated in the two conferences.

In essence, the conclusions from both documents included the following:

- Appraisers must instill *trust* in their profession.
- Appraisers must instill *reliability* in their services.
- Appraisers must *develop competitive services.*
- Appraisers must *ensure federal and state regulations are maintained.*

How does the appraisal profession implement the changes necessary to achieve these goals? Valuation professionals must determine how to effectively mine data throughout their markets, understand their customer's needs, and provide meaningful analysis based on their experience. They must embrace technologies that will provide cutting-edge and unique market information, but they must also provide personalized market insight. Appraisers must continue to develop the knowledge base and skill set to position themselves as the most knowledgeable players at the property and neighborhood levels. In this way, appraisers can provide keen insights to assist in the decision-making process.

Ultimately, the answer is to truly change the appraisal paradigm. The IVM technology should sift, analyze, and push data to appraisers. IVMs should suggest and present alternative information. The model should extract adjustments from the market and allow the appraiser to interpret that data. Finally, the IVM should provide all the essential data necessary for whole-market analysis, enabling all stakeholders to have a more complete understanding of the subject property and its market.

Identifying and Implementing Innovation

Why have residential appraisers failed to implement and adapt to the profusion of technology and data that are presently available to provide a more sophisticated understanding of the dynamics of the residential marketplace? Reasons include the lack of the necessary skill sets, software, and automated tools among appraisers, the reluctance of clients to compensate appropriately for more sophisticated services, the absence of individuals connecting the dots, and the isolated nature and work-from-home aspects of the business for many practitioners. Another reason often cited by veterans in the industry is the virtually universal use of prescribed appraisal report forms that drive the mortgage origination side of the industry. The residential appraisal marketplace has been characterized by many as mere "form-filling." Ever since the introduction of fixed-format appraisal report standards by Fannie Mae more than 25 years ago (the so-called "Green Hornet" form), appraisers have migrated to a single-solution appraisal report format–the Uniform Residential Appraisal Report (URAR) form.

An unintended consequence of the growth of Fannie Mae and Freddie Mac is that the appraisal report forms promulgated by these entities have grown to become the accepted norm for the market. If a lender wanted to sell its mortgage loans to either organization, the lender would be required to use the forms prescribed for residential appraisal. Gradually, appraisers came to use the URAR form (or Fannie Mae Form 1004) as a solution for virtually all appraisal assignments, whether for mortgage work or any other use. As a result, the form itself became the appraisal process.

More than any other single factor, the introduction of this form has provided the basis for the commoditization of the residential appraisal process and product. Indeed, software developers who provide the software that automates the process of filling out and generating the URAR form have made significant gains during this time period to provide operational shortcuts to make the process of completing the forms easier and more efficient. As a result, the analysis that once characterized the residential marketplace was largely lost. Much of the energy that the industry could have channeled into innovative valuation tools went instead to making the process of filling out the 1004 form even more efficient.

In response to concerns that the property-specific nature of the URAR prevented a thorough understanding of the greater marketplace in which a single property existed, Fannie Mae announced plans in 2009 to require additional information from residential appraisers. The new addendum, labeled the *1004MC* form (for "market conditions"), was required in mid-2009 and immediately created a stir within the appraisal community. The main concern among appraisers was directly related to their inability to fully implement the relatively straightforward requirements of the 1004MC, chief among them a requirement to provide sales velocity, absorption, and predominant price points for the neighborhood surrounding a given property.

Many appraisers have in fact been analyzing markets, but the results of that analysis has not been communicated to clients because the nature of the form limits the information provided to the client. The client sees only the results of the analysis as reported in the form, not the data, analysis, and support that serves as the basis of the analysis. Appraisers will not be able to provide transparency without a change in the form itself.

How then does this dynamic change? How does innovation enter such a market? Where does it come from, and which stakeholders would be the most successful in introducing a product into mainstream use? The answer is no more complex for the residential appraisal marketplace than any other mainstream market. The status quo is sufficient for most stakeholders, who often perceive no compelling need to change.

So how do appraisers move beyond the form and focus on interactive data analysis? The KW401 team members concluded that a proactive effort of investing the time, effort, and financial resources was the only certain course to bringing effective innovation into the market. Working with a team of vendors, practitioners, and educators, an application development team was assembled, and many KW401 participants became the nucleus of a broad-based development effort.

The Development of a Meaningful Solution

In 2006, a group of industry stakeholders, including subject matter experts, appraisers, academics, statisticians, and software developers, came together to create a prototype solution that would serve to move the appraiser away from entering data into a form and instead concentrate on analyzing the data that was pulled into the report and aggregated from a variety of public and private data sources.

The first step was to identify the type of data that was required by the appraiser. Data sources included local Realtor associations and their multiple listing services (MLS). Public record data was included along with predictive data from an econometrics research firm, flood map data, single-property aerial imagery, and finally global aerial imagery. All of this data was integrated through a Web service platform that was tightly integrated into the prototype.

Development began on an interactive software application spearheaded and funded by Jeff Bradford, CEO of Bradford Technologies, an original member of the KW401 group. The product, code-named CompCruncher, was designed to integrate data and analytics in a robust and comprehensive fashion. The idea was to empower appraisers with the most sophisticated tools possible and create operational efficiencies that would improve the valuation process and the bottom line.

The goals included transparency, whole-market analysis, and the integration of sophisticated regression analysis at the core of the application. Another goal was to meld an appraiser's experience, judgment, and local knowledge with the ability to quickly parse and understand data. The most important core goal of the product, however, was that the appraiser be in control at all times of the data, the analysis, and the outcome. Since the product was intended to provide a USPAP-compliant process, the development group felt that it was critical that appraisers feel confident in the valuation process and the product produced.

Characteristics of a Successful Solution

It would be logical to assume that the work of a valuation professional augmented by more robust statistical analysis and techniques could ultimately result in a more supportable valuation with the analytical support of sophisticated data analysis. Rather than attempting to build the best models by eliminating the appraiser component, empowering the valuation professional with the mathematical rigor of appropriate statistics would be a more fruitful solution. Ultimately, integrating the strengths of regression with the inherent skill set of appraisers would seem to yield the best of both techniques.

The concept of applying advanced analytics to residential property information is the critical driver to a better understanding of the market. The concept of advanced analytics itself basically means collecting and organizing raw data, applying statistical analysis to it, and deriving meaningful outcomes that permit effective decisions to be made regarding adjustments and the underlying adequacy of the data, and supporting valuation in the marketplace. A hands-on approach that delivers practical applications is the only acceptable route toward equipping appraisers with the skill sets that will be required for more rigorous analysis of data. Taking the theoretical and making it practical will provide a best-practices perspective that can demonstrate the costs and benefits of using advanced techniques effectively.

Discussions with industry experts reveal that the availability of an interactive, appraiser-driven product would open a floodgate of acceptance by appraisers and additionally satisfy lender demands for a more advanced understanding of market dynamics. Valuation professionals will be able to leverage this technology towards other valuation processes, positioning themselves to provide different and diverse services to clients. With recent market declines in most parts of the country and the rise of two-tier markets (foreclosures and short sales as opposed to arm's-length transactions), ghost foreclosure markets, seller concessions, and other market-related influences on value, appraisers and their clients recognize that a solid, supportable value opinion has merit and must be the ultimate goal of appraisers to regain credibility with their clients.

What are the characteristics of a tool set that can serve to enhance an appraiser's capacity to analyze the profusion of data that exists in the market? The key questions identified by the application development team were

- What should an application do?
- What type of functionality should it have?
- What features should a successful and meaningful application have?

Based on these goal-oriented questions, the development team further identified the environment that the tool set would exist within. Key features of the future environment included the following:

- Lenders will require additional analysis from appraisals.
- Regression analysis will become the coming standard.
- Appraisers will have more data than ever before.
- Appraisers will need tools to analyze this data.
- New tools will require new training and skills.
- Appraisers will have a need to understand what will be required.

Standards Used in CompCruncher

The CompCruncher application was built with the assistance and guidance of appraisers, reviewers, and USPAP experts who considered all of the applicable elements that were required to support a credible value. The entire body of knowledge was used. One of the chief architects was the coauthor of two existing foundational books, *A Guide to Appraisal Valuation Modeling* and *Practical Applications in Appraisal Valuation Modeling*. Other guidelines considered in the implementation and design included guidelines from all relevant white papers on testing and the recommendations of the International Association of Assessing Officers (IAAO) and the Collateral Assessment Technologies Committee (CATC) of the Real Estate Information Providers Association (REIPA).

Looking at Accepted Standards and the Body of Knowledge

In constructing a solution, USPAP experts were consulted and incorporated in the development process. Ultimately, it was critical that an appropriate scope of work be developed and implemented to assure appraisers of the efficacy of the process developed. Assessment experts were also consulted and experienced industry practitioners were surveyed with respect to the deficiencies in existing tool sets.

Exhibit 12.1 Source Documents

Uniform Standards of Professional Appraisal Practice
Standards 1 and 2 (Appraisal Development and Reporting)
Standard 6 (Mass Appraisal)
Advisory Opinion 18 (Use of an Automated Valuation Model)

Joint Industry Task Force on Automated Valuation Models
Standards and Testing Guidelines

International Association of Assessing Officers
Standard on Ratio Studies
Mass Appraisal of Real Property
Standard on Automated Valuation Models

Appraisal Institute
A Guide to Appraisal Valuation Modeling
Practical Applications in Appraisal Valuation Modeling
The Appraisal of Real Estate, 13th Edition

Exhibit 12.1 lists the resources consulted by the development team to ensure that the Collateral Valuation Report (CVR) produced by the CompCruncher application met the generally accepted industry standards for technology, data, and statistical processes.

Competence

Due to the nature and extent of the regression analytics that the developers chose to embed in the CompCruncher application, the appraiser had to be trained in a manner sufficient to understanding the various statistical measures outlined in the CompCruncher application. The statistical measures defined within the analysis allow the appraiser to understand the data and draw certain conclusions based on the accuracy of the data, the amount and quality of the data, the measures of statistical significance, and the accuracy of the analysis applied.

Given the previous lack of familiarity with regression techniques, the appraisers using the CompCruncher application needed to feel comfortable that they had undergone sufficient training and had a strong understanding of the statistical measures underlying the regression component of the CompCruncher application. The appraisers would have to understand in a general manner the method and manner of analysis. While appraisers were not required to be statisticians, they would need to be aware of the basic guidelines pertaining to the use of the statistical analysis tool set. It was critical that appraisers feel comfortable in analyzing small market data sets and understand the analysis and methodology in a manner sufficient to render a credible estimate of value in tandem with the other data and analysis presented in the application and report. The final value conclusion is the appraiser's own and is based on the appraiser's knowledge and experience in the field of appraisal. The data and analysis in this report, whether through direct information or through derived statistical information, aids the appraiser in understanding the dynamics of the market and the neighborhood.

The development team recognized that most appraisers did not have a working knowledge of regression, and thus determined that one of the requirements for using CompCruncher would be a working knowledge and demonstrated competency in basic statistics to ensure an understanding of the underlying processes of the application. An educational curriculum was developed and deployed with a rigorous testing process that would be required before a user could operate the application.

CompCruncher Data Analysis and Value Conclusions

Using CompCruncher, an appraiser would be able to deliver a CVR, which is in essence a summary appraisal report, significantly faster than a traditional 1004. The amount of data, analysis, and transparency in the process would be enhanced, and all relevant stakeholders would be served in the process.

The series of screen shots on the following pages (Exhibits 12.2-12.14) provides a sequential introduction to the functional aspects of the CompCruncher application. Each screen shot has been annotated with the functional elements connected to that screen.

How accurate is CompCruncher and the Collateral Valuation Report (CVR) that it produces? At its core CompCruncher is as accurate as the appraiser driving the application. In this respect the CVR is similar to traditional appraisal in that the accuracy of both appraisal types is in some part a function of how accurately the appraiser performs the analysis. In the 1004 report, for example, the appraiser is entirely responsible for the analysis, and must extract data from the market, consider comparable sales, apply adjustments to those sales, and conclude a value that is supported by the relevant data considered. The same process must be considered when evaluating the accuracy of an appraiser using CompCruncher to derive a value for a given property. One of the key differences however, is that the CompCruncher application enhances the process for the appraiser by bringing unprecedented amounts of data to the appraiser's desktop, including public records, MLS data, GIS images, flood maps, census data, and forecast data, to provide a comprehensive base of data to begin the analysis. In addition analytics are embedded in the application that empower the appraiser.

The four components of the analytics that are applied are

1. Comprehensive data analysis of the delineated neighborhood (similar to the 1004MC)

Exhibit 12.2 Features—Screen 1

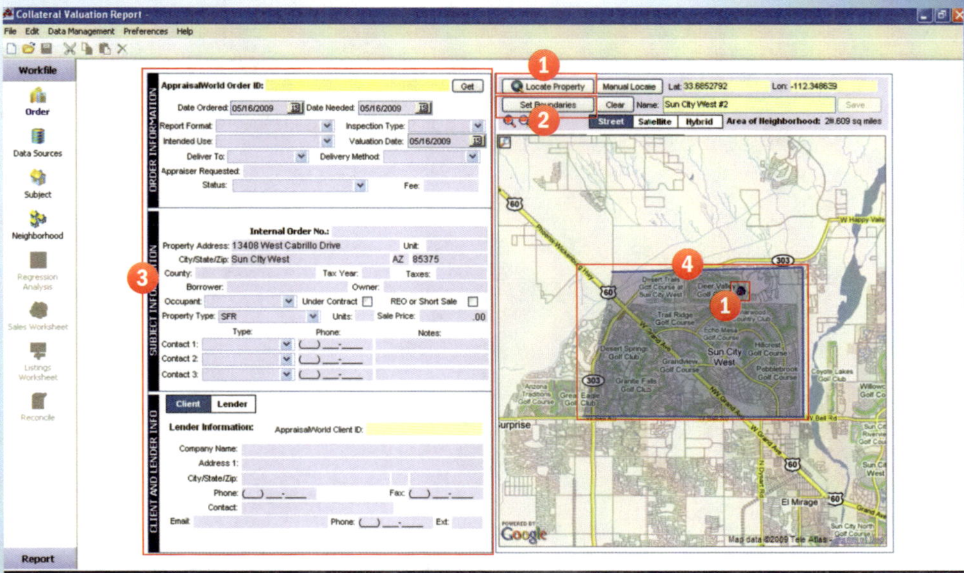

1. CompCruncher locates the subject property and displays the subject.
2. The appraiser defines and draws the subject market area based on local market expertise.
3. Boundaries can be saved and modified. If saved, they can be easily used the next time an appraisal is completed in this area. They can ultimately be shared with others, increasing the knowledge base of the appraiser.
4. All appraisal reporting features, client information, contact information, and other relevant data is provided to the appraiser.

Exhibit 12.3 Features—Screen 2

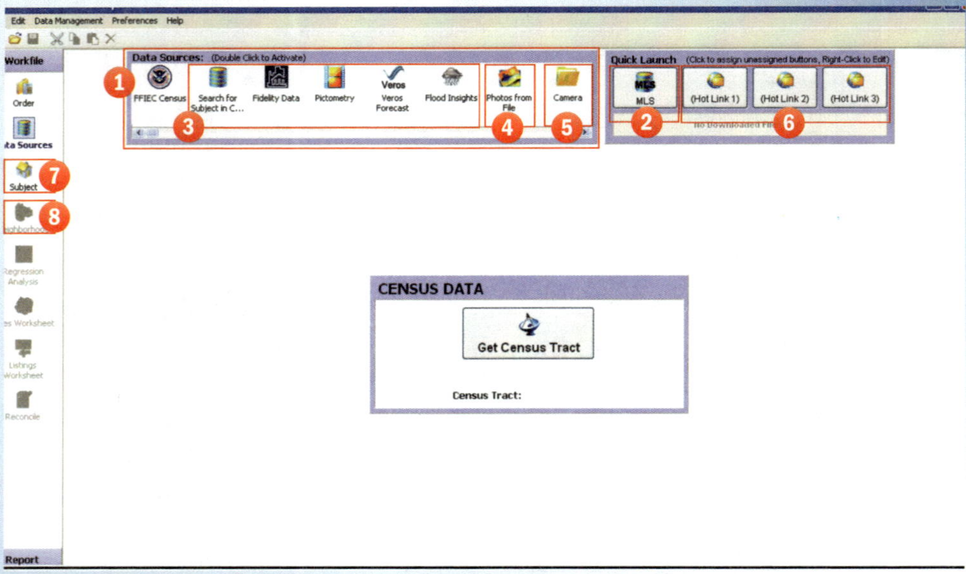

1. All web services are integrated and deployed through CompCruncher. These include
   - Census data
   - Internal database search functionality
   - Fidelity/LPS public record data: 500 sale records in subject market area
   - Pictometry imagery
   - Veros forecasting
   - Flood map–Flood Insights
   - Mapping–Microsoft
   - Aerial imagery–Google
2. Appraisers can import MLS data directly into the application from their own MLS; data can be appended into an internal database.
3. Other proprietary, enhanced, or internal databases can be used by the appraiser.
4. Photographs can be imported into CompCruncher.
5. Direct import from digital camera sources is also available.
6. Hot links on this page allow direct access to frequently used data sources, such as the assessor's office, etc.
7. The appraiser can see if the subject property is contained within the database from any data source.
8. The appraiser can integrate general neighborhood-level property data from other properties in the vicinity to complete the appropriate subject data record.

Exhibit 12.4 Features—Screen 3

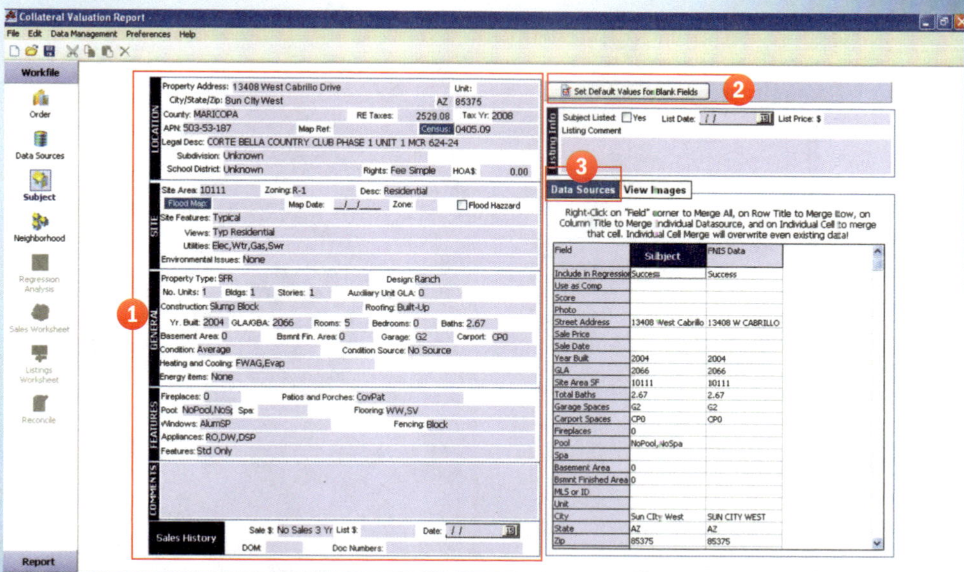

1. The appraiser can compare data on the subject obtained through differing data sources and use the most reliable data. In addition, data from the sources can be combined to create a more complete data record.
2. The data can fill in blank fields not contained within public record from general data sources available for the neighborhood. This would include utilities, etc.
3. If data is available from other sources, the appraiser may complete the appropriate fields directly.

Exhibit 12.5 Features—Screen 4

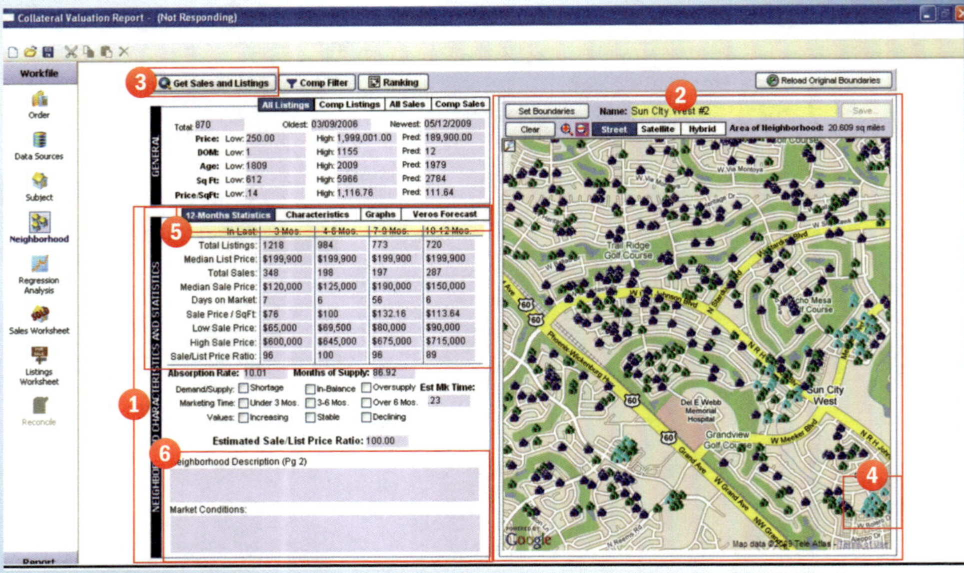

1. The neighborhood and market conditions section of CompCruncher completes all of the relevant data requirements for the 1004MC form. All of the data is automatically imported and calculated, and the relevant metrics are displayed for the user's examination and analysis. Note: The user does none of the calculation. Rather, all of the user's time can be spent examining the data and determining the current market conditions in the subject neighborhood.
2. All sales data is imported into CompCruncher, and the sales are displayed on the map with an appropriate color code within the previously defined neighborhood.
3. All of the listings are imported into CompCruncher and then displayed.
4. Non-SFR properties such as condos and townhomes are displayed in a different color code so that their impact can be evaluated by the appraiser later in the application.
5. Numerous tabs allow the user to see other relevant data, which will be demonstrated on the pages that follow.
6. The appraiser can make appropriate comments in two sections of this page: one is a neighborhood description and the second is a discussion of current market conditions based on the appraiser's examination of the market conditions metrics displayed in the matrix.

Exhibit 12.6 Additional Features—Screen 4

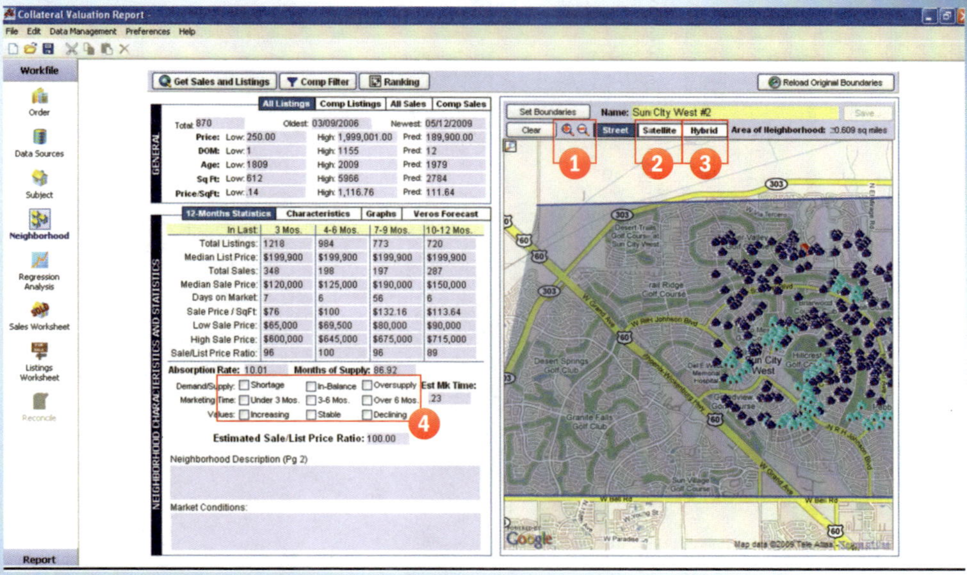

1. The appraiser can zoom in and out to gain a greater understanding of the neighborhood.
2. The appraiser can create a hybrid of both imagery sources, overlaying one on the other.
3. The appraiser can also bring up aerial imagery to discern neighborhood-level geographic features, boundaries, and the like.
4. The appraiser can check appropriate boxes that delineate the nature of the neighborhood and whether the market is declining, stable, or improving, similar to the 1004MC.

Exhibit 12.7 Additional Features—Screen 4

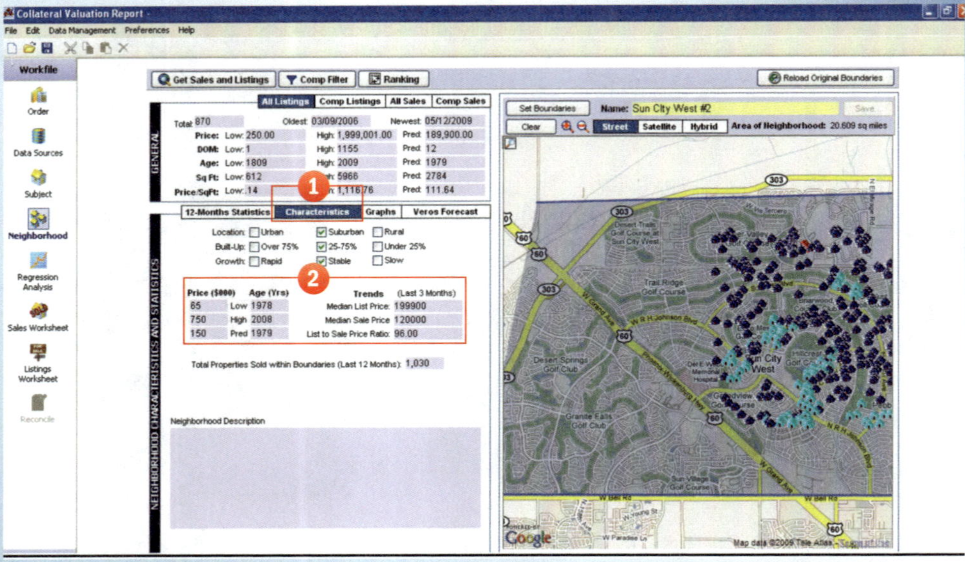

1. The appraiser, by clicking on a tab, can look at additional neighborhood-level information and can check boxes that pertain to the nature of the neighborhood development.
2. Additional information is provided automatically that further describes the nature of the housing stock in the subject neighborhood.

Exhibit 12.8 Additional Features—Screen 4

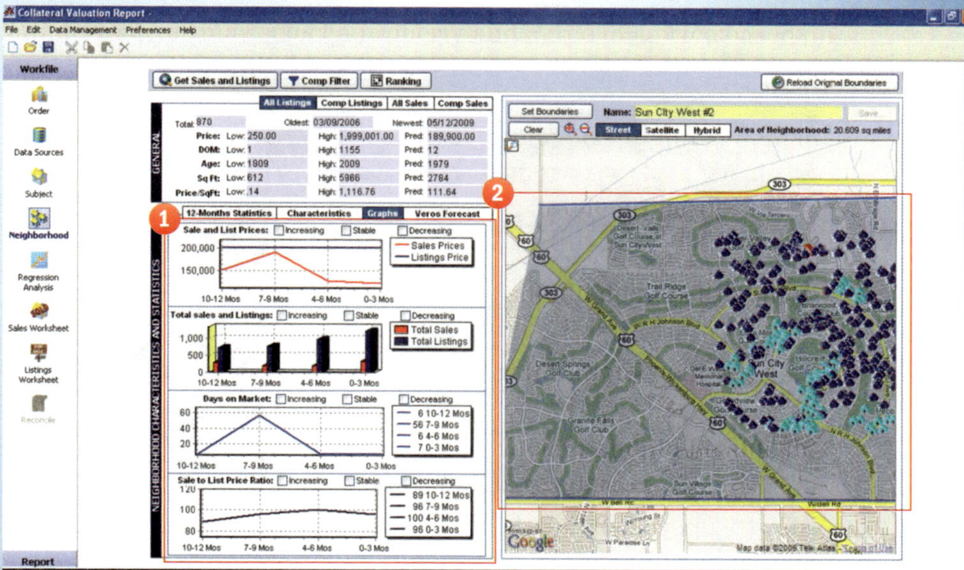

1. CompCruncher provides relevant graphs that relate to a comparison of market trends, sales-to-listings ratios, sales velocity, listings, and other information that assists in understanding the market dynamics.
2. All of the graphed information is drawn from and is consistent with the matrix.

Exhibit 12.9 Additional Features—Screen 4

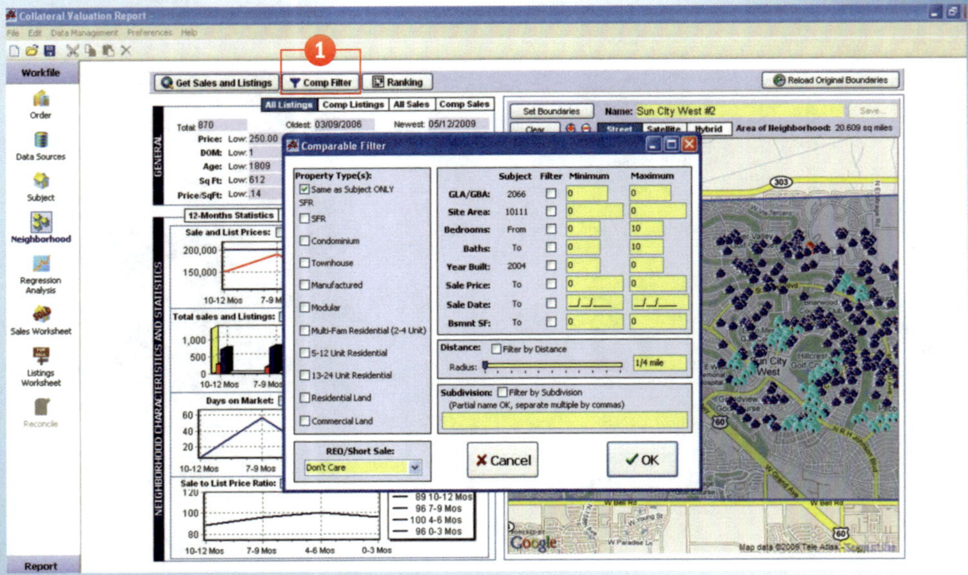

1. The appraiser can create data filters for the overall information, including date, size, physical characteristics, location, etc., so that the most relevant information is sent automatically to CompCruncher. In addition, this information can be used later in the application to run regression analysis and import comparable sales into the sales grid.

Exhibit 12.10 **Features—Screen 5: Sales/Listings Data Analysis**

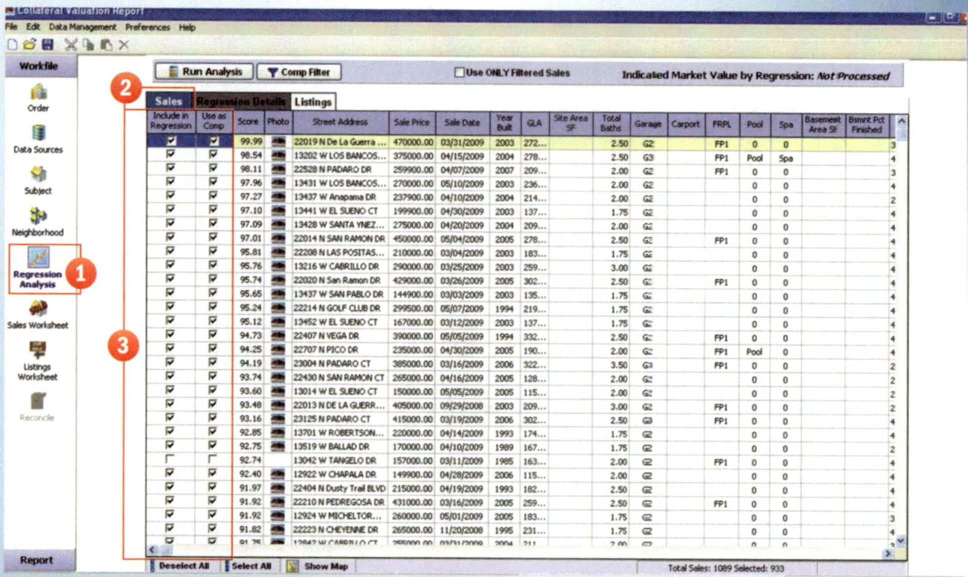

1. The core valuation engine of the CompCruncher application is the Regression Module, which enables an appraiser to properly perform sophisticated market analysis.
2. Sales are provided to the appraiser to examine and analyze. This is consistent with traditional regression in that an examination of the data allows an appraiser to examine the completeness of the data record, omissions that could impact data, and similar issues.
3. All data can be arrayed, sorted, and examined to determine its appropriateness for use in regression.

Exhibit 12.11 Features—Screen 5: Regression

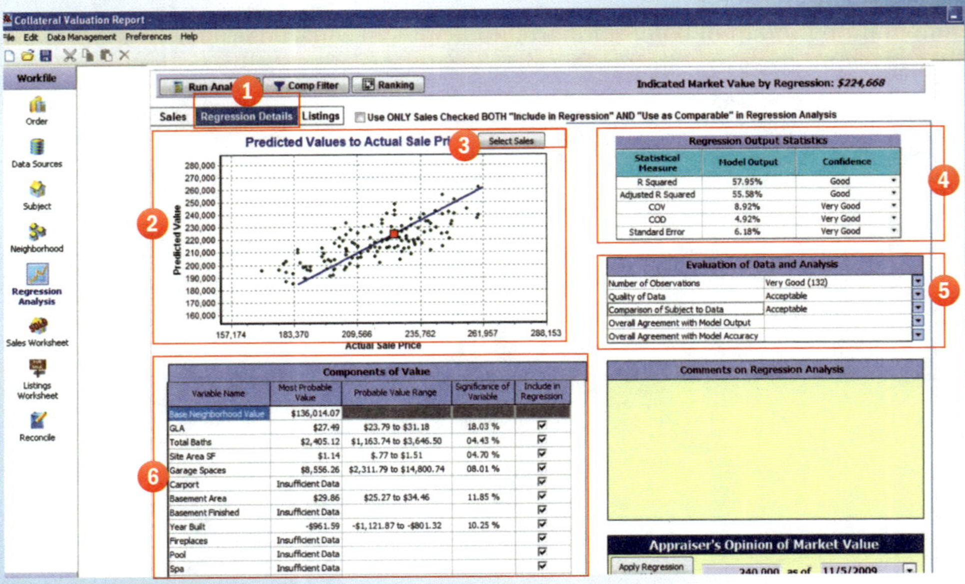

1. The most sophisticated feature of CompCruncher is the regression analysis output.
2. The scatter plot displays the regression line itself, demonstrating the accuracy of the individually predicted values for the data set and the actual sale prices. This visual representation of regression provides a visual understanding of the overall accuracy of the regression process. The appraiser can "see" the data points in a manner that permits their evaluation and allows the appraiser to discern different market segments, overall markets, and other relevant information that could later lead to a further filtering of the relevant data.
3. Within the regression scatter plot, the appraiser has the ability to create a polygon that can either concentrate the data the appraiser wishes to use or cut out and trim information that is determined to lack relevance in the analysis.
4. In the grid at the upper righthand side of the screen, the appraiser can examine the model output and determine (based on expertise from training and experience) if the model is properly examining the overall data. Measures of model accuracy, central tendency, and average error are provided.
5. The grid immediately below the model output asks the appraiser to provide an opinion of the adequacy of the data sample, the amount of sales, and other relevant responses, with a series of questions and pick lists.
6. The grid under the scatter plot contains a listing of all of the relevant property components and their indicated contributions to value for the subject. The information displayed includes the value contribution of gross living area (GLA), bathrooms, site area, year of construction/age, basement, basement finish, pools, spas, and fireplaces and the affect of changing market conditions (time adjustment). The actual data points plus the normal distribution around the most likely data points are provided for the appraiser's analysis and opinion.

Exhibit 12.12 Features—Screen 6: Sales Comparison

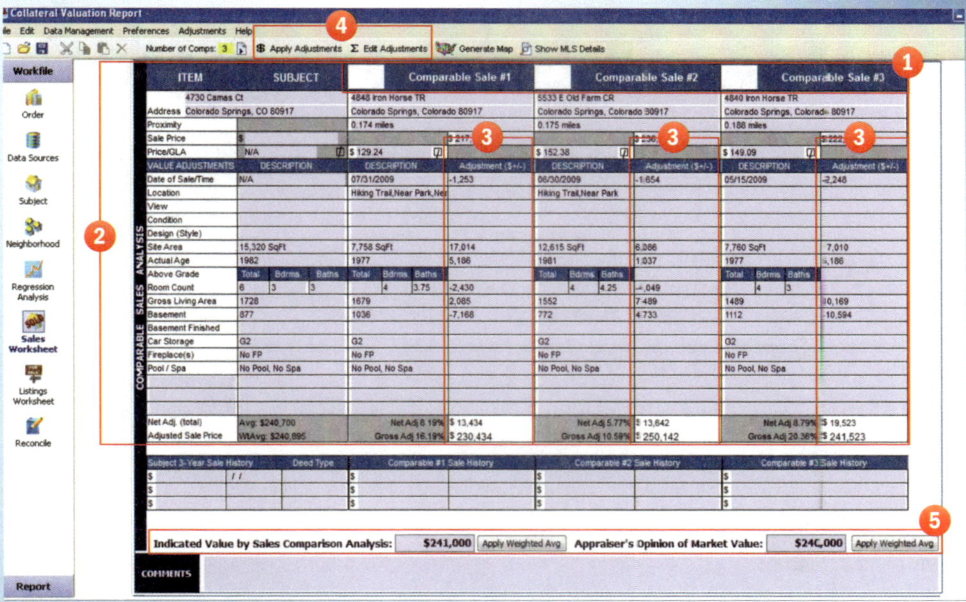

1. All of the relevant sales examined by the appraiser have been scored and ranked within the application. The score measures the similarity of a given sale to the subject property. All physical and locational characteristics as well as date of sale are considered in this ranking.
2. The three most relevant sales are pushed into a traditional adjustment grid. The appraiser may choose these sales or select others that appear most appropriate.
3. The appraiser can examine the adjustments suggested from the regression analysis and apply personal preferences as to their appropriateness. At this stage, the adjustments can be modified and rounded as well.
4. The appraiser can apply the adjustments into the grid automatically. Additional editing can be performed directly within the grid if desired.
5. The appraiser can choose to conclude a value from direct sales comparison or apply a weighted average.

Exhibit 12.13 Features—Screen 7: Listings Comparison

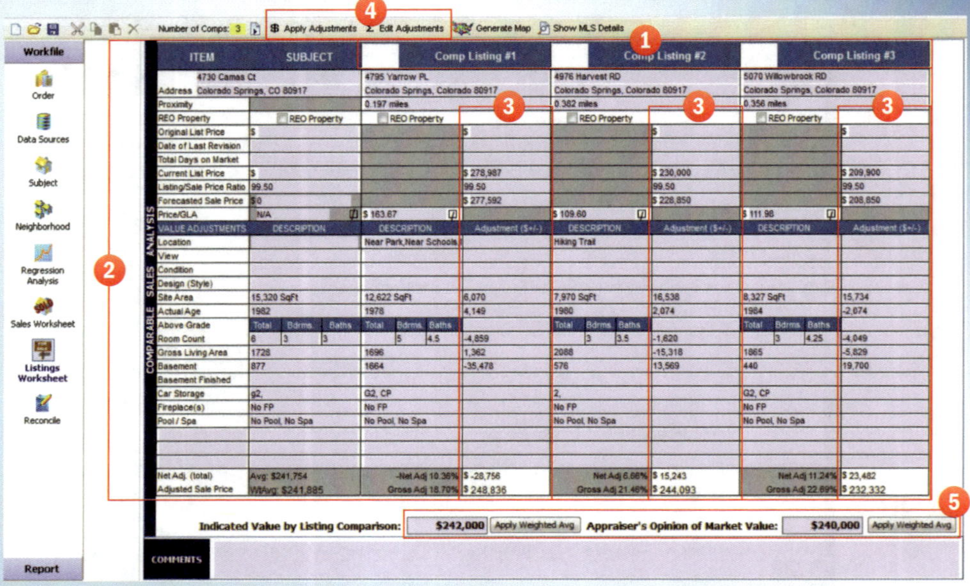

1. All of the relevant listings examined by the appraiser have been scored and ranked within the application. The score measures the similarity of a given listing to the subject property. All physical and locational characteristics as well as the date of sale are considered in this ranking.
2. The three most relevant listings are pushed into a traditional adjustment grid. The appraiser may choose these listings or select others that appear more appropriate.
3. The appraiser can examine the adjustments suggested from the regression analysis and apply personal preferences as to their appropriateness. At this stage the adjustments can be modified and rounded as well.
4. The appraiser can apply the adjustments into the grid automatically. Additional editing can be performed directly within the grid if desired.
5. The appraiser can choose to conclude a value from direct listing comparison or apply a weighted average.

Exhibit 12.14 Features—Screen 8: Reconciliation

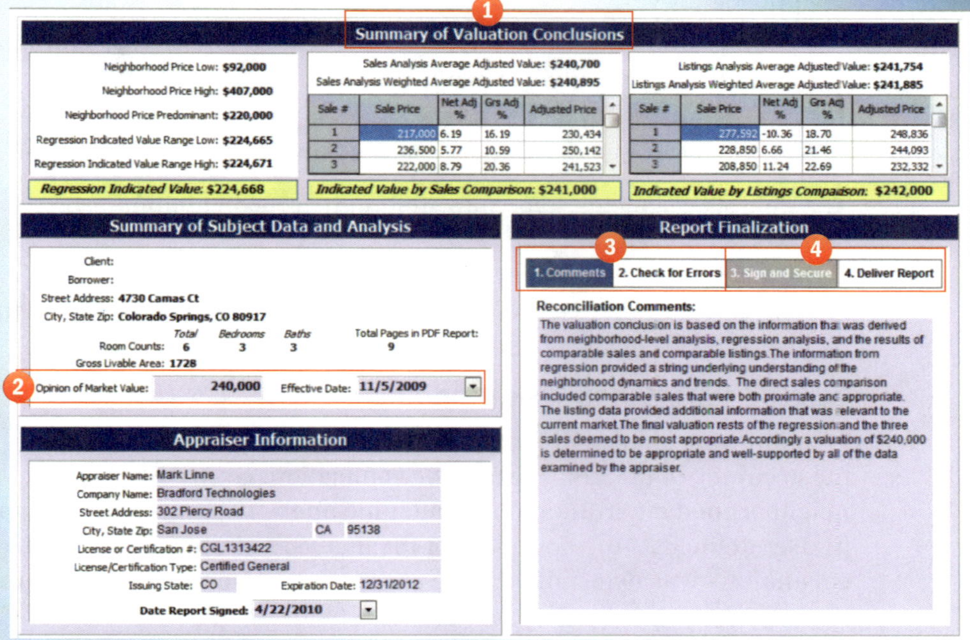

1. All of the data considered within the application is provided to the appraiser in one spot, so that all of the various conclusions can be considered at once. Neighborhood high and low sales and the regression analysis, sales comparison analysis, and listings analysis conclusions are all provided.
2. The appraiser can examine all of this data in tandem and, after considering the strength of each data set, conclude a final value.
3. The appraiser can then automatically scan the report for errors. If an error is found by the application, the appraiser can then go to that section and correct or complete the information.
4. The appraiser can then digitally sign the report, create a PDF, and send it to the client. All data is paperless, digitally retained in a secure digital file format with signature authentication and with full transparency. This entire process creates a report in 35-45 minutes that is USPAP-compliant, fully transparent, and credible.

2. A regression analysis model confirmed and modified by the appraiser
3. Comparable sales ranked for characteristic similarity, proximity, and currency of sale by CompCruncher and the appraiser
4. Comparable listings ranked for characteristic similarity, proximity, and currency of listing by CompCruncher and the appraiser

The appraiser considers the information in a reconciliation process that weighs all relevant data.

The final value selected calls upon the appraiser's local area knowledge and expertise as well as personal judgment on the strengths and weaknesses of any one given data set or indication of value.

Testing Comparison Methodology

A significant range of differing techniques could be used to gauge the accuracy of the CompCruncher application. Given the smaller neighborhood environment in which the analysis occurs after the appraiser delineates the boundary of the analysis, simplicity of statistical analysis was determined to be appropriate. The ultimate measure of accuracy comes from a consideration of how well the analytics compare to a benchmark such as sale price.

The sale price of a given property is the "gold standard" of its most likely market value. The industry standard for such measures, as indicated by the Collateral Assessment and Technologies Committee, was previously described on the committee's Web site (http://catcommittee.org) as follows:

> Q. What is a reasonable range for values to fall within compared to the sale price?
>
> A. The true value of a property is what a buyer and seller agree upon in the absence of fraud or duress. It is widely accepted that each purchase transaction is unique, resulting in natural variances in actual purchase prices for properties that are identical. While every lender will have to establish a reasonable variance tolerance according to its own lending guidelines, industry literature suggests that a range of plus or minus 6%-12% is acceptable.

The IAAO, in its benchmark Standard on Automated Valuation Models, states that a sample of properties should have predicted valuation outcomes that are within 95%-105% of the actual sale prices in order to be credible.

Raw Data Regression Accuracy

The raw regression values are a starting point from which the appraiser analyzes the strength of the regression relationship. The results are consistent as sample size increases. Larger tests in Denver, in which actual sale prices were used as the measurement benchmark, resulted in overall initial values, prior to the appraiser's fine-tuning of the raw regression outcome, that were on average within 8.1% of the actual sale price, based on absolute values. The computed sales ratio (predicted price divided by sale price) resulted in a value of 1.01, which means that on average the predicted values for the sample of 165 properties were within 1% of the actual sale price. (This type of analysis does not consider absolute value differences and so will have a different outcome given that some predicted values will be higher and some predicted values will be lower than the sale price.) It is important to note that the IAAO's AVM standards for regression of this type have an acceptable range of 0.95 to 1.05. The performance of CompCruncher falls within the range for performance measures in all tests performed.

Accuracy of the Analytic Process

Once the regression analysis has been performed, the appraiser completes the analytic process with the following steps:

1. Consider the "best" comparable sales and apply adjustments
2. Use the "best" sales listings and apply adjustments
3. Use personal local knowledge and judgment to reconcile neighborhood data, regression statistics, sales, and listings to a final value conclusion

In blind-test comparisons with actual 1004 appraisals performed by appraisers in four cities–Denver, Phoenix, Minneapolis, and Houston–appraisals performed at the desktop level using CompCruncher of properties that had previously been appraised with a full interior field inspection resulted in values that were within ±6.6% of appraised values.

Smaller samples were employed using actual appraisers to see how the outcome compares to both sales and previously performed appraisals that included an interior inspection of the subject property. In instances in which an exterior field inspection accompanied the CompCruncher process, concluded values were within 2% of the actual sale price in a field test in Phoenix. This further enhancement of the accuracy is based on the appraiser's ability to further refine a value based on observing the subject property and the neighborhood properties. Results from the smaller test environments are illustrated in Exhibit 12.15.

Additional Test Parameters

Large tests were also performed in which sale samples were first examined in their "raw" state (i.e., applying regression before an ap-

Exhibit 12.15 Results of Small-Sample Test

A	B	D	F	G	H
Test	Property Address	:omp Crunche Value	Appraisal Value	Value%	Difference
1	**4635 E Lafayette Blvd** Phoenix, AZ	473,000	440,000	7.50%	Sale at $500k
2	**20632 North Stonegate** Sun City West, AZ	250,000	275,000	-9.09%	
3	**15532 East Greystone Drive** Fountain Hills, AZ	1,100,000	915,000	20.22%	Unknown GLA unfinished lower level
4	**13408 West Cabrillo Drive** Sun City West, AZ	268,000	260,000	3.08%	
5	**3430 N. Mtn Ridge Unit 60** Mesa, AZ	305,000	312,000	-2.24%	
6	**1345 E. Desert Fern Trail** Casa Grande, AZ	77,000	90,000	-14.44%	Strange Little House
7	**9412 North Broken Bow** Fountain Hills, AZ	400,000	360,000	11.11%	
8	**22528 North Pandano** Sun City West, AZ	254,000	259,900	-2.27%	Actual Sale Price-No Appraisal
9	**23309 North Las Positas Drive** Sun City West, AZ	260,000	280,000	-7.14%	Actual Sale Price-No Appraisal

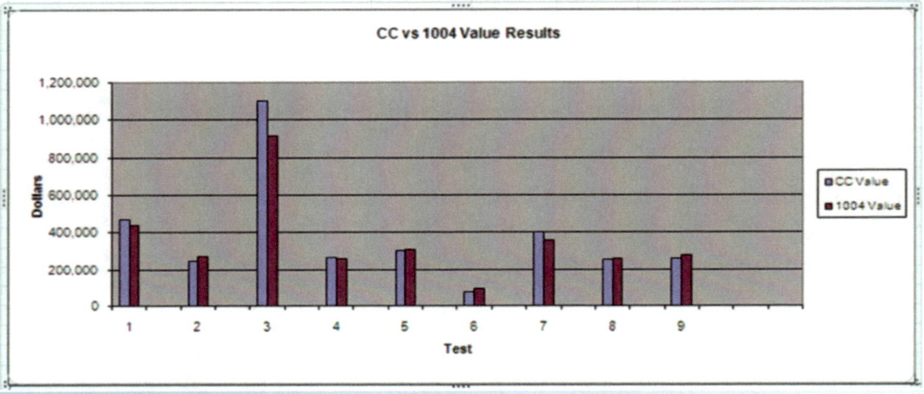

praiser examines the outcome to apply appraisal judgment) and then afterwards as appraisers apply a critical eye toward improving the accuracy of the predicted valuation outcome.

In the representative smaller test sample in Denver shown in Exhibit 12.16, the overall performance comparing sales to predicted values resulted in a variance of less than 8% initially, based on the results of the raw regression analysis performed on the data set. Once the appraiser examined the regression and used personal judgment and appraisal expertise, the overall deviation from sale prices dropped to less than 5%.

Tests ranging from 150 to over 300 properties have also been performed. The detailed outcomes of one such test, in which the raw regression results are displayed on a property-by-property basis, are shown in Exhibit 12.17. This test included 164 properties that sold in the 12 months prior to August 2009.

Exhibit 12.16 **Variance of Sale Price and Predicted Value**

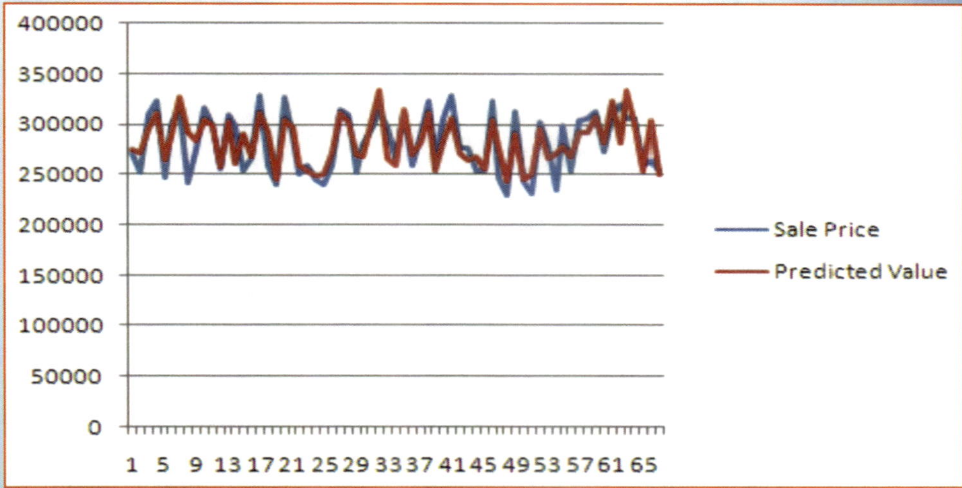

The following data analysis considers a representative neighborhood in Denver, drawn from data extracted during August 2009. This data set is the first run of a regression analysis process based on a neighborhood drawn by an appraiser. Note that the regression model has not been altered by the appraiser. Any effort by the appraiser comes later in the application/appraisal process.

Exhibit 12.17 **Comparison Between Sale Prices and Raw CompCruncher Predicted Values**

Sale Price	Predicted Value	Difference	Sales Ratio	% Difference	Absolute Value
385000	383464	1536	1.00	0.40%	0.004005591
370000	376146	-6146	0.98	-1.63%	0.0163394
412000	343821	68179	1.20	19.83%	0.198297952
315000	360369	-45369	0.87	-12.59%	0.125895957
348000	310234	37766	1.12	12.17%	0.121733917
389000	384657	4343	1.01	1.13%	0.011290578
375000	355603	19397	1.05	5.45%	0.054546784
412500	360614	51886	1.14	14.39%	0.143882378
379900	375067	4833	1.01	1.29%	0.012885698
477000	402797	74203	1.18	18.42%	0.184219346
290000	302791	-12791	0.96	-4.22%	0.04224366
353000	368278	-15278	0.96	-4.15%	0.041484965
450000	394682	55318	1.14	14.02%	0.140158406
320000	318025	1975	1.01	0.62%	0.006210204
330000	346561	-16561	0.95	-4.78%	0.047786681
424900	412658	12242	1.03	2.97%	0.029666213
314000	328358	-14358	0.96	-4.37%	0.043726664
390000	328672	61328	1.19	18.66%	0.186593321
297000	336606	-39606	0.88	-11.77%	0.117662787
438000	390885	47115	1.12	12.05%	0.120534172

Exhibit 12.17 **Comparison Between Sale Prices and Raw CompCruncher Predicted Values *(continued)***

Sale Price	Predicted Value	Difference	Sales Ratio	% Difference	Absolute Value
341500	301732	39768	1.13	13.18%	0.13179908
305000	330713	-25713	0.92	-7.78%	0.077750194
277900	282282	-4382	0.98	-1.55%	0.015523484
255000	277302	-22302	0.92	-8.04%	0.080424952
343000	371258	-28258	0.92	-7.61%	0.076114185
260000	298761	-38761	0.87	-12.97%	0.129739156
307500	307122	378	1.00	0.12%	0.001230781
301500	343480	-41980	0.88	-12.22%	0.122219634
500000	423453	76547	1.18	18.08%	0.180768586
355000	365436	-10436	0.97	-2.86%	0.028557668
378000	369559	8441	1.02	2.28%	0.022840737
277500	267746	9754	1.04	3.64%	0.036430049
389000	392594	-3594	0.99	-0.92%	0.009154495
325000	335681	-10681	0.97	-3.18%	0.031818899
320000	372640	-52640	0.86	-14.13%	0.141262344
274500	325439	-50939	0.84	-15.65%	0.156523957
266700	300264	-33564	0.89	-11.18%	0.111781632
310000	366598	-56598	0.85	-15.44%	0.154387094
260000	303531	-43531	0.86	-14.34%	0.143415335
365000	381398	-16398	0.96	-4.30%	0.042994457
412000	391856	20144	1.05	5.14%	0.051406639
265000	307391	-42391	0.86	-13.79%	0.137905794
267500	296855	-29355	0.90	-9.89%	0.098886662
311000	314309	-3309	0.99	-1.05%	0.010527856
304500	304271	229	1.00	0.08%	0.000752619
367230	327221	40009	1.12	12.23%	0.122269048
325000	342084	-17084	0.95	-4.99%	0.04994095
380000	373658	6342	1.02	1.70%	0.01697274
467000	470924	-3924	0.99	-0.83%	0.008332555
380000	381292	-1292	1.00	-0.34%	0.003388479
306700	315955	-9255	0.97	-2.93%	0.029292146
330000	313467	16533	1.05	5.27%	0.052742394
299500	257460	42040	1.16	16.33%	0.163287501
324000	300952	23048	1.08	7.66%	0.076583641
482650	426709	55941	1.13	13.11%	0.131098711
260000	275544	-15544	0.94	-5.64%	0.056412043
301000	322411	-21411	0.93	-6.64%	0.066409025
274900	305970	-31070	0.90	-10.15%	0.101545903
373300	312589	60711	1.19	19.42%	0.194219886
430000	393534	36466	1.09	9.27%	0.092662896
290000	308869	-18869	0.94	-6.11%	0.061090624
280000	272763	7237	1.03	2.65%	0.026532191
316500	321119	-4619	0.99	-1.44%	0.014384076
405000	424964	-19964	0.95	-4.70%	0.046978097
317500	323367	-5867	0.98	-1.81%	0.018143472
358500	391047	-32547	0.92	-8.32%	0.083230405
440000	411120	28880	1.07	7.02%	0.07024713
365000	388122	-23122	0.94	-5.96%	0.059574051
435500	412242	23258	1.06	5.64%	0.056418317
260000	277803	-17803	0.94	-6.41%	0.064084981

Exhibit 12.17 Comparison Between Sale Prices and Raw CompCruncher Predicted Values *(continued)*

Sale Price	Predicted Value	Difference	Sales Ratio	% Difference	Absolute Value
330000	330257	-257	1.00	-0.08%	0.000778182
279000	294367	-15367	0.95	-5.22%	0.052203542
417500	357747	59753	1.17	16.70%	0.167025859
315000	362063	-47063	0.87	-13.00%	0.129985665
459000	421642	37358	1.09	8.86%	0.08860123
304900	293748	11152	1.04	3.80%	0.037964514
317000	316476	524	1.00	0.17%	0.001655734
300000	285053	14947	1.05	5.24%	0.052435863
391000	348723	42277	1.12	12.12%	0.121233759
305000	290783	14217	1.05	4.89%	0.048892129
308000	344945	-36945	0.89	-10.71%	0.107104031
340000	314770	25230	1.08	8.02%	0.080153763
279850	272561	7289	1.03	2.67%	0.026742637
490000	413893	76107	1.18	18.39%	0.183880858
279900	334540	-54640	0.84	-16.33%	0.16332875
350000	388898	-38898	0.90	-10.00%	0.100021085
272000	268321	3679	1.01	1.37%	0.013711189
319500	322439	-2939	0.99	-0.91%	0.009114902
429000	422211	6789	1.02	1.61%	0.016079638
385000	389118	-4118	0.99	-1.06%	0.010582908
425000	378673	46327	1.12	12.23%	0.122340383
254918	239166	15752	1.07	6.59%	0.065862204
325000	309613	15387	1.05	4.97%	0.049697526
299900	303818	-3918	0.99	-1.29%	0.012895878
254500	249538	4962	1.02	1.99%	0.019884747
295000	299056	-4056	0.99	-1.36%	0.013562677
256000	247240	8760	1.04	3.54%	0.03543116
310000	353315	-43315	0.88	-12.26%	0.122595984
399500	408689	-9189	0.98	-2.25%	0.022484089
395000	436762	-41762	0.90	-9.56%	0.095617293
315000	314279	721	1.00	0.23%	0.00229414
388000	446364	-58364	0.87	-13.08%	0.130754272
317500	351711	-34211	0.90	-9.73%	0.097270202
445000	398412	46588	1.12	11.69%	0.116934229
445100	431581	13519	1.03	3.13%	0.031324363
320500	310012	10488	1.03	3.38%	0.033830948
311000	303561	7439	1.02	2.45%	0.024505783
329000	296982	32018	1.11	10.78%	0.107811248
275000	315109	-40109	0.87	-12.73%	0.127286114
349900	427337	-77437	0.82	-18.12%	0.181208274
392500	366369	26131	1.07	7.13%	0.071324266
310000	296237	13763	1.05	4.65%	0.046459423
420000	362233	57767	1.16	15.95%	0.159474703
271040	306065	-35025	0.89	-11.44%	0.114436476
300000	247295	52705	1.21	21.31%	0.213126024
371500	367284	4216	1.01	1.15%	0.011478856
305000	350454	-45454	0.87	-12.97%	0.129700332
264950	298318	-33368	0.89	-11.19%	0.111853794
255095	278458	-23363	0.92	-8.39%	0.083901342
290000	303336	-13336	0.96	-4.40%	0.043964449

Exhibit 12.17 Comparison Between Sale Prices and Raw CompCruncher Predicted Values *(continued)*

Sale Price	Predicted Value	Difference	Sales Ratio	% Difference	Absolute Value
377500	401937	-24437	0.94	-6.08%	0.060798085
475000	462607	12393	1.03	2.68%	0.026789478
366500	366843	-343	1.00	-0.09%	0.000935005
397000	448295	-51295	0.89	-11.44%	0.114422423
397000	359151	37849	1.11	10.54%	0.105384643
319000	299184	19816	1.07	6.62%	0.066233488
306000	292947	13053	1.04	4.46%	0.044557548
265000	349519	-84519	0.76	-24.18%	0.24181518
292500	333051	-40551	0.88	-12.18%	0.121756127
500000	428095	71905	1.17	16.80%	0.167965054
445000	391171	53829	1.14	13.76%	0.137609894
325000	318228	6772	1.02	2.13%	0.02128034
357900	384094	-26194	0.93	-6.82%	0.068196848
370000	445403	-75403	0.83	-16.93%	0.16929163
350000	306058	43942	1.14	14.36%	0.143574094
496000	440359	55641	1.13	12.64%	0.126353725
385000	436775	-51775	0.88	-11.85%	0.118539294
315000	345854	-30854	0.91	-8.92%	0.089211054
485000	437219	47781	1.11	10.93%	0.109283906
260000	257557	2443	1.01	0.95%	0.009485279
260000	237075	22925	1.10	9.67%	0.096699357
316000	333420	-17420	0.95	-5.22%	0.052246416
415000	361348	53652	1.15	14.85%	0.148477368
392000	392023	-23	1.00	-0.01%	5.867E-05
275000	311478	-36478	0.88	-11.71%	0.117112605
457000	409068	47932	1.12	11.72%	0.117173673
254000	236453	17547	1.07	7.42%	0.074209251
255000	243643	11357	1.05	4.66%	0.046613283
325000	370597	-45597	0.88	-12.30%	0.123036614
298000	363015	-65015	0.82	-17.91%	0.179097282
438000	411040	26960	1.07	6.56%	0.065589724
298000	340370	-42370	0.88	-12.45%	0.124482181
270000	292371	-22371	0.92	-7.65%	0.076515797
261000	301886	-40886	0.86	-13.54%	0.135435231
375000	348705	26295	1.08	7.54%	0.075407579
285300	253563	31737	1.13	12.52%	0.12516416
400000	348776	51224	1.15	14.69%	0.146867904
263000	311461	-48461	0.84	-15.56%	0.155592514
480000	408591	71409	1.17	17.48%	0.174768901
494731	468532	26199	1.06	5.59%	0.055917205
449900	486250	-36350	0.93	-7.48%	0.074755784
494524	479442	15082	1.03	3.15%	0.031457403
485139	456927	28212	1.06	6.17%	0.061742904
460430	489796	-29366	0.94	-6.00%	0.059955573

Overall Sales Ratio	1.00098
Standard Deviation	0.0562
Average Absolute Deviation	0.0816
Average Difference	0.09769%
Median	0.0745

The outcome of this initial regression run shows that CompCruncher can derive an initial value that is equal to or better than traditional vendor-built AVM accuracy. What distinguishes this regression run is that the appraiser can then examine the output and model accuracy as well as the appropriateness of the individual component outcome and can then refine and specify a value for a given subject property that is more accurate and applicable to a single property than any AVM could ever be.

Even in a raw state (not having been modified by an appraiser), the results in Exhibit 12.17 demonstrate a sales ratio of 101% (compared to the acceptable IAAO range of 95%-105%). The median differentiation is 7.45%, and the average difference is 10%. These statistics are meaningful by any measure but also demonstrate the accuracy that can result from small market models in which the appraiser has defined the neighborhood specifically. The small market model built within the CompCruncher process is unique from the standpoint that all AVMs built in the United States for either private sector or assessment purposes are usually quite large and cover 10,000 to over 100,000 properties. It is usually not cost effective for vendors to apply regression at the same granular level as is applied within CompCruncher. Accordingly, AVMs are built on the multi-neighborhood or metropolitan level in most cities, which results in economies of scale but with a clear loss in accuracy.

CompCruncher's Collateral Valuation Report is the result of data analysis at its most granular level–i.e., the neighborhood that has the most direct effect on the subject. Regression has been constructed to consider the most consistently relevant variables, such as gross living area, number of bathrooms, site size, age, and sales trends. An appraiser examines each of the variable outcomes critically to determine if they make appraisal sense.

In the final analysis, the Collateral Valuation Report produced by an appraiser using the CompCruncher application will generate an accurate and supportable value with robust data, better analysis, and product transparency.

Conclusion

New tools and technology such as CompCruncher that interact with and enhance an appraiser's ability to analyze the market and deliver meaningful information to clients have the potential to change the face of valuation. These tools can be leveraged in the current market environment to help define the nature of how the industry can move forward. Keep in mind the following:

- The data exists.

- The analytics are identified.
- There is a clear need for more sophisticated analysis.
- The industry must encourage new products to come to market.
- Better solutions must be implemented.
- Ultimately appraisers must get involved to restore their own credibility and stay relevant.

Maintaining relevance requires that the clients served by appraisers recognize and accept the special skill set and market knowledge that appraisers bring to the collateral valuation and risk mitigation process. If appraisers are to remain relevant, they must provide values to their clients and leverage their market knowledge to provide innovative solutions. Only then will they survive and prosper.

The blueprint for the future is being developed now. It is up to the profession to examine how to meet the needs of clients in innovative and ultimately more satisfying ways. In this way, valuation professionals can truly achieve a lasting and meaningful change in how they are perceived by the real estate industry and others for whom they provide professional services.

Section III

Pushing the Envelope: Advanced Case Studies from Academia

13 Advanced Statistical Analysis of Real Estate Data Using Geoda

Françoise Vermeylen, Sumeeta Srinivasan, PhD, and
Michelle M. Thompson, PhD

The Spatial Hedonic Pricing Models

Recent research indicates that empirical models of hedonic pricing have begun to take into account the potential bias and loss of efficiency that can result when spatial effects such as spatial autocorrelation and spatial heterogeneity are ignored in the estimation process. This spatial autocorrelation and spatial heterogeneity is due to the fact that housing of similar type will be expected to cluster spatially and have similar characteristics. Ignoring this phenomenon, besides providing biased estimates, will often lead to an underestimation of the standard errors (see Exhibit 13.1).

Using OLS (ordinary least squares) regression to estimate housing prices assumes that the residuals of the model are independent and identically distributed with a zero mean and constant variance. The residuals of an OLS model for housing prices can be expected to have spatial autocorrelation because houses of similar prices tend to cluster together for a variety of other reasons besides crime, public school quality, and similar issues that are usually not measurable by observed variables. Using spatial statistics, researchers have found that there is a relationship between house prices and negative envi-

Exhibit 13.1 **Housing Prices and Median Housing Value**

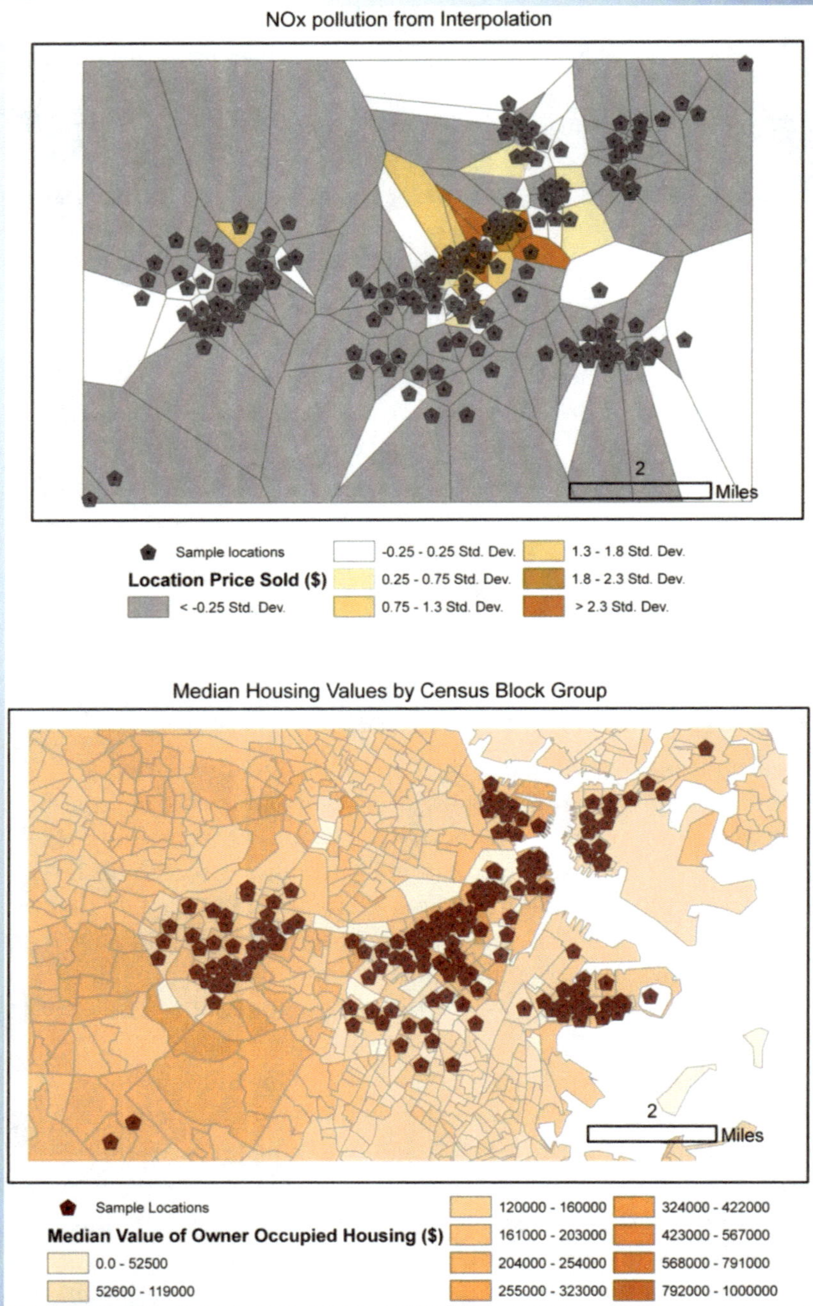

Data Sources: MassGIS (2007); Census (2000)

ronmental characteristics, confirming that nearby point-source pollutants depress house price.[1]

The availability of geographical information systems (GIS) software for estimating such models has improved the ability of hedonic models to include local information about the amenities that are available.[2] Bastian et al. use GIS to measure recreational and scenic amenities and specify their hedonic model with these GIS-derived measures.[3] Anselin et al. have suggested the incorporation of the spatial dependence in cross-sectional data into model specification, estimation, and testing.[4] The incorporation of spatial dependence in the models has allowed theoretically sound estimation of spatial hedonic models that relate housing price to amenities. Tu et al. show that when the correlations between hedonic residuals are used to assign housing units into clusters, the precision of price prediction is improved by 17.5%.[5]

In this chapter, we use GIS to estimate the effects of several negative environmental characteristics and amenities including proximity to highways, open spaces, and air pollution on housing price. Furthermore, we use spatial statistics to demonstrate the spatial autocorrelation of the prices and other land value–related variables.

Data

In this chapter, we use a combined set of data from various sources. This data set includes sale transactions and their corresponding latitude and longitude, information regarding the number of rooms, square footage of the house, and amenities such as the availability of a garage and swimming pool among others. These data elements had to be cleaned to create a subset because many observations did not have any values associated with them.

The latitude and longitude values for the housing transactions were georeferenced to match up with other local Massachusetts

---

1. D. M. Brasington and D. Hite, "Demand for Environmental Quality: A Spatial Hedonic Analysis," *Regional Science and Urban Economics*, vol. 35, no. 1 (2005): 57-82.
2. L. Anselin, "GIS Research Infrastructure for Spatial Analysis of Real Estate Markets," *Journal of Housing Research*, vol. 9, no. 1 (1998): 113-134; and Soren T. Anderson and Sarah E. West, "Open Space, Residential Property Values, and Spatial Context," *Regional Science and Urban Economics*, vol. 36, no. 6 (2006): 773-789.
3. Chris T. Bastian, Donald M. McLeod, Matthew J. Germino, William A. Reiners, and Benedict J. Blasko, "Environmental Amenities and Agricultural Land Values: A Hedonic Model Using Geographic Information Systems Data," *Ecological Economics: The Journal of the International Society of Ecological Economics*, vol. 40, no. 3 (2002): 337-349.
4. Luc Anselin, *Spatial Econometrics: Methods and Models* (Dordrecht; Boston: Kluwer, 1988); L. Anselin (1998); and Luc Anselin and Nancy Lozano-Gracia, "Errors in Variables and Spatial Effects in Hedonic House Price Models of Ambient Air Quality," *Empirical Economics*, vol. 34, no. 1 (2008): 5-34.
5. Yong Tu, Hua Sun, and Shu-Ming Yu, "Spatial Autocorrelations and Urban Housing Market Segmentation," *Journal of Real Estate Finance Economics*, vol. 34, no. 3 (2007): 385-406.

data from MassGIS (2007) and the Census 2000 data by block group. MassGIS data included Massachusetts Highway Department data that showed the location of highways and major roads, open space data from the Department of Conservation and Recreation, and public transit data from the Massachusetts Bay Transportation Authority (MBTA).

The census data included variables for population density, number of workers using transit, median owner-occupied housing values, and median income. The Massachusetts Department of Environmental Protection (DEP) data for pollution was obtained from 11 monitoring sites. This data was interpolated to show the magnitude of NO_2 pollution in $\mu g/m^3$ in Boston across the study area. (See Exhibits 13.2, 13.3, and 13.4.)

Spatial Statistics Results

Spatial dependence is measured by spatial autocorrelation, which is a property of data located on a map that arises whenever a spatial pattern occurs in the values, as opposed to a random pattern that indicates no spatial autocorrelation. To measure the spatial pattern (i.e., spatial association and spatial autocorrelation), some standard global and local spatial statistics have been developed, which include Moran's I, Geary's C, Getis, and Local Indicators of Spatial Autocorrelation (LISA) statistics. We use Geoda, the spatial statistics software developed by Anselin et al.[6], to measure Moran's coefficients and LISA statistics to locate local clusters of similar values. Moran's coefficients for the variables (Exhibit 13.5) indicate global spatial clustering in the price, assessed land value, per capita income, and median

Exhibit 13.2 Variables Used in the Analysis

Variable Name	Description
Price	Price that the property sold for ($)
Rooms	Number of rooms
Baths	Number of bathrooms
Fireplace	Number of fireplaces
Density	Population density per dry square mile of the census block group in 2000
VL_Med_Own	Median value of owner-occupied housing in the census block group in 2000
Income per capita	Income per capita in census block group in 2000
Open_dist	Average distance to nearest park in meters
Hwy_dist	Average distance to nearest highway in meters
T_dist	Distance to nearest subway stop in meters
Transit	Percentage of workers in the census block group who commute to work by transit in 2000
NO_2	Maximum NO_2 measured in the location in the year 2004 $\mu g/m^3$

6. Luc Anselin, "Under the Hood: Issues in the Specification and Interpretation of Spatial Regression Models," *Agricultural Economics: The Journal of the International Association of Agricultural Economists*, vol. 27, no. 3 (2002): 247.

Exhibit 13.3 Sampled Locations and Transportation Characteristics

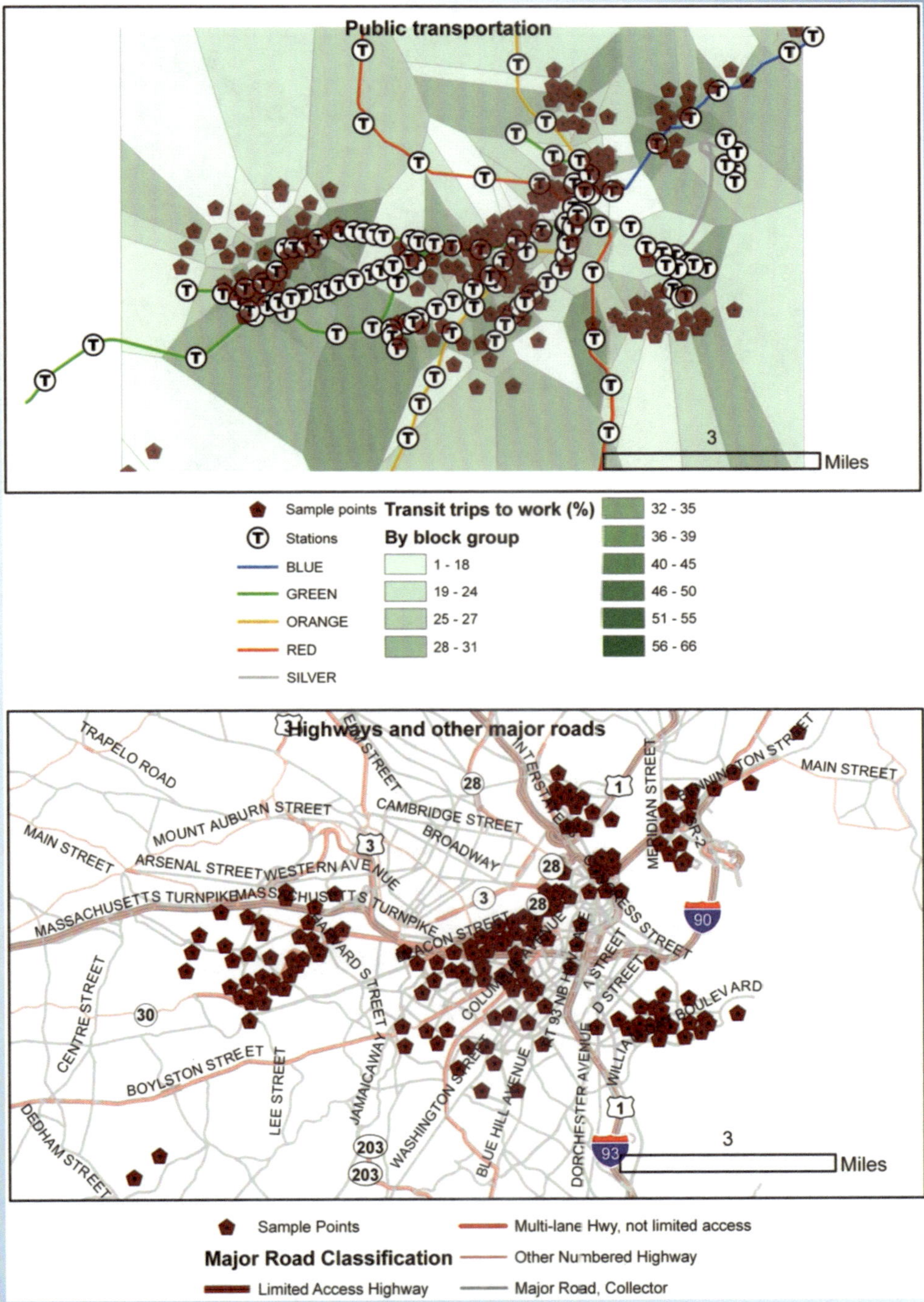

Data Sources: MassGIS (2008); Census (2000)

Exhibit 13.4 Open Space and Pollution

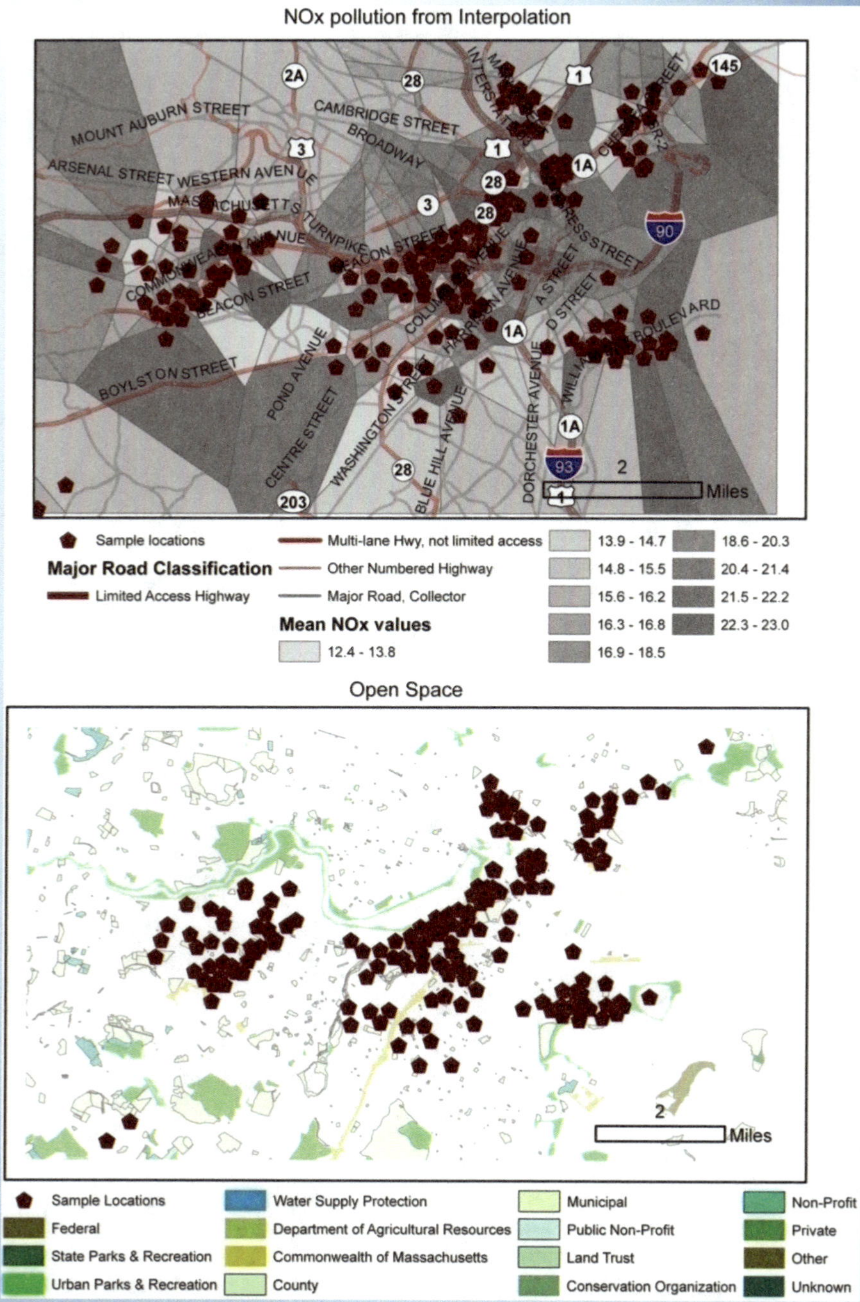

Data Sources: MassGIS (2007); DEP (2004); MHD (2007)

value of owner-occupied housing. Also, several amenities, such as distance to the transit stop, open space, and number of baths, indicated some spatial clustering. Most of the data points are within the urban boundaries of Boston, so the pollution indicator did not show much variation or clustering within the study. Exhibits 13.6 and 13.7 indicate considerable local clustering of high and low land values and prices. The central parts of Boston have clusters of high land values and prices that show local clustering.

Exhibit 13.5 Moran's Coefficients for Variables Used in the Analysis

Variable Name	Moran's I	*p* Value after 99 Permutations
Price	0.58	0.01
Assessed land value	0.56	0.01
Rooms	0.2	0.01
Baths	0.31	0.01
Fireplace	0.21	0.01
Density	0.29	0.01
VL_Med_Own	0.38	0.01
Income per capita	0.52	0.01
Open_dist	0.34	0.01
Hwy_dist	0.13	0.02
T_dist	0.43	0.01
Transit	0.36	0.01
NO_2	0.04	0.13

Spatial Regression Results

Besides spatial dependence in the data, there can be spatial heterogeneity, which indicates that the underlying process being studied may vary systematically over space. Biased coefficients will result from using regression and other econometric models that ignore the spatial variation in the relationships being modeled. Regression models that do account for this spatial autocorrelation are called *spatial regression models.* Three spatial autoregressive models are commonly referred to in the literature:

1. The conditional autoregressive model (CAR)

Exhibit 13.6 LISA Clustering of Land Values in Boston

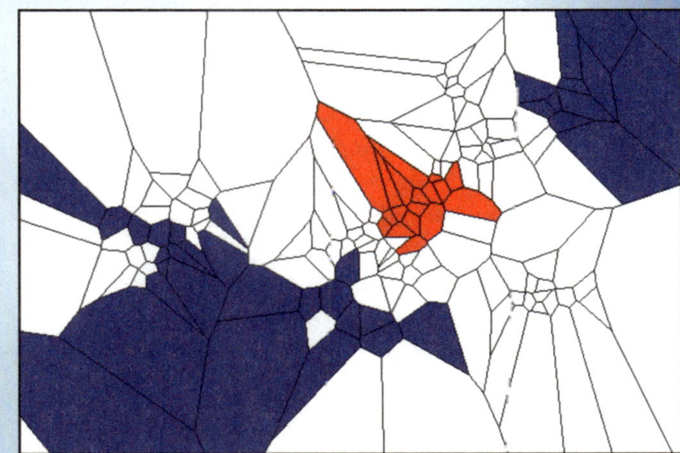

Red indicates high values surrounded by high values and blue indicates low values surrounded by low values.

Exhibit 13.7 **LISA Clustering of House Prices in Boston**

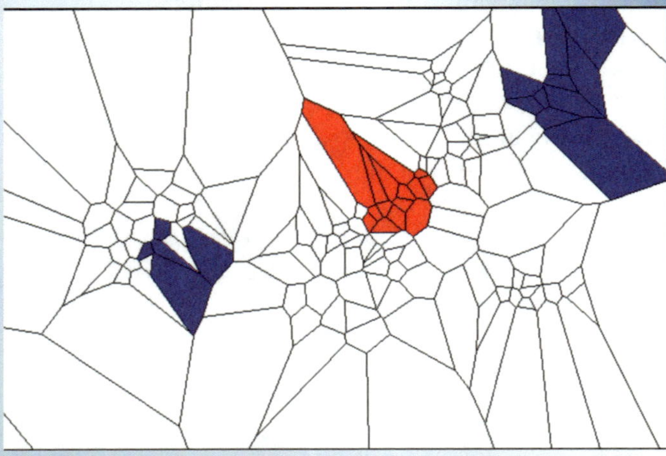

Red indicates high values surrounded by high values and blue indicates low values surrounded by low values.

2. The simultaneous autoregressive model (SAR)
3. The moving average (MA) process model[7]

Spatial dependence in data indicates that one observation at location *i* depends on other observations at other locations $j \neq i$. In general, the CAR model is appropriate for situations with first-order dependency or relatively local spatial autocorrelation, and the SAR and MA models are more suitable where there is second-order dependency or a more global spatial autocorrelation.

Two special cases of the SAR model are the "spatial lag" or mixed regressive spatial autoregressive model and the "spatial error" model.[8] The spatial lag model incorporates a spatially lagged dependent variable on the right-hand side of the regression model.[9] This method addresses the issue of spatial dependence but assumes that spatial dependence only affects the dependent variables. In the case of the spatial autoregressive error model, the parameter λ is a coefficient of the spatially correlated errors, similar to the serial correlated errors in time series models. The data points are considered interdependent through unmeasured variables that are correlated across space or measurement error that is correlated in space. Theoretically, this type of spatial dependence could be eliminated with proper explanatory variables and the correct geographical units for the data.

An OLS regression of the Boston data set indicated clustering of the residuals. The Moran's I for the residuals was 0.48, and the map indicates clusters where the model is under- or over-predicting prices (Exhibit 13.8). The inclusion of a spatial lag or error variable

7. Michael Tiefelsdorf, "Modeling Spatial Processes—The Identification and Analysis of Spatial Relationships in Regression Residuals by Means of Moran's I," *Lecture Notes in Earth Sciences*, no. 87 (Berlin; New York: Springer Verlag, 2000).
8. Anselin (1998).
9. Anselin (2002).

improved the R^2 statistic. However, the lag and error variables were not themselves significant, indicating that the model does not account fully for spatial variation in the dependent variables. Furthermore, the signs of the coefficients indicate problems. For example, proximity to highways would be expected to be a negative amenity, but the spatial lag model indicates a small positive amenity. Likewise proximity to the subway or park lowered the price for the OLS, spatial lag, and error model. Adding the surrounding prices as a lagged dependent variable suggested more consistent coefficients, but the model indicated that higher pollution increased housing prices in Boston.

Clearly, the size of the data set and tremendous variation in housing prices without much variation in the amenities resulted in a lack of robustness in this price model. Also, this model does not account for the fact that some center city locations in Boston, which may be relatively polluted, are valued highly as places to buy housing. Furthermore, we have not differentiated between the qualities of open spaces within Boston. Some open spaces are more highly valued than others. For example, proximity to the Franklin Park Zoo with its location near a high crime cluster could be valued differently than proximity to the Boston Common, which is located in the center of downtown Boston, a difference that is not incorporated in the model.

This data set was confined mostly to within the city of Boston and showed little variation in terms of some of the negative amenities like high pollution and crime, and public school test scores. Another set of regressions for an older data set[10] with more variation indicated

Exhibit 13.8 **OLS Regression Estimation Derived Residuals for Housing Price (2000 data)**

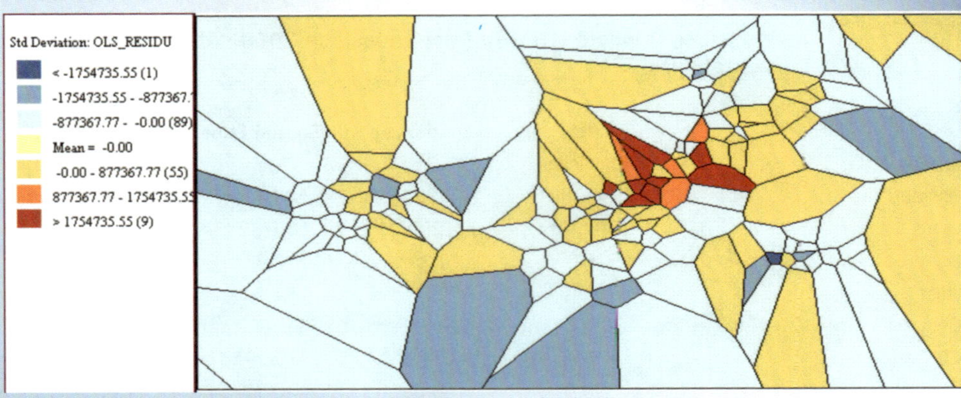

10. R. Kelley Pace and Otis W. Gilley, "Using the Spatial Configuration of the Data to Improve Estimation," *Journal of Real Estate Finance and Economics*, vol. 14, no. 3 (1997): 333-340; and CMU, Corrected data for Boston housing prices (2008): http://papers.ssrn.com/sol3/papers.cfm?abstract_id=9272/. Last accessed: March 2008.

that a spatial regression with lagged variables would be useful in the hedonic pricing. The positive and significant Moran's I (0.45) in combination with the mapped residuals for the OLS (Exhibit 13.9) estimated model indicate that the residuals are positively autocorrelated. Including a lagged variable improved the R^2 and reduced the Moran's I of the residuals.

The Lagrange Multiplier (LM) for both lag and error based estimation was significant. However, the robust LM for lag was not significant, which suggested that the model was best estimated as an error-based spatial regression. The results of the regression are shown in Exhibit 13.11. As would be expected, higher crime and pollution significantly decrease housing value. Low pupil-to-teacher

Exhibit 13.9 **OLS Regression Estimation Derived Residuals for Median Values of Owner-Occupied Housing in USD 1000 (based on CMU data)**

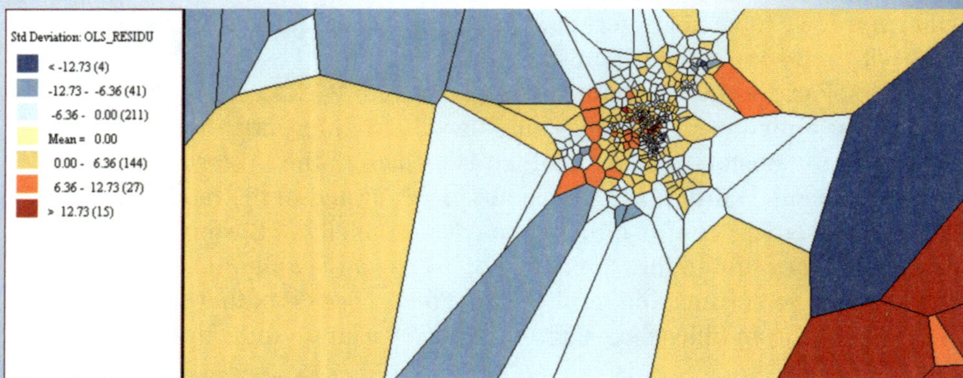

Exhibit 13.10 **Regressions Predicting House Price (based on 2000 data for a small subset of data)**

Variable	OLS Coefficient	Spatial Lag	Spatial Error	Lagged Dependent Variable
Constant	-1792369**	-1342311**	-541624**	-1316811**
Rooms	179720**	141259.2**	146732**	141162**
Hwy_dist	61	2.3**	-7.9**	-14.3**
T_dist	192	70.3**	81.6**	66.5**
Open_dist	80515**	47969.9**	28599**	-110.6**
NO_2	-166	-129.5**	-100.9**	40546.6**
Lag or error variable	N/A	0.65	0.72	1.03**
R^2	0.32	0.64	0.64	0.63
Adjusted R^2	0.30	-	-	0.64
N = 168				

* Indicates significance at 0.1 ** Indicates significance at 0.05

Exhibit 13.11 Regressions Predicting Median Values of Owner-Occupied Housing in USD 1000 (based on CMU data)

Variable	OLS Coefficient	Spatial Lag	Spatial Error	Lagged Dependent Variable
Constant	15.8**	-10.6*	18.6**	-17.7**
Average numbers of rooms per dwelling	6.3**	5.2**	6.5**	4.9**
Per capita crime	-0.2**	-0.13**	-0.12**	-0.11**
Accessibility to radial highways per town (Index)	0.1*	0.1*	0.2*	-0.1*
Pupil-teacher ratios per town	-1.2**	-0.5*	-0.8**	-0.3*
Nitric oxides concentration (parts per 10 million) per town	-18.5**	-7.2**	-43.5**	-4.2
Lag or error variable	-	0.57**	0.83**	0.72**
R^2	0.56	0.69	0.78	0.70
Adjusted R^2	0.55	-	-	0.70
$N = 442$				

* Indicates significance at 0.1   ** Indicates significance at 0.05

ratios also increase housing values on average. The significance of the error variable suggests that there are other variables that are not accounted for in this model. Except for pollution, all the other coefficients remain stable across the regression models.

Conclusions

Until recently few statistical techniques were available in predictive models to assess and incorporate the potential spatial dependence of measurements taken. The increased computer capacity available now with any personal computer, coupled with the fast development of easy-to-implement statistical methodology to address these issues, provide much-improved tools to assess associations and to build more accurate predictive models in the presence of spatial dependence. These tools are especially useful in the field of real estate where it has always been recognized that a large part of the price of housing is influenced merely by location and cannot be captured with any measured variable.

Using housing data, this chapter highlighted the bias that occurs when the models do not control for the spatial dependence. Spatial autocorrelation as a measure of spatial dependence and the appropriate measurements to represent it, such as Moran's I and LISA statistic, were introduced. Several spatial regression models (CAR, SAR, and MA) that incorporate this spatial dependence were then discussed. The pros and cons of each of these are illustrated with the example data set. Not one of these models will surpass the others in accuracy and efficiency at all times, but choosing one over the other will depend on where the spatial dependence occurs and how much is being captured by the available information. The chapter also showed that these techniques will only perform satisfactorily with the availability of a sufficiently large data set. At this point in time, however, there is no excuse not to use these types of techniques if a user wants to model a response that exhibits some form of spatial dependence.

REFERENCES

Anderson, Soren T., and Sarah E. West. "Open Space, Residential Property Values, and Spatial Context." *Regional Science and Urban Economics*, vol. 36, no. 6 (2006): 773-789.

Anselin, L. "GIS Research Infrastructure for Spatial Analysis of Real Estate Markets." *Journal of Housing Research*, vol. 9, no. 1 (1998): 113-134.

Anselin, Luc. *Spatial Econometrics: Methods and Models*. Dordrecht; Boston: Kluwer, 1988.

___. "Under the Hood: Issues in the Specification and Interpretation of Spatial Regression Models." *Agricultural Economics: The Journal of the International Association of Agricultural Economists*, vol. 27, no. 3 (2002): 247.

Anselin, Luc, and Nancy Lozano-Gracia. "Errors in Variables and Spatial Effects in Hedonic House Price Models of Ambient Air Quality." *Empirical Economics*, vol. 34, no. 1 (2008): 5-34.

Bastian, Chris T., Donald M. McLeod, Matthew J. Germino, William A. Reiners, and Benedict J. Blasko. "Environmental Amenities and Agricultural Land Values: A Hedonic Model Using Geographic Information Systems Data." *Ecological Economics: The Journal of the International Society of Ecological Economics*, vol. 40, no. 3 (2002): 337-349.

Brasington, D. M., and D. Hite. "Demand for Environmental Quality: A Spatial Hedonic Analysis." *Regional Science and Urban Economics*, vol. 35, no. 1 (2005): 57-82.

CMU. Corrected data for Boston housing prices (2008): http://lib.stat.cmu.edu/data sets/boston_corrected.txt/. Last accessed: March 2008.

Pace, R. Kelley, and Otis W. Gilley. "Using the Spatial Configuration of the Data to Improve Estimation." *Journal of Real Estate Finance and Economics*, vol. 14, no. 3 (1997): 333-340.

Tiefelsdorf, Michael. "Modelling Spatial Processes–The Identification and Analysis of Spatial Relationships in Regression Residuals by Means of Moran's I." *Lecture Notes in Earth Sciences*, no. 87. Berlin; New York: Springer Verlag, 2000.

Tu, Yong, Hua Sun, and Shu-Ming Yu. "Spatial Autocorrelations and Urban Housing Market Segmentation." *Journal of Real Estate Finance Economics*, vol. 34, no. 3 (2007): 385-406.

14

Using Econometrics and Geographic Information Systems for Property Valuation

A Spatial Hedonic Pricing Model

Richard Bernknopf, PhD, Kevin Gillen, PhD, Susan Wachter, PhD, and Anne Wein, PhD

OVERVIEW

The hedonic pricing function is an econometric model for estimating property values. Typically, an econometric model for land values is estimated from variables that characterize properties. Here we use the hedonic pricing function with and without spatial explanatory variables that include distances to location amenities such as parks, central and secondary business districts, urban growth boundaries, and environmentally sensitive locations. Using a geographic information system (GIS), it is possible to measure distance-related explanatory and location variables for economic applications.

In this chapter, after a review of the literature, we apply the hedonic pricing function to explain and estimate land values in Miami-Dade County, Florida, where environmental regulation across the county and land preservation near national parks are contentious issues. We demonstrate that environmental regulations and location amenities have a significant effect on land values. Furthermore, we demonstrate that including explanatory variables of distances, measured in a GIS, improves the model's predictive accuracy in explaining the spatial variability of the price of land.

Introduction

The hedonic pricing function is a statistical regression model based on the behavior of buyers and sellers who participate in actual real estate transactions. This property valuation model relates the market price of a property to the values of its individual attributes, such as lot size, number of bedrooms, zoning classification, and environmental amenities. Differences in property prices are used to deduce what individual attributes are worth to participants (either negatively or positively) in a real estate market. In addition to a conventional set of site characteristics, we added independent variables derived in a geographic information system (GIS) analysis to examine the influence of location characteristics and the measured distances of a parcel to specific amenities.

While different approaches to real estate appraisal exist such as sales, income, and cost analyses, this chapter presents a statistically rigorous method as an alternative. The chapter includes a description of the hedonic pricing function, a review of past applications, and then the application for land valuation in Miami-Dade County, Florida. The strengths and weaknesses of using the hedonic pricing function for predicting real estate prices are described in the chapter summary. The application discussed in this chapter is a hedonic pricing function for land that has been integrated into a decision support tool (DST) that ranks ecologic and economic land values to compare development and preservation choices. The DST assists users in comparing residential development value with its ecological impact on Biscayne and Everglades National Parks in South Florida.

The addition of the GIS explanatory variables almost doubles the explanatory power of the econometric model for estimating land prices across Miami-Dade County. The incorporation of GIS-derived spatial explanatory variables into the hedonic pricing function in Miami-Dade County shows that spatial factors (location, location, location) do have a significant influence on the statistical accuracy and robustness of land value estimation.

Background

The hedonic pricing function is an indirect or inferential method of valuation that explains the price of a heterogeneous market good, such as land, in terms of its valuable characteristics both within and between market segments. The method of estimation is statistical regression analysis where the property sale transaction price is correlated with the parcel's characteristics to describe the market value of the parcel as a function of the property's physical characteristics and location amenities.[1]

Various applications of the hedonic pricing function to estimating housing and land values involve characteristics such as property listings, urban amenities, agricultural productivity, environmental impacts, and natural and anthropogenic hazards. Studies of these problems with econometric methods have, on occasion, included spatial explanatory variables such as a county boundary, an urban development boundary, or a linear distance to a central business district (CBD) or a park. Following a review of hedonic pricing models in urban and rural applications, we apply the method to valuing land in Miami-Dade County.

Some of the land in the county currently zoned for open space, agriculture, or recreation or currently vacant is likely to be subject to development pressure and to be converted to other higher-value land uses in the coming years. However, there seems to be strong public interest, based on many newspaper articles, books, and public

---

1. Sherwin Rosen, "Hedonic Prices and Implicit Markets: Product Differentiation in Pure Competition," *Journal of Political Economy*, vol. 82, no. 1 (1974): 34-55; and C. Redfearn, "Land Markets and Terrorism: Uncovering Perceptions of Risk by Examining Land Price Changes Following 9/11," in Harry Ward Richardson, Peter Gordon, and James Elliott Moore II, eds., *The Economic Impacts of Terrorist Attacks* (Cheltenham, U.K.; Northampton, MA: Edward Elgar, 2005), 152-169.

discourse, in preserving some of these parcels in their current state to help minimize the effects of development on the natural environment, specifically the areas bordering the Biscayne and Everglades National Parks. These environmentally sensitive parcels may help protect endangered species, critical habitats, and hydrologic processes that provide ecological goods and services to the region.

Within this context, the hedonic pricing function provides a useful analytical tool because it can be used to focus on appraisal (individual) and policy (aggregate) issues.[2] For appraisal purposes, studies establish values for specific physical and location characteristics. For example, the price of a farmland parcel is indicative of characteristics such as productivity, soil erosion, rural amenities, access to water, and urban density. House sales are typically analyzed to determine the price effects of location, structure characteristics, view quality, or other intangible goods that affect the market price. This information is important in estimating the value of a parcel for purchase or sale. The regression equation coefficients estimated in the econometric model can be used to establish market values for characteristics of a property like community flood mitigation.

These analyses also can provide economic information used in a policymaking process. When location characteristics are influenced by public policies, hedonic studies can provide estimates of some of the influences that a policy could have on parcel price. The terms in the hedonic equation that represent a parcel characteristic, such as open space, can positively influence price as a reflection of an economic benefit. Alternatively, a regulation such as an urban growth boundary could be seen as a negative outcome because that regulation can limit high-density development to a confined area and inhibit urban growth. This information makes it feasible to discuss and explicitly quantify the implicit tradeoffs associated with the positive or negative ramifications to a community of a proposed "smart growth" policy.

This chapter is divided into a description and explanation of the general hedonic pricing model, a review of applications, and an application of the model in a case study. In the application, the hedonic pricing function is incorporated into a decision support tool (DST). The DST ranks both ecologic and economic land values for application in development and preservation choices in Miami-Dade County. In the DST, economic land valuation considers classification of a parcel, land use, parcel characteristics such as size, environmental regulations that affect a parcel's potential use, and measured distances of

2. J. Miranowski and M. Cochran, "Economics of Land in Agriculture," in G. Carlson, D. Zilberman, and J. Miranowski, eds., *Agricultural and Environmental Resource Economics* (New York: Oxford University Press, 1993), 392-440.

a parcel to specific amenities. We incorporated spatial explanatory variables into the land market valuation and compared the estimated land price (per square foot) with and without these variables. As will be shown, inclusion of the spatial variables improved econometric estimation. In particular, the spatial variables were significant predictors, as shown by measurable improvement in the goodness of fit. The strengths and weaknesses of using the hedonic pricing function for predicting real estate prices are described in the chapter summary.

The Hedonic Pricing Function

The hedonic pricing function describes the relationship between the market price of a property and its characteristics or the services the property provides. The technique differentiates positive and negative characteristics of land parcel price.[3] The literature attributes the hedonic pricing function to multiple origins.[4] The method distinguishes between sources of utility (described as an assembly of independent variables such as the number of bedrooms, air quality, soil conditions, earthquake liquefaction potential, proximity to airports, and others) and a traded commodity (which is a dependent variable of market price).[5] Hedonic pricing statistically divides the total price of a market good into the portions that are attributable to characteristics not separately for sale themselves.[6] In this way, otherwise indistinguishable characteristics, commodities, and externalities are segregated and quantified.

The Generic Hedonic Price Model

A hedonic price function is a strategy to model an individual's demands for a type of heterogeneous good. There are many different versions of the same basic commodity–say, land–and the individual generally consumes one type of property.[7] The hedonic price function matches demanders and suppliers with the heterogeneous commod-

---

3. Timothy J. Bartik and V. Kerry Smith, "Urban Amenities and Public Policy," in Edwin S. Mills, ed., *Handbook of Regional and Urban Economics, Volume II: Urban Economics* (Lausanne; Amsterdam; New York: North-Holland, 1987), 1207-1254.
4. A. Court, "Hedonic Prices Indexes with Automotive Examples," in *The Dynamics of Automobile Demand* (New York: General Motors Corporation, 1939), 99-117; Z. Griliches, "Hedonic Price Indexes for Automobiles: An Econometric Analysis of Quality Change," *The Price Statistics of the Federal Government, General Series No. 73* (New York: National Bureau of Economic Research, 1961), 137-196; Kevin J. Lancaster, "A New Approach to Consumer Theory," *Journal of Political Economy*, vol. 74, no. 2 (1966): 132-157; and Rosen (1974).
5. D. Zilberman and M. Marra, "Agricultural Externalities," in G. Carlson, D. Zilberman, and J. Miranowski, eds., *Agricultural and Environmental Resource Economics* (New York: Oxford University Press, 1993), 221-267.
6. Nick Hanley, Jason F. Shogren, and Ben White, *Environmental Economics in Theory and Practice* (New York: Oxford University Press, 1997).
7. Bartik and Smith (1987).

ity so that demand equals supply in a real estate market. Households as demanders and firms (landlords) as suppliers of properties are assumed to be price takers, i.e., no individual can influence the market by their actions. The hedonic pricing function provides an estimate of the highest bids of demanders wanting the good and the lowest offering prices of suppliers making the good available.

A demander selects a property to maximize utility subject to the available budget and existing prices for properties in the market. The supplier chooses a property type and number of units to offer to maximize profits. The hedonic price function is the transformation of characteristics to dollars. A property's value is equal to the present discounted value of all future services provided by that property to its current owner.

Several additional assumptions are associated with the implementation and use of the hedonic pricing model. Some of the most important ones are

- The property characteristics must vary continuously and over a considerable range of values.
- The prices must be interpretable as value in use and not solely in terms of investment or speculative value.
- Consumers must have similar perceptions of the property characteristics.
- The market studied must be open to migration, meaning that buyers can choose to buy somewhere else at no disadvantage.
- The data must document the whole market under consideration, and the data set must represent a random sample of that market.
- The data must represent most, if not all, of the characteristics that influence buyers' decisions.[8]

The hedonic price model in Equation 1 illustrates a statistical/econometric estimation technique that correlates sale price with a series of independent (explanatory) variables:

$$\ln p_{it} = \alpha + \sum_{i=1}^{k} \beta_i \chi_i + \sum_{i=2}^{j} \gamma_i t_i \qquad (1)$$

$$t_i = \begin{cases} 0 \text{ if time period} \neq i \\ 1 \text{ if time period} = i \end{cases}$$

where

$\ln p_{it}$ = the natural logarithm of the arm's-length transaction price of property i during time period t

8. R. Bernknopf, R. Champion, E. Schwerer-Duffie, E. Graffy, P. Ng, and R. Taketa, "Final Report to TRPA: Tahoe Constrained Optimization Model, Proof of Concept, Initial Results, and Application to Lake Tahoe Management," *USGS Administrative Report* (2003).

The hedonic price function is best estimated with closed sales of parcels of property (built or vacant) rather than with assessed values because assessed or appraised values are one step removed from a buyer's actual willingness to pay for properties as revealed by market prices.[9] The variable x_i represents the location characteristics of the property that are measured continuously (e.g., square feet of land parcel and linear distance to the central business district) and measured categorically (e.g., land use). The variable t_i is a dummy variable indicating year the transaction took place, and α, β_t, and γ_i are the coefficients to be estimated. In particular, the series of coefficients γ_i is the price index for real estate over time.

Estimation of Equation 1 from available data yields regression coefficients of the model that can be used to determine the marginal price associated with each characteristic of a property, holding all other characteristics constant. If characteristic x_i measures parcel elevation, then the marginal price per foot due to height above sea level is the partial derivative of the price of the property with respect to the elevation characteristics. Taking the natural logarithm of the left (price) side of the equation suggests that the variables on the right side have multiplicative effects on price. That is to say, if a real estate market had high prices in 2005, such that a large coefficient is associated with that year's index variable, then that factor should tend to inflate the effective prices of all the relevant characteristics in that year. On the other hand, taking the log of some of the other explanatory variables assumes that these variables do not interact with themselves. Each additional bedroom in a house costs the same amount. Interaction would mean that the cost of each additional room would increase exponentially.

Applying the Hedonic Pricing Method

Hedonic valuation is performed in two stages. The first stage of the analysis measures the market price trends for the characteristics contained in the specified equation.[10] The coefficients of the statistical regression equation are interpreted as the "marginal prices" of the characteristics, i.e., the sum buyers would have had to pay to acquire an additional unit of a characteristic. For example, a unit of a characteristic could be the extra amount (marginal value) a buyer of a two-bedroom home would have to pay for an otherwise comparable home with three bedrooms in the property market. The second stage of the

9. Robert McFadden, Presentation at the 2nd World Congress of Environmental and Resource Economists, Monterey, CA (2002).
10. W. Patrick Beaton and Marcus Pollock, "Economic Impact of Growth Management Policies Surrounding the Chesapeake Bay," *Land Economics*, vol. 68, no. 4 (1992): 434-453.

analysis incorporates the marginal prices from the first stage with additional data on buyer incomes and tastes to specify a market demand function and characterize the market segment. This chapter focuses on the first stage of the hedonic pricing function used to estimate the market price for a property. To derive theoretically correct estimates of the total costs and benefits of management and regulatory choices, the second stage would be necessary.[11]

A Review of Hedonic Price Function Applications

The economics literature abounds with applications of the hedonic pricing function in urban settings with location amenities[12] and in rural contexts for agricultural land production.[13] Hedonic pricing functions also have been used to assess how people value more intangible property characteristics such as exposure to air and water pollution, natural hazards, and potential terrorist activities. The following review of the literature is a short summary of studies that are relevant to the land valuation example in this chapter. This review is not exhaustive, nor is it intended to be. It simply highlights the vast literature on the subject.

The hedonic pricing method has been used extensively to estimate economic values for ecosystem goods and services that directly affect residential property market prices.[14] The foundation for these types of studies is the purchaser's willingness to pay for property that includes the quality of the local environment as part of the consumer's decision.[15] In general, the price of a house is related to the characteristics of the house and property, the characteristics of the neighborhood and community, and the environmental amenities. That is, all other things being equal, houses and properties in neighborhoods with clean air would be expected to command higher prices than comparable dwellings in neighborhoods with polluted air.

---

11. Rosen (1974).
12. Bartik and Smith (1987).
13. Miranowski and Cochran (1993).
14. Freeman, A. Myrick III, *The Measurement of Environmental and Resource Values: Theory and Methods* (Washington, D.C.: Resources for the Future, 1993); Hanley et al. (1997); and John B. Loomis and Gloria E. Helfand, *Environmental Policy Analysis for Decision Making* (Dordrecht, Netherlands; Boston: Kluwer Academic Press, 2001). Ecosystem function is an ecosystem characteristic related to the set of conditions and processes whereby an ecosystem maintains its integrity (such as primary productivity, food chain, and biogeochemical cycles). Ecosystem functions include such processes as decomposition, production, nutrient cycling, and fluxes of nutrients and energy. Ecosystem services are the benefits people obtain from ecosystems. These include provisioning services such as food and water; regulating services such as flood and disease control; cultural services such as spiritual, recreational, and cultural benefits; and supporting services such as nutrient cycling that maintain the conditions for life on Earth (Millennium Ecosystem Assessment, 2003).
15. Thomas H. Tietenberg, *Environmental Economics and Policy*, 2nd ed. (Reading, MA: Addison-Wesley, 1998).

By statistically comparing the market values of similar properties in neighborhoods with different levels of air quality, the analysis divides the aggregate property values into distinct values of the characteristics. With adequate data on properties from several locations in a community where the air quality varies, a regression coefficient (the marginal price of this characteristic) for air quality can be estimated.[16]

As reflected by market-based housing prices, this type of information is useful in analyzing the societal benefits of an air quality improvement program. The coefficient provides the current willingness to pay for a one-unit change in air quality. The value of an incremental change in air quality is approximated by calculating the change in house price with the change in air quality resulting from the program. The hedonic pricing function works well for small changes in air quality.[17] To accurately estimate larger or nonmarginal changes in environmental quality with the hedonic model, estimation of the demand for air quality is required.

In addition to air quality, the hedonic method has been used in a variety of applications to estimate the economic value associated with changes in environmental quality for water pollution, noise, and other forms of negative environmental impacts. Extensive literature exists on the estimation of the value households living in urban areas place on improving air and water quality.[18] Brookshire et al. compared the effect of differences in sulfur dioxide and particulate levels in the atmosphere and their impact on house prices in Los Angeles.[19] Results showed that a house in a location that has the best air quality was worth significantly more than a house in an area with poor air quality. If the analysis reveals a premium paid for clean air by consumers, this premium can serve as a measure of the value of clean air to a population. Also, the hedonic pricing function has been used to assess scenic, regional, and environmental amenities such as aesthetic views or proximity to recreational sites.[20]

Besides the application of the hedonic pricing function to environmental issues, the technique has been used in a variety of studies concerning the effect of natural hazards and the possible threat of terrorism on property prices. To test the application of the expected

16. Loomis and Helfand (2001).
17. Ibid.
18. Ibid.
19. David S. Brookshire, William D. Schulze, Mark A. Thayer, and Ralph C. d'Arge, "Valuing Public Goods: A Comparison of Survey and Hedonic Approaches," *The American Economic Review*, vol. 72, no. 1 (1982): 165-177.
20. Kevin J. Boyle, S. Lawson, H. Michael, and R. Bouchard, *Lakefront Property Owner's Economic Demand for Water Clarity in Maine Lakes*, Miscellaneous Report 410 (Orono, ME: Maine Agricultural and Forest Experiment Station, 1998).

utility model for self-insurance, the hedonic price function was applied to earthquake hazards. Brookshire et al. documented a housing price gradient for safety in Los Angeles and San Francisco for property near known earthquake faults.[21] That is, individuals are willing to buy properties located in more hazard-prone areas, if they cost less. Brookshire et al. used a California regulation based on the 1974 Alquist-Priolo Act that delineates an earthquake Special Study Zone (SSZ) to identify parcels close to known earthquake faults for further geotechnical analysis. Within the SSZ, residential housing prices in San Francisco were negatively and significantly correlated. The presence of the SSZ, which represents a spatial explanatory variable, however, was not significant in the Los Angeles sample. This difference could be due to the number and locations of earthquake faults in the L.A. basin (faults are ubiquitous in the L.A. Basin as opposed to the San Francisco Bay region, where faults are more sparsely distributed).

In an analysis of volcano and earthquake hazards, Bernknopf et al. used a hedonic function to demonstrate the negative and significant impact of hazard warning announcements by the U.S. Geological Survey (USGS) on real estate properties at the Mammoth Lakes, California, ski resort during the 1980s.[22] Bernknopf et al. found that short-term recreational visits to the area were unaffected because consumers felt that skiing was riskier than exposure to natural hazards. However, for real estate investment decisions, which are longer-term commitments, Bernknopf et al. found that property prices were affected negatively in Mammoth Lakes, in comparison to the other ski resorts in the western United States, where no price response was detected.

Recently, the hedonic pricing function has been applied to identify the effects of terrorism attacks on property prices.[23] Like the environmental and natural hazards applications described above, proximity to the site (or potential site) of terrorist attacks among other site characteristics, is considered an external effect on the real estate market. In the Redfearn application, housing values are expected to vary as a function of the proximity to the externality. The closer a property is to a target such as an airport the greater the property would be discounted by the market to a lower value. This discount is a penalty of lower

---

21. D. Brookshire, M. Thayer, J. Tschirhart, and W. Schulze, "A Test of the Expected Utility Market: Evidence from Earthquake Risks," *Journal of Political Economy*, vol. 93, no. 2 (1985): 369-389.
22. R. Bernknopf, D. Brookshire, and M. Thayer, "Earthquake and Volcano Alerts: An Economic Evaluation of Risk Perception Changes," *Journal of Environmental Economics and Management*, vol. 18, no. 1 (1990): 35-49.
23. Redfearn (2005); and V. K. Smith and D. Hallstrom, "Designing Benefit-Cost Analyses for Homeland Security Policies," in Harry Ward Richardson, Peter Gordon, and James Elliott Moore II, eds., *The Economic Impacts of Terrorist Attacks* (Cheltenham U.K.; Northampton, MA: Edward Elgar, 2005), 170-195.

value for the risk of potential damage incurred by the target's effect on the surrounding neighborhood. If consumers believe this spatial relationship to be true, the effect has been isolated and the value of urban property could diminish in high population density areas.

This application of the hedonic price function tested whether there is a housing price gradient around potential terrorist targets following the September 11, 2001, attacks. The results showed that consumers perceived no threat from terrorism on property values, as indicated by absence of an impact on local price indexes, sales volume, or the implicit price of proximity to potential targets in Los Angeles before and after the September 11, 2001, attacks.[24]

The research undertaken by Smith and Hallstrom identified ways to design a benefit-cost analysis of homeland security policies based on the risk of owning property in a hazardous location.[25] Smith and Hallstrom suggest using the analogy of risk-related information from regional-scale natural hazards as a gauge of how consumers would respond to changes in the risks to security. Among the various approaches to valuing security policies, the hedonic pricing function has been proposed as a way to measure the impact of a specific policy on the choice of risk reduction. The analysis is based on relative property prices before and after a specific event to estimate the market value of a property. By using a regional-scale natural hazard as an analogy to a terrorist attack, the scale of destruction is assumed to be consistent with the affect of a wide-ranging, regional-scale terrorist attack on a city's infrastructure.[26]

If a major natural catastrophe were to hit a large U.S. city, the size of the losses could be as large if not greater than the September 11, 2001, attacks. A study by Brookshire and Chermak asserted that natural hazards provide a reasonable parallel for estimating losses of an attack by applying benefits transfer methods.[27] In situations of hurricane and flooding, if the zoning criteria accurately represent the geographic extent of the likely hazard and the severity of the damaged structures, then locations designated as inside a flood zone should reflect a reduction in value. Areas outside these zones should not be affected in the same way.

That said, specific natural hazard events do not affect a region uniformly. Some storm affected locations are "near misses" that suf-

---

24. Redfearn (2005).
25. Smith and Hallstrom (2005).
26. Ibid.
27. D. Brookshire and J. Chermak,"Benefit and Informational Transfers," in S. Navrud, R. Ready, and O. Olvar, eds., *Environmental Value Transfer: Issues and Methods*, volume 9 in *The Economics of Non-Market Goods and Resources* series (Dordrecht: Kluwer Academic Publishers, 2007).

fer limited wind and water damage but are not affected by the higher intensities of a particular storm.[28] This variation in damage across locations serves as the data for the model because physical flood damage did not occur. Yet these places (designated as within a particular flood zone) are considered to be part of the same market as those places that suffered losses. The Smith and Halstrom model suggests that, although the physical damage and need for reconstruction will be less, the "near miss" locations will still reflect perceived changes in economic value. They assert that these "near misses" in a large damaging natural hazard application are similar to a failed terrorist attack and that risk information from the "near miss" is incorporated into and reflected by the housing market. Data from Lee County, Florida, a "near-miss" location during Hurricane Andrew, is used in the model. In this case, Federal Emergency Management Agency–designated flood zones are included in the model to capture baseline subjective risk beliefs. The distinction of being inside or outside of a predicted flood zone is included in the hedonic pricing model. The effect of the risk information was statistically significant in the model and confirmed that the "near miss" information reduced property values for homes in the region prone to coastal hazards.[29] This result suggests that the perceived risk of a hazard in a region, even if it is not realized on the ground, still influences the real estate market. Furthermore, this example suggests that a hedonic type model could offer a way to measure the perceived and real trade-offs required to evaluate some types of homeland security policies.

The hedonic price function has been formulated to estimate the value of agricultural land as reflected by its production function and ownership status. Agricultural output is a function of labor, technique, equipment, and land. One important characteristic of land that impacts farm productivity and affects land value is the rate and cumulative amount of soil erosion. A second important characteristic is geographic location, in terms of distance, absence of nuisances (e.g., insects and other pests), and presence of exceptional attractions (e.g., access to transportation).[30] Other spatially heterogeneous land characteristics that affect productive capacity such as agrochemical uptake can influence the parcel value.[31] These characteristics affect profits and land price.[32] Miranowski and Hammes have applied the hedonic pricing

---

28. Smith and Hallstrom (2005).
29. Ibid.
30. Miranowski and Cochran (1993).
31. Zilberman and Marra (1993).
32. Raymond B. Palmquist, "Land as a Differentiated Factor of Production: A Hedonic Model and Its Implications for Welfare Measurement," *Land Economics*, vol. 65, no. 1 (1989): 23-28.

function to explain variations in agricultural land values in terms of soil quality characteristics and the impacts of soil erosion.[33] Palmquist and Danielson considered the effects of farmland improvements, urban density, urban growth, and rural amenities on land values.[34]

Ownership of the land presents additional considerations, specifically owner-operator or landlord-tenant situations. Miranowski and Cochran summarize the analytical framework in an owner-operator agricultural land price problem that supports the empirical estimation of land price as a function of spatial heterogeneity in soil quality and quantity characteristics.[35] For a landlord-tenancy arrangement, estimation of the land rental price is determined by the actions of demanders and suppliers in a particular land market. Market participants influence the rental price by choosing specific land characteristics they prefer when they contract for a parcel. They cannot affect the equilibrium or market-clearing price. The rental price is established by all of the demanders and suppliers interacting in the farmland lease market. The renter is assumed to maximize variable profits derived from the value of the outputs less the costs of inputs other than land.[36] While approaching the rental market problem by using the hedonic price function is possible, we found a land price model by Miranowski and Cochran[37] related more closely to the other models reviewed and the following case study.

Building on the owner-operator agricultural land price problem, Miranowski and Cochran developed a county-level hedonic model from location and soil characteristics data to estimate farmland value.[38] The empirical model included explanatory variables for

- Physical characteristics including soil erodability, soil depth, and average rainfall
- Economic activities including the portion of total farmland devoted to cropland, average agricultural extension expenditure for the county, and real estate tax mill rate on farmland
- Spatial variables and demographics including radial distance to the nearest metropolitan area and population density per square mile in the county

---

33. John A. Miranowski and Brian D. Hammes, "Implicit Prices of Soil Characteristics for Farmland in Iowa," *American Journal of Agricultural Economics*, vol. 66, no. 5 (1984): 745-789.
34. R. Palmquist and L. Danielson, "A Hedonic Study of the Effects of Erosion Control and Damage on Farmland Values," *American Journal of Agricultural Economics*, vol. 71, no. 1 (1989): 55-62.
35. Miranowski and Cochran (1993).
36. Palmquist (1989) and Miranowski and Cochran (1993).
37. Miranowski and Cochran (1993).
38. Ibid.

Based on their analysis, Miranowski and Cochran made the following conclusions:

1. As topsoil depth increases, or soil lost to erosion decreases, there is a positive effect and land value increases.
2. An increase in erosion potential decreases land value.
3. The geographical proximity of a parcel to a metropolitan area was valued more highly for agricultural land use relative to all other land types.
4. The effective property tax rate had a larger-than-expected effect on the price of agricultural parcels.
5. Help from the public sector through increased agricultural extension activities increases land value.
6. Counties with higher population densities should increase in land value, thereby suggesting a possibility of a price premium due to the scarcity of agricultural land.

Furthermore, the relative share of cropland to all land was significant in reflecting differences in cropland and noncropland values.

In sum, the review of select literature shows that hedonic models establish an interpretation of land and property values based on both land characteristics and nonland characteristics. In the case study below, the hedonic pricing function is applied to undeveloped land in a model that compares the hedonic pricing function with and without GIS-measured distances.

A Case Study in Land Valuation

Miami-Dade County covers approximately 1,950 square miles in southeast Florida (Exhibit 14.1). In July 2006, the county population was approximately 2.4 million and the population density was about 1,230 people per square mile, according to the U.S. Census (2007). County statistics available for 2005 report more than 928,700 housing units and about 74,260 nonfarm establishments, and at that time private nonfarm employment exceeded 858,000. Currently, development is encroaching on lands believed to be critical for the survival of national parks and refuges.[39] In this context, the case study demonstrates the use of the hedonic pricing method to understand how the value of undeveloped land throughout the county could be affected by parcel characteristics, neighborhood characteristics, environmental regulations, location constraints, and GIS-measured distances.

39. D. Hallec, U.S. Department of Interior, National Park Service, personal communication (2006).

Exhibit 14.1 Location of Study Site in Miami-Dade County, Florida (Labiosa et al., 2008)

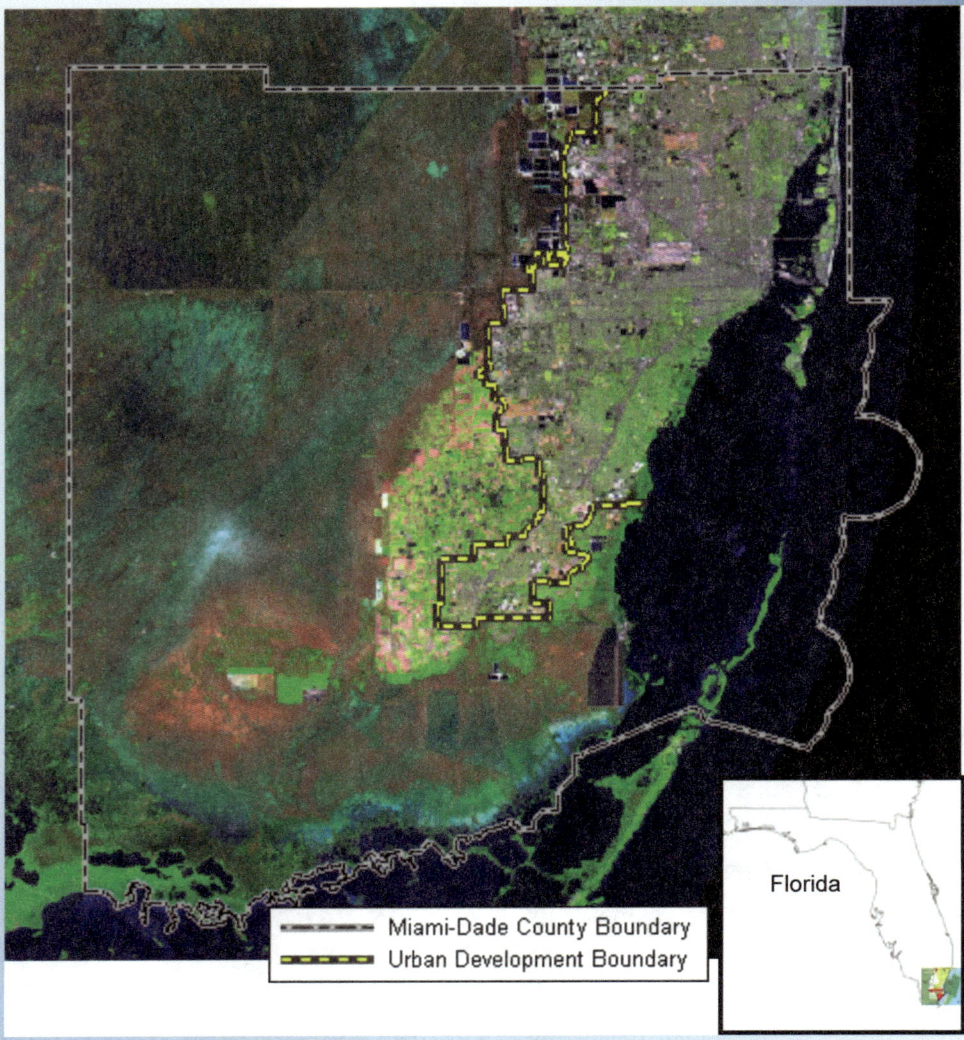

Research was conducted to estimate a hedonic price function for land for use in a decision support tool. The land valuation relates market price to land use/land cover characteristics indicative of conservation and development decisions in Miami-Dade County. The primary objective of the USGS project was to develop an integrated ecological and socioeconomic land use evaluation model for resource managers to use in reconciling the need to maintain the ecological health of South Florida parks and refuges with the demand for land to maintain regional economic development and community growth. With increasing pressures for higher density development in the agricultural lands outside of the urban development boundary, there is considerable stress on parkland and private lands adjacent to protected areas.[40] The research project made use of contributions from conservation ecology, landscape ecology, decision science, real estate economics, environmental economics, urban planning, GIS analysis, and Web technologies.[41]

State, county, and regional environmental regulations and management policies that restrict development are designed to protect and preserve environmentally sensitive lands. These same regulations and policies influence land values by altering the amount and density of land that can be developed, which is further complicated by land use planning, zoning, and covenants, codes, and restrictions of subdivisions and homeowners associations. There has been considerable debate about the future direction of development and urban expansion in Miami-Dade County, the effects that this growth will have on lands that have been protected, and what additional lands could be protected for present and future generations. For example, the *Miami Herald* reported the following:

> The most daunting threat to the Everglades is the runaway development that is still wiping out its wetlands and stressing its aquifers. The Miami-Fort Lauderdale-West Palm Beach conurbation has become America's sixth-largest metropolitan area, obliterating almost every patch of green space between the Atlantic and the perimeter levee. Postwar Everglades suburbs such as Coral Springs, Hialeah, Miami Gardens, Miramar, Pembroke Pines, and Sunrise have all attracted 100,000 residents, and are approaching build-out.[42]

Another opinion of the situation published a few years later was

> Westward sprawl has become the area's hottest political issue. Miami-Dade County has already approved two developments outside its "urban service boundary"–one built by Governor Bush's former business partner–and is now embroiled in a battle over proposals to shift the entire boundary west and south.[43]

A statement in the Miami-Dade County Comprehensive Plan, which was written with a different tone concerning growth and the environment, provides insight to the public context for the debate:

> The overarching aims of the Economic Element are to expand and further diversify the Miami-Dade economy, provide employment for all who want to work, and increase income and wealth. More specifically, the Element provides a set of goals and associated objectives and policies that will enhance

---

40. Ibid.

41. W. Labiosa, R. Bernknopf, P. Hearn, D. Hogan, D. Strong, L. Perlstine, A. Wein, K. Gillen, and S. Wachter, "The South Florida Ecosystem Portfolio Model: An Integrated Ecological, Economic, and Geographic Land Use Planning Tool," USGS Open-file Report, unpublished manuscript (2008).

42. Noah Bierman and Tim Henderson, "South Florida's Sprawl Quickly Nearing Limit; Western Cities Near Built-Out," *Miami Herald* (July 10, 2003).

43. M. Grunwald, *The Swamp: The Everglades, Florida, and the Politics of Paradise* (New York: Simon and Schuster, 2006), 363.

> Miami-Dade County government's contribution to the economic development of the area. The Element will serve as the general policy framework for economic development decisions and it will be the guide for operational activities, which influence economic development.[44]

Part of the debate concerning the effectiveness of development policies that alter otherwise unregulated development is the economic effect of

1. Creating an urban development boundary (UDB), shown as the thick yellow-and-black line in Exhibit 14.1.
2. Requiring land development that is contiguous to existing development, as indicated by the adjacent location of a vacant parcel with developed land illustrated in Exhibit 14.2.
3. Allowing development on land that does not require drainage for development.
4. Avoiding ecologically sensitive lands. (Exhibit 14.2 depicts protected areas located near vacant parcels that are adjacent to a developed area.)

In addition to federal and state regulations, the overlapping mechanisms for regulating the development of land include county or municipal zoning designations, subdivision regulations, and building codes. These mechanisms can dictate such practices as type of use, parcel and road layout, density, height, mitigation measures, and setbacks from human development and environmental landscape features. In general, zoning determines coarser-scale patterns and design of land use, preservation and connectivity, and hence development potential. In some jurisdictions, county or municipal officials regulate site design through stormwater management, erosion control, and best management practices. In addition to shaping the community's character, aesthetic qualities, and additional best management practices (beyond county or municipal levels), subdivision regulations determine the extent to which property owners can subdivide their property, designate minimum lot sizes, and require infrastructure for development. Building codes set minimum standards for construction, and relate more to the exteriors and interiors of structures and their associated systems (i.e., plumbing, electrical).

Environmental Protection Programs

The Miami-Dade County government has realized that the region's unique environmental circumstances require special attention.[45] As a result, many policies and regulations have been included in Miami-

44. http://www.miamidade.gov/planzone/cdmp.asp/.
45. Ibid.

Exhibit 14.2 **Land Use Map in the USGS Ecosystem Portfolio Decision Support Tool (Labiosa et al., 2008)**

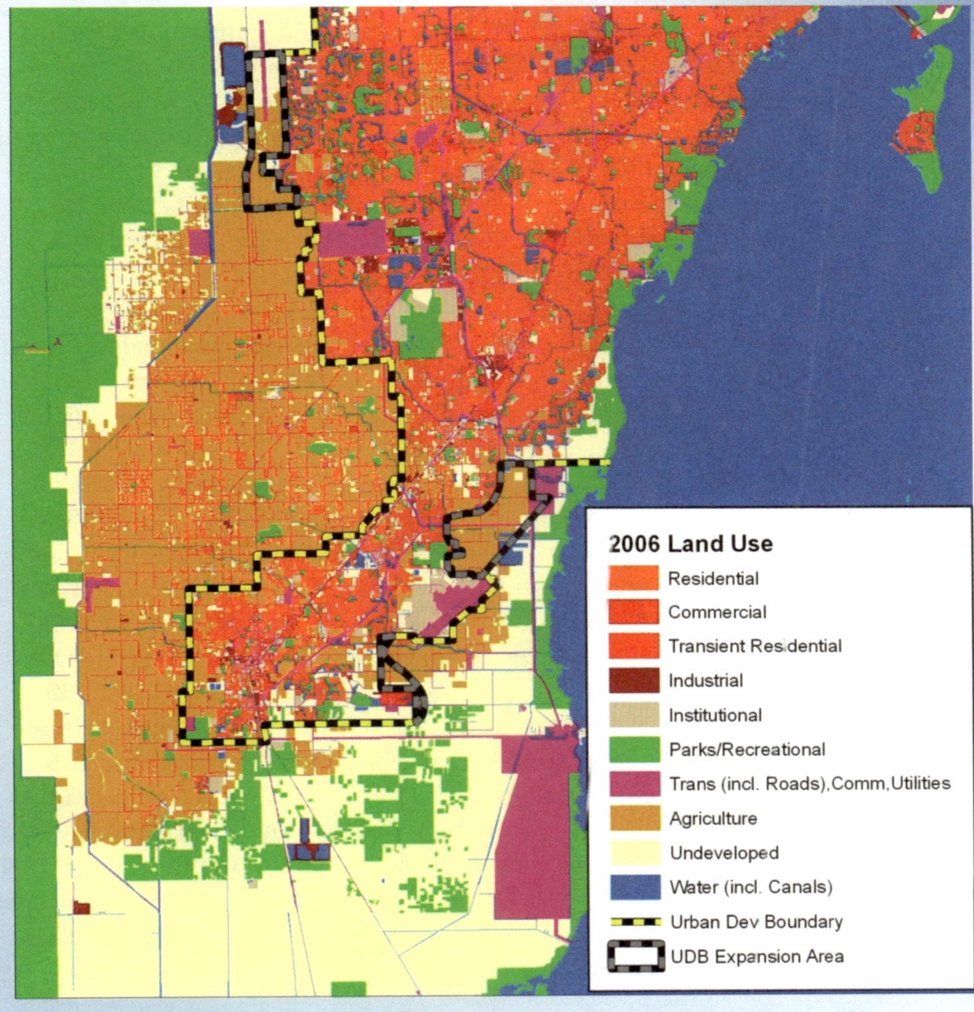

Dade County's Comprehensive Plan to protect fragile natural resources and ensure the region's future sustainability. In general, the County protects environmentally sensitive lands by

- Establishing an urban development boundary.
- Requiring development to be contiguous to existing development.
- Only allowing development on dry land. Land can not be drained for development.
- Adhering to Florida's state concurrency requirements.
- Administering the Environmentally Endangered Lands Program.

Each of these protection strategies is described briefly below.

Urban Development Boundary

The primary purpose of the UDB is to promote contiguous development rather than scattered, patchy development. This helps to provide efficient delivery of public services and infrastructure, protect environmentally sensitive land and agriculture from urban encroachment, and promote infill development and redevelopment. The UDB is included in the County Comprehensive Plan to distinguish the area where urban development may occur through the year 2015 from areas where development should not occur. The UDB is always under intense scrutiny and pressure to be changed or altered to suit the needs of different groups. At its heart, the appropriate designation of boundaries for urban containment can help provide protection and a regulatory mechanism for preserving sensitive landscapes that are unsuitable for intensive development.

The land use planning literature suggests that the UDB should have an influence on land values. Theoretical models consider the impact of land use policy to preserve farmland on property values.[46] According to Nelson, urban land value must rise because urban containment would result in more intensive use and because benefits of urban containment and greenbelt preservation should be capitalized in the land market. Programs such as the Williamson Act in California, established in 1965, are intended to preserve the agricultural or open space lands by decreasing their property tax assessment.[47] If urban containment is effective, the land market will value proximity to the boundary. Near the boundary, property owners can take advantage of the views and privacy of the greenbelt beyond the UDB, which is a rural amenity effect.[48] (The example later in the chapter evaluates whether land values outside the UDB are systematically lower than land values inside the boundary.)

Contiguous Urban Development Requirement

Both Miami-Dade County and the State of Florida have policies in place to prevent leapfrog or patchy development. Several policies encourage local decision-makers to deny development projects that are

---

46. Arthur C. Nelson, "Demand, Segmentation, and Timing Effects of an Urban Containment Program on Urban Fringe Land Values," *Urban Studies*, vol. 22, no. 5 (1985): 439-443.

47. Department of Conservation, State of California, http://www.conservation.ca.gov/dlrp/lca/Pages/Index.aspx/.

48. D. Hellerstein, C. Nickerson, J. Cooper, P. Feather, D. Gadsby, D. Mullarkey, A. Tegene, and C. Barnard, *Farmland Protection: The Role of Public Preferences for Rural Amenities, Agricultural Economic Report No. 815* (Washington, D.C.: U.S. Department of Agriculture, Economic Research Service, 2002).

not contiguous to other development projects. This policy requirement complements the purpose of the UDB to concentrate development in certain areas.

No Drainage Policy

Miami-Dade County's environmental regulation prohibits draining wetlands for development.

State and County Concurrency Requirements

Concurrency has its roots in state legislation called the *Local Government Comprehensive Planning and Land Development Regulation Act*, which was adopted by the Florida legislature in 1985 and which amended Chapter 163 of the Florida Statutes. The act mandated that specific levels of service standards be adopted for roadways, mass transit, water, sewer, solid waste, local recreation open space, and drainage, and that each of these services be defined and addressed in local comprehensive plans. Also, the act further stated that no development orders can be issued when the adopted levels of service would not be met.

Environmentally Endangered Lands Program

The Environmentally Endangered Lands (EEL) Program has been established to acquire, preserve, enhance, restore, conserve, and maintain threatened natural forest and wetland communities located in Miami-Dade County for the benefit of present and future generations.[49] The purpose of the EEL Program has been to acquire and protect environmentally endangered lands that, if not acquired, would threaten the environmental integrity of the existing resource or that, if acquired, would enhance the environmental integrity of the resource with the primary objective of maintaining and preserving their natural resource values for present and future generations.

A Hedonic Model for Land Valuation in Miami-Dade County

In applying a hedonic pricing model to property in Miami-Dade County, we estimated land value as a function of parcel square footage, zoning and environmental regulations, location and distance to amenities, and year and season of property transaction. In this section, we demonstrate that, consistent with planning concepts and theory in the literature, land use controls have an effect on potential land values inside and outside the UDB and around other environmentally sensitive and protected land. Furthermore, an important contribution

49. Ord. No. 04-214, §§ 1, 5, 12-2-04.

to the spatial variation in price that was included in the model was the measured distances from a parcel to a variety of amenities.

Study Area, Data, and Variables

Property data was aquired from the Miami-Dade County tax roll. We obtained comprehensive information for every unique parcel amounting to 541,184 observations. Exhibit 14.3 summarizes the real estate statistics for Miami-Dade in the data set. Of the total number of parcels, undeveloped land parcels in the county accounted for approximately 51,000 observations. There were about 28,500 observations of arm's-length transactions. The final data set consisted of about 24,000 observations after removing statistical outliers[50] and observations with missing or incorrect data.

Exhibit 14.3 lists statistics for parcels inside the UDB and non-UDB (outside the UDB). There is a considerable disparity between parcels located inside and outside of the UDB for lot size, sale price, assessed value, and price per square foot. A parcel located inside the UDB is slightly less than 13.5% the size (square feet) and has an average sale price that is 20 times greater and a median sales price that is over 60 times greater than a parcel outside the UDB. The assessed values in Exhibit 14.3 show the same order of magnitude difference as the sale price (i.e., the average and median assessed price are 19 and 54 times greater respectively) inside the UDB relative to outside the UDB. This price differential between parcels inside and outside

Exhibit 14.3 **Summary Statistics for Properties in Miami-Dade County, Florida, in 2006**

	# of Parcels	sq. ft.	Sale Price	Assessed Value	$/sq. ft.
Total parcels	541,184				
Total vacant	50,850				
Vacant inside UDB	30,739	26,399	$14,117	$25,238	
Vacant non-UDB	20,111	18,739	$9,991	$19,454	
Average size inside UDB	71,208	74,850			
Average size non-UDB	523,590	557,902			
Average sale price inside UDB			$1,592,882		$21.28
Average sale price non-UDB			$583,168		$1.05
Median sale price inside UDB			$240,000		$3.21
Median sale price non-UDB			$25,000		$0.05
Average assessed price inside UDB				$252,664	$3.38
Average assessed price non-UDB				$100,457	$0.18
Median assessed price inside UDB				$81,445	$1.09
Median assessed price non-UDB				$9,615	$0.02

50. Outliers were defined as observations that lay in the 2.5% tails of any variable's distribution.

the UDB has a significant influence on predicting the value of vacant land in the hedonic model.

Depending on consumer preferences, each variable was anticipated to have either a positive or negative effect on price in the following ways:

- Lot size was expected to be negatively correlated with price because as a parcel size increases, a buyer's marginal willingness to pay will decline on a price per square foot basis while the total price of the property increases.
- Land zoning influences were expected to be positive for agricultural lands and negative for recreational parcels. We hypothesized that recreation zoning is worth less because it is perceived to have a lower probability of conversion to a higher (developed) land use. We expected the opposite for agriculturally zoned parcels.
- "Transacted in Winter Months" was expected to be negatively correlated with price because real estate transactions traditionally are less likely in those months even in Florida.

GIS provided the important capability to use the location of amenities and measures of distance as explanatory variables that identify land use and land cover, and to visualize parcel characteristics. The land use map shown in Exhibit 14.2 illustrates where a parcel is located relative to the location of some spatial amenities. The GIS data was used to represent development restrictions and boundaries based on environmental regulations and flood zones.

We expected properties located inside the UDB (shown as the yellow-and-black boundary line in Exhibit 14.1) and properties that were contiguous to existing development (shown, for example, as the areas within gray-and-black boundaries that have been proposed for urban expansion) to have positive effects on economic value. Locations within the UDB should be positively correlated with price because land within the UDB is supported by significant public infrastructure and greater development density is permitted inside the boundary. A parcel contiguous to development satisfies a county regulatory requirement (e.g., the environmental protection programs discussed earlier in the chapter) and should be worth more than parcels located away from existing development. We expected rural and urban amenities such as proximity to the ocean and environmentally sensitive parcels to have positive effects on land values. Environmentally endangered land was identified, publicly listed for preservation, and purchased as part of a county land acquisition program.[51] An EEL listing requires agreement by the owner. Parcels marked for

51. Code of Ordinances, Miami-Dade County, Florida, Chapter 24-50.

EEL designation would lower the probability of conversion to developed land because there was an expressed desire to protect the land.

The spatial variables included distances to central business districts, highways, parks, and waterways. We anticipated both positive and negative impacts on the sale price depending on the variable. The variable *Distance to the Miami CBD* was the Euclidean distance (in miles) from a parcel to the Miami central business district (CBD). We expected that the farther a parcel is from the city center, the greater the negative effect on land price. The *Distance to Nearest Secondary CBD* variable was expected to be positively correlated to price as a convenience to residents. However, the primary CBD may have a dominating effect. *Distance to Nearest Highway* (in miles) was hypothesized to have a negative correlation with price, suggesting properties closer to highways are more valuable.

Distance to Park (in miles) was a parcel's Euclidean distance to a park other than a national park and was hypothesized to be inversely correlated with price. We anticipated that the distance to a park (including a national park) to be negatively correlated with price, i.e., the closer the parcel is to a park, the more valuable the parcel is to the buyer.

We hypothesized *Distance to Canal* (in miles) to be positively correlated with price, i.e., price should increase the farther the parcel is from a canal. Some canals can contain contaminated water that can influence the value of nearby properties. We also realized that there could be canals that are desirable (e.g., for navigation). The variable *Distance to Nearest Inland Body of Water* (in miles) should be positively correlated with price because these bodies of water are caused by extraction of construction materials that then fill with undesirable mineralized water. Parcels west of Canal L31 were hypothesized to be less valuable than parcels east of the canal because this particular canal separates parcels that are protected from flooding from those that are not.[52] A high potential for flooding was expected to have a negative effect on parcel price because of the no drainage rule, which does not allow for the dewatering of a parcel. Likewise, the *Coastal Flooding* variable was expected to have a negative coefficient because these parcel values would be discounted in response to the coastal flooding hazard. Oceanside properties were hypothesized to have a positive correlation with land price because locations near the ocean are desirable for views and access.

All of the explanatory variables were evaluated for inclusion in the model. Exhibit 14.4 lists the dependent and all of the explanatory

52. C. Dwyer, Miami-Dade County Department of Planning and Zoning, personal communication (2007).

Exhibit 14.4 Variables and Definitions in the Hedonic Price Function for Miami-Dade County

Variable	Definition
price_sqft	Lot price in dollars per square foot
AREA	Lot area in square feet
PERIMETER	Lot exterior perimeter in feet
LOT_SIZE	Lot size of each parcel in square feet
xlot_sqft	Lot square footage is the size of each parcel in square feet
Winter	Real estate transaction occurred during the winter months, = 1 if transacted in winter months
year_19XX	A parcel sale occurs in a specific year, = 1 if transaction occurs in the given year
Recreational*	**Recreational encompasses all parcels designated as zoned "recreational" in the Miami-Dade County Land Use map, = 1 if zoned "recreational"**
Agricultural	**Agricultural encompasses all parcels designated as zoned "agricultural" in the Miami-Dade County Land Use map, = 1 if zoned "agricultural"**
Flood_zone	**= 1 if coastal flood zone**
EEL_Private	**A parcel is designated as Environmentally Endangered Land if it has ecologically desirable characteristics that the landowner and the county have agreed to not develop, = 1 if parcel is private EEL purchase**
*Dist_CBD***	*Distance to CBD is the linear distance in miles from the Miami Central Business District to the parcel using a GIS*
Dist_Canal	*Distance to canal is a parcel's distance in miles to the nearest canal*
Dist_HWY	*Distance to highway is a parcel's distance in miles to the nearest state or federal highway*
Dist_inland	*Distance to inland water is a parcel's distance in miles to the nearest inland body of water*
Dist_Park	*Distance to park is a parcel's distance in miles to a local park*
Dist_Biscayne	*Distance to Biscayne National Park is a parcel's distance in miles to Biscayne National Park*
Dist_Everglades	*Distance to Everglades National Park is a parcel's distance in miles to Everglades National Park*
Dist_UDB	*Distance to UDB is a parcel's distance in miles to the Urban Development Boundary*
UDB_Qtr_Mi	*Parcels identified to be within ¼ mile of the Urban Development Boundary*
Dist_Ocean	*Distance to ocean is a parcel's distance in miles to the Atlantic Ocean*
Dist_MjrRd	*Distance to major road is a parcel's distance in miles to the nearest major local road*
Dist_Scndry_CBD	*Distance to CBD is the distance in miles from a secondary central business district to the parcel*
Canal_L31_West	*L31 West Canal = 1 if parcel is west of Canal L31*
Dist_CERP	*Distance to a Comprehensive Everglades Restoration Project (an investment in restoring degraded everglades habitat) is the linear distance in miles from the parcel*
Dist_EEL	*Distance to a Environmentally Endangered Land parcel is the linear distance in miles from the parcel*
Oceanside	*Oceanside is an oceanfront property identified in GIS*
EEL_Size	*Lot size of an Environmentally Endangered Land parcel*
EEL_UDB	*EEL parcel inside the UDB, = 1 if an EEL parcel is inside the UDB*
Canal	

Exhibit 14.4	**Variables and Definitions in the Hedonic Price Function for Miami-Dade County *(continued)***
Variable	**Definition**
UDB*	***A parcel is designated as inside the Miami-Dade County's Urban Development Boundary, = 1 if parcel is within Urban Development Boundary***
Contiguous	***A parcel is designated as contiguous for development if it is located to an existing developed parcel, = 1 if parcel is contiguous to development***
Zone_A	***Zone A is the flood insurance rate zone that corresponds to the 1% annual chance floodplains that are determined in the Flood Insurance Study by approximate methods of analysis. Mandatory flood insurance purchase requirements apply.***
Zone_AE	***Zone AE is the flood insurance rate zone that corresponds to the 1% annual chance floodplains that are determined in the Flood Insurance Study by detailed methods of elevation analysis. Mandatory flood insurance purchase requirements apply.***
Zone_AH	***Zone AH is the flood insurance rate zone that corresponds to the areas of 1% annual chance of shallow flooding with a constant water-surface elevation (usually areas of ponding) where average depths are between 1 and 3 feet. Mandatory flood insurance purchase requirements apply.***
Zone_VE	***Zone VE is the flood insurance rate zone that corresponds to areas within the 1% annual chance coastal floodplain that have additional hazards associated with storm waves. Base Flood Elevations derived from the detailed hydraulic analyses are shown at selected intervals within this zone. Mandatory flood insurance purchase requirements apply.***
Zone_X500	***Zone X is the flood insurance rate zone that corresponds to areas outside the 1% annual chance floodplain, areas of 1% annual chance sheet flow flooding where average depths are less than 1 foot, areas of 1% annual chance stream flooding where the contributing drainage area is less than 1 square mile, or areas protected from the 1% annual chance flood by levees. No Base Flood Elevations or depths are shown within this zone. Insurance purchase is not required in these zones.***
Flood_zone_inland	***Inland flood zone is measured as the proximity to the ocean-caused coastal flooding, which is within a half mile of the shoreline***

* Bold font indicates a nonspatial regulatory variable.

** Italic font indicates a spatial variable measured in a GIS.

*** Bold italic font indicates a spatially measured regulatory variable.

variables for each parcel in the county that could be included in the hedonic price function. Exhibit 14.5 contains the salient statistics for the dependent and the explanatory variables.

The relationships of a parcel and its characteristics to other nearby parcels and their characteristics can help explain spatial price differentials. In combination, the visualization of land use in Exhibit 14.2 and the characteristics listed in Exhibit 14.5 provide a "picture" of the real estate market for land in Miami-Dade County. In this way, GIS adds a new dimension to the economic valuation of properties using the hedonic pricing function.

Model Specification

In our model, we applied the log-semilog formulation of the hedonic price function. Several variants of the statistical regression were run

Exhibit 14.5 **Statistics for the Dependent and Explanatory Variables in the Hedonic Price Function for Miami-Dade County**

Variable	*N*	Min.	Max.	Mean	Standard Deviation
AREA	24224	19.869	97208597	194207.9	1110742
price_sqft	24127	0.010099	993.6655	21.66898	64.22064
PERIMETER	24224	29.571	58257.65	1291.744	1904.62
LOT_SIZE	24224	0.5	7907838	26170.98	175847.1
lot_sqft	24224	8	94050396	199601.3	1138525
Recreational	**24224**	**0**	**1**	**0.000991**	**0.031461**
Agricultural	**24224**	**0**	**1**	**0.012178**	**0.109682**
Flood_zone	**24224**	**0**	**1**	**0.634412**	**0.481605**
EEL_Private	**24224**	**0**	**1**	**0.051148**	**0.220303**
Dist_Canal	*24224*	*0*	*5.82*	*1.079939*	*0.959757*
Dist_HWY	*24224*	*0*	*10.38*	*1.331828*	*1.755592*
Dist_Park	*24224*	*0*	*12.77*	*2.010266*	*2.33851*
Dist_Biscayne	*24224*	*0*	*24.8*	*11.23411*	*5.039312*
Dist_Everglades	*24224*	*0*	*27.81*	*9.765871*	*7.371954*
Dist_UDB	*24224*	*0*	*13.74*	*1.513842*	*2.697697*
UDB_Qtr_Mi	*24224*	*0*	*1*	*0.0534*	*0.2248*
Dist_Ocean	*24224*	*0*	*24.08*	*6.8902*	*5.816657*
Dist_MjrRd	*24224*	*0*	*2.75*	*0.124106*	*0.337726*
Dist_Water	*24224*	*0*	*5.95*	*0.536313*	*0.604201*
Dist_CBD	*24224*	*0.035*	*41.07*	*15.36*	*9.32*
Dist_Scndry_CBD	*24224*	*0*	*15*	*3.076701*	*2.916524*
Canal_L31_West	*24224*	*0*	*1*	*0.157117*	*0.363918*
Dist_CERP	*24224*	*0*	*15*	*5.17301*	*3.848523*
Dist_EEL	*24224*	*0*	*15*	*2.860882*	*2.840016*
Oceanside	*24224*	*0*	*1*	*0.129004*	*0.335212*
EEL_Size	*24224*	*0*	*27593706*	*36423.03*	*452665.1*
EEL_UDB	*24224*	*0*	*1*	*0.008132*	*0.089814*
Canal	*24224*	*0*	*1*	*0.040208*	*0.196451*
UDB	***24224***	***0***	***1***	***0.605***	***0.489***
Contiguous	***24224***	***0***	***1***	***0.110964***	***0.314094***
Zone_A	***24224***	***0***	***1***	***0.160213***	***0.366811***
Zone_AE	***24224***	***0***	***1***	***0.192454***	***0.394236***
Zone_AH	***24224***	***0***	***1***	***0.27803***	***0.448038***
Zone_VE	***24224***	***0***	***1***	***0.003715***	***0.060841***
Zone_X500	***24224***	***0***	***1***	***0.058909***	***0.235458***
Flood_zone_inland	***24224***	***0***	***1***	***0.123803***	***0.329363***
year_1970	24224	0	1	0.000165	0.012849
year_1971	24224	0	1	0.003179	0.056291
year_1972	24224	0	1	0.004046	0.063477
year_1973	24224	0	1	0.012219	0.109866
year_1974	24224	0	1	0.017792	0.132198
year_1975	24224	0	1	0.01193	0.108575

Exhibit 14.5 **Statistics for the Dependent and Explanatory Variables in the Hedonic Price Function for Miami-Dade County *(continued)***

Variable	*N*	Min.	Max.	Mean	Standard Deviation
year_1976	24224	0	1	0.017834	0.132349
year_1977	24224	0	1	0.018246	0.133844
year_1978	24224	0	1	0.016595	0.127751
year_1979	24224	0	1	0.014985	0.121496
year_1980	24224	0	1	0.026626	0.160993
year_1981	24224	0	1	0.021797	0.146022
year_1982	24224	0	1	0.015481	0.123456
year_1983	24224	0	1	0.014283	0.118659
year_1984	24224	0	1	0.014614	0.120003
year_1985	24224	0	1	0.015687	0.124264
year_1986	24224	0	1	0.015481	0.123456
year_1987	24224	0	1	0.014325	0.118828
year_1988	24224	0	1	0.01643	0.127125
year_1989	24224	0	1	0.018205	0.133695
year_1990	24224	0	1	0.016058	0.125703
year_1991	24224	0	1	0.017132	0.129765
year_1992	24224	0	1	0.015357	0.122969
year_1993	24224	0	1	0.018907	0.136199
year_1994	24224	0	1	0.022746	0.149096
year_1995	24224	0	1	0.018081	0.133248
year_1996	24224	0	1	0.020558	0.141902
year_1997	24224	0	1	0.025471	0.157553
year_1998	24224	0	1	0.029475	0.169137
year_1999	24224	0	1	0.033025	0.178706
year_2000	24224	0	1	0.039837	0.195579
year_2001	24224	0	1	0.042272	0.201213
year_2002	24224	0	1	0.059693	0.236922
year_2003	24224	0	1	0.092512	0.289753
year_2004	24224	0	1	0.11245	0.315926
year_2005	24224	0	1	0.118271	0.322936
year_2006	24224	0	1	0.028236	0.165651

to test its robustness to different model specifications, different data reconciliation choices, and different ways of representing urban development and environmental regulations.[53] The hedonic price function for Miami-Dade County combined elements from all of the

53. The hedonic pricing literature addresses but does not resolve the question of functional form. A general principle of statistics asserts that it is best to have a logical, maximum form in place *a priori* so that the available data can be devoted to fitting and testing that form rather than to finding a form. To counteract the danger of "overfitting" data and coming up with meaningless or skewed results, the reliability of the model can be tested.

types of applications identified and described in the literature review. Integration of these elements with distance and location explanatory variables along with other parcel characteristics provided a way to achieve more robust estimates of land value. As part of the development of the valuation model, we were able to evaluate whether the location variables derived in a GIS are intuitive and increase the explanatory power to help improve the model.[54]

The hedonic price function for land in Miami-Dade County is

$$\ln p_{i\tau} = \alpha + \beta_1 H + \beta_2 Y + \beta_3 Z + \gamma_t T \qquad (2)$$

where

$\ln p_{i\tau}$ = the natural log of the transaction price per square foot of property i during time period τ, ($\tau = 1, \ldots, n$)

$H = (h_1, \ldots, h_i)$ and $Y = (y_1, \ldots, y_i)$ are property characteristics with h_i measured continuously, e.g., square feet of land parcel, y_i measured categorically, land use

$Z = (z_1, \ldots, z_i)$ are variables measured continuously as linear distances (in miles), e.g., distance to the central business district

T are fixed effects variables indicating whether the transaction takes place during time period t

α, β_j, and γ_τ are the coefficients to be estimated. In particular, the series of coefficients γ_τ are the price indices.

The model was estimated with and without the spatial explanatory variables for the Miami-Dade County data.

Model Results

Two regression equation models were estimated, and they performed as expected. The dependent variable for both models was ln(*price*/*ft*^{2}), and the estimation method to explain the variation in land price was ordinary least squares. The results for Model 1, a nonspatial version of the hedonic price function, are listed in the four left columns in Exhibit 14.6 that include property characteristics, land zoning, and sale year. Model 2 results, a spatial version of the hedonic price function, are contained in the columns on the right side of Exhibit 14.6 (Model 2 includes Model 1 variables). Addition of the spatial explanatory variables (see Exhibit 14.4 for which variables are coded as spatial) in Model 2 enhanced the nonspatial model with distance measurements from a parcel to a variety of amenities and destinations and with delineation of environmental and growth regulations and standards. The explanatory power of Model 2 was more than twice that of Model 1, i.e., the adjusted R^2 rises from about 0.34 to about 0.77. Furthermore, a significant number of the GIS-measured explan-

54. Krister Sandberg, "Hedonic Prices, Economic Growth, and Spatial Dependence," PhD thesis, (2004); and Baifu Xu, "An Hedonic Analysis of Southwestern Louisiana Wetland Prices Using GIS," M.S. thesis (2007).

Exhibit 14.6 **Regression Results for the Hedonic Price Nonspatial and Spatial Models***

Model	Model 1: Excludes Spatial Explanatory Variables; adjusted R^2 = 0.3352				Model 2: Includes Spatial Explanatory Variables; adjusted R^2 = 0.7683			
Variable	**Estimate**	**Error**	**t Value**	**Pr > |t|**	**Estimate**	**Error**	**t Value**	**Pr > |t|**
Intercept	-1.29076	0.99461	-1.3	0.1944	-1.09794	0.58801	-1.87	0.0619
Lot_sqft	-2.53E-07	1.13E-08	-22.34	< 0.0001	-6.91E-09	6.76E-09	-10.23	< 0.0001
Flood_zone	0.79381	0.21058	3.77	0.0002	-0.82635	0.1265	-6.53	< 0.0001
Recreational	-.24303	0.40674	-0.6	0.5502	-0.68378	0.23946	-2.86	0.0043
Agricultural	-1.0501	0.11666	-9.0	< 0.0001	0.24256	0.06964	3.48	0.0005
EEL_private	-1.88146	0.14145	-13.3	< 0.0001	-0.70475	0.08686	-8.11	< 0.0001
Dist_CBD					-0.07466	0.00165	-45.26	< 0.0001
Dist_Canal					0.0278	0.01041	2.67	0.0076
Dist_Water					0.14813	0.01834	8.08	< 0.0001
Dist_HWY					-0.06415	0.00776	-8.27	< 0.0001
Dist_Everglades					-0.00558	0.00234	-2.38	0.0172
Dist_Scndry_CBD					0.07062	0.00704	10.03	< 0.0001
Canal_L31_West					-0.66027	0.04186	-15.77	< 0.0001
Dist_EEL					-0.08386	0.00647	-12.96	< 0.0001
Oceanside					0.96204	0.02917	32.98	< 0.0001
UDB_Qtr_Mi					-0.29458	0.03487	-5.87	< 0.0001
UDB					1.85972	0.02935	63.36	< 0.0001
Contiguous					0.40938	0.0187	21.9	< 0.0001
Year_1981	0.13933	0.99836	0.14	0.889	1.0468	0.58775	1.78	0.0749
Year_1983	0.77995	1.00034	0.78	0.4356	1.00454	0.58889	1.71	0.0881
Year_1984	0.98356	1.00021	0.98	0.3254	1.27435	0.5888	2.16	0.0304
Year_1985	1.03367	0.99986	1.03	0.3012	1.11276	0.5886	1.89	0.0587
Year_1986	1.26586	1.00005	1.27	0.2056	1.3149	0.58871	2.23	0.0255
Year_1987	1.17613	1.0003	1.18	0.2397	1.29243	0.5888	2.19	0.0282
Year_1988	1.3414	0.99958	1.34	0.1796	1.42426	0.58846	2.42	0.0155
Year_1989	0.93626	0.99909	0.94	0.3487	1.36329	0.58816	2.32	0.0205
Year_1990	1.21352	0.99975	1.21	0.2248	1.42193	0.58855	2.42	0.0157
Year_1992	1.55093	1.0001	1.55	0.121	1.32121	0.58874	2.24	0.0248
Year_1993	1.9065	0.99892	1.91	0.0563	1.46483	0.58804	2.48	0.013
Year_1994	1.7666	0.9982	1.77	0.0768	1.4684	0.58765	2.5	0.0125
Year_1995	1.97575	0.99914	1.98	0.048	1.46483	0.58819	2.49	0.0128
Year_1996	2.19656	0.99857	2.2	0.0278	1.57375	0.58783	2.68	0.0074
Year_1997	2.16248	0.99781	2.17	0.0302	1.63725	0.58738	2.79	0.0053
Year_1998	2.44851	0.99737	2.45	0.0141	1.72018	0.58713	2.93	0.0034
Year_1999	2.42212	0.99707	2.43	0.0151	1.71521	0.587	2.92	0.0035
Year_2000	2.49168	0.99665	2.5	0.0124	1.72573	0.58671	2.94	0.0033
Year_2001	2.73561	0.99653	2.75	0.0061	1.78815	0.58666	3.05	0.0023
Year_2002	2.70334	0.99596	2.71	0.0066	1.84433	0.5863	3.15	0.0017
Year_2003	2.96231	0.99547	2.98	0.0029	2.05272	0.58602	3.5	0.0005
Year_2004	3.45777	0.99532	3.47	0.0005	2.38655	0.58594	4.07	< 0.0001
Year_2005	3.69459	0.99528	3.71	0.0002	2.62069	0.58592	4.47	< 0.0001
Year_2006	4.07573	0.99754	4.09	< 0.0001	3.06471	0.58724	5.22	< 0.0001

* All variables are significant at least at the 10% level.

atory variables were statistically significant (Pr < 0.01) to improve the hedonic valuation of the Miami-Dade County land market.

The Model 1 variables that performed as expected were lot size and environmentally endangered land. Lot size measured as lot square footage and EEL had the expected signs and were statistically significant.

However, other explanatory variables did not behave intuitively, including land zoned for recreation and agriculture and coastal flooding potential. We found the zoning variable for recreation had the intuitive sign but was statistically insignificant, while the agriculture zoning for land use was statistically significant but had a nonintuitive sign, although less negative than recreation. The flood zone variable was nonintuitive in that the value of a property was found to increase when subject to coastal flooding. One of the best predictors was the transaction year fixed effect for transactions that occurred between 1981 and 2006. Those variables were found to be positively related for market conditions that prevailed in that year. Overall, the result for Model 1 was disappointing because of the low explanatory power and the lack of evidence for the importance of the zoning variables as drivers in the real estate market for conversion of land parcels to higher uses.

The results for Model 2 produced a substantial improvement over Model 1 due, in part, to the change from nonintuitive results for specific explanatory variables in Model 1 to intuitive results in Model 2 and the dramatic increase in explanatory power. As in Model 1, the lot size measured as square footage had the intuitive sign and was statistically significant. The remaining parcel characteristic variables performed as hypothesized in Model 2. These variables were the recreation and agriculture zone variables, which had opposite signs and were significant. Land zoned for recreational use was negatively correlated with price, and land zoned for agricultural use was positively correlated, suggesting that agricultural lands were higher valued by the market. Also, parcels subject to coastal flooding changed signs to have a negative impact on price. This could suggest that the variable of Model 1 could have been confounded by distance to the ocean (*dist_ocean*), which was isolated in Model 2 and not found to be significant.

The 11 spatial and two spatial regulatory variables that were derived and measured using GIS added a new dimension to the hedonic price function. The effects of the two spatial regulatory variables (*Parcel is Located Within the UDB* and *Parcel is Contiguous to Development*) were positively correlated with price as expected. The effect was quite large and can be observed in Exhibit 14.7 for land parcels near the UDB. As the map shows, the predicted price from the equation for all vacant parcels displays how the price declines if the parcel is located outside the UDB (in Exhibit 14.7, the UDB is the

Exhibit 14.7 **Map of Predicted 2006 Land Prices for Miami-Dade County Based on Model 2 (n = 50,850)**

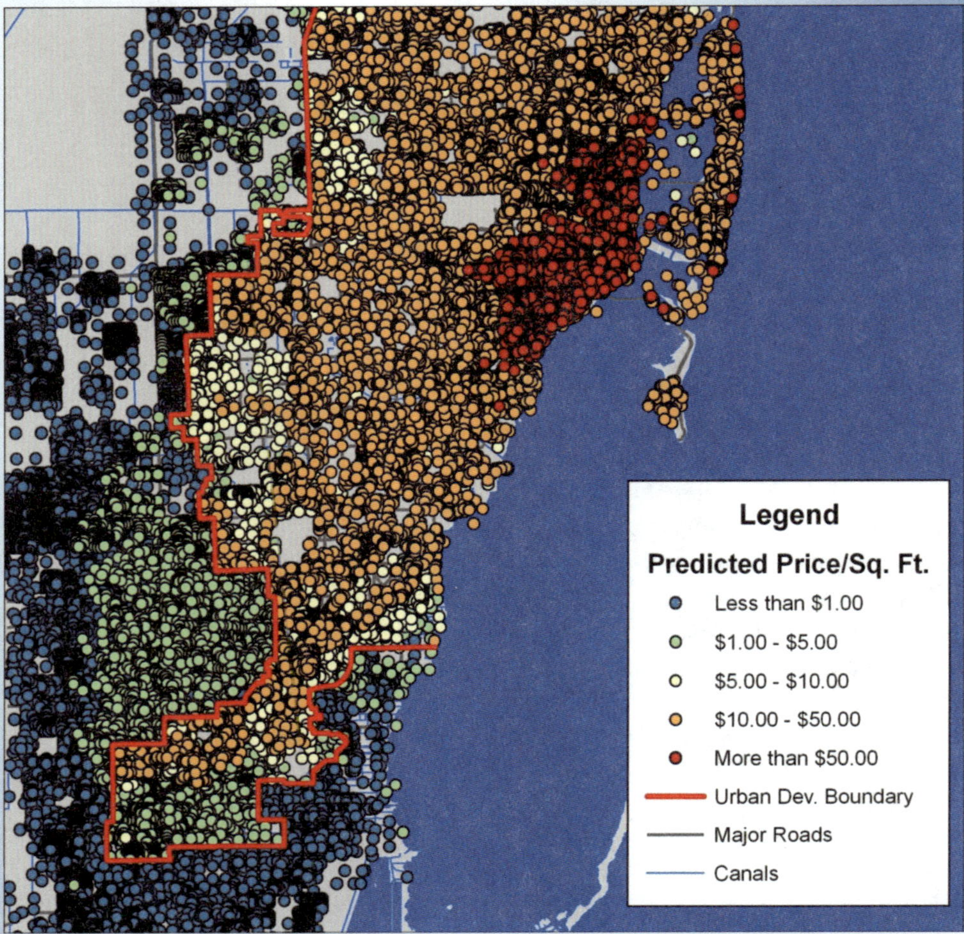

Parcel price circles are not plotted to scale.

red boundary line). Like Model 1, the EEL was negatively correlated with price and is significant. The price was being discounted because the parcel had been designated environmentally endangered by the county, e.g., a critical habitat.

Turning to the 11 spatial variables, the *Oceanside Properties* variable performed as expected. That is, oceanfront properties were positively correlated with price. The variables *Parcels West of Canal L31* was intuitive and statistically significant due to the fact that parcels west of the canal were not flood protected. *Distance to Nearest Inland Body of Water* was positively correlated with land price. This variable confirmed that distance from these potentially contaminated bodies of water matters.

The variable *Distance to Miami CBD* was negatively correlated and was significant. The influence of this variable can be seen in Exhibits 14.7 and 14.8. As the map in Figure 14.7 shows, the predicted price from the equation can be plotted to show that priced declined if the parcel was located farther away from the CBD. The spatial effect is confirmed on the graph of the bid-rent curve in Exhibit 14.8 that shows a steep decline in price with increasing distance from the Miami CBD.

The variable *Distance to Canal* was ambiguous because not all canals are undesirable. *Distance to Nearest Highway* (in miles) was intuitive and was negatively correlated with price. Distances to the two national parks were negatively correlated with price. However, only the *Distance to Everglades National Park* was statistically significant. *Distance to Nearest Secondary CBD* was positively correlated with price and intuition. The variable *UDB_Qtr_Mi* was negatively correlated with price. The variable was ambiguous because properties within one-quarter mile and inside the UDB would be valued more

Exhibit 14.8 **Miami Bid-Price Gradient. Miami-Dade County Land Price Indices as a Function of the Distance from the CBD**

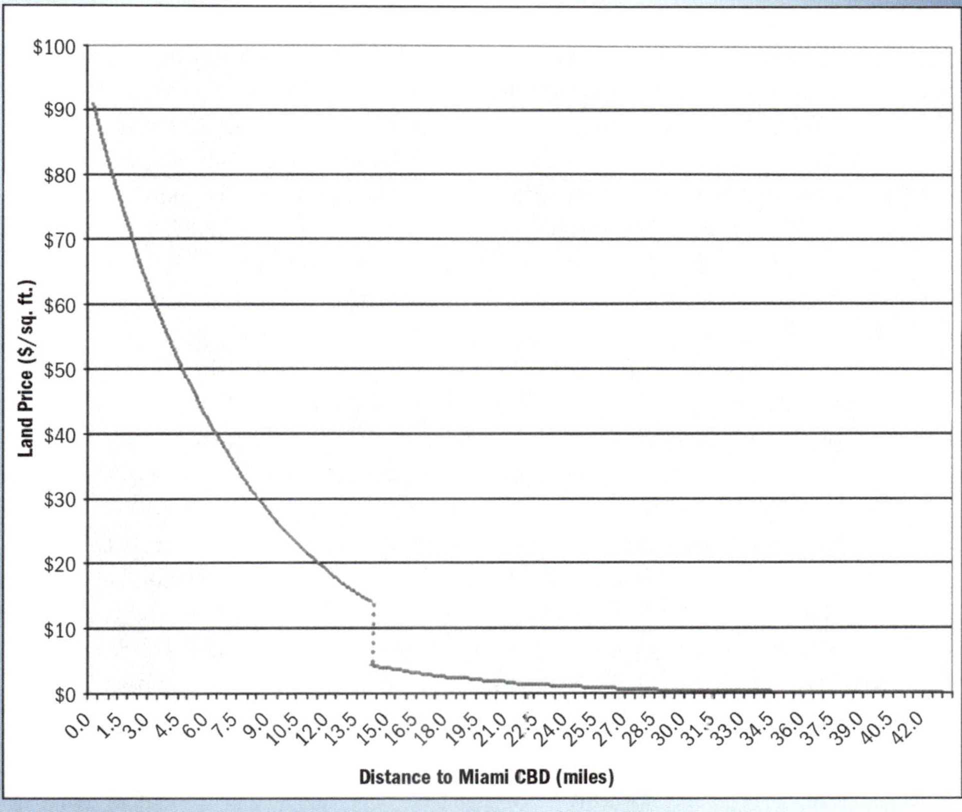

highly than properties outside the regulatory boundary. *Distance to Park* was positively correlated with price but statistically insignificant.

The results of Model 2 included explanatory variables that included parcel, regulatory, spatial, spatial regulatory framework (e.g., hazards and amenities), and GIS-measured distances. Each of the explanatory variables had different effects on land price. Parcel size measured in total square feet contributed a very small negative decrease to price per square foot of -0.000007, which translated to a value of -0.0007%.[55] A parcel location inside the UDB (+183%) and contiguous development (+42%) increased land price by a combined 225%. An agricultural zoning increased price by 24%, while a recreation zoning designation (-69%) and an EEL listing (-70%) reduced land price by 138%. This demonstrated the trade-offs of different zoning characteristics in the land market.

Environmental amenities, as expressed through regulations, had a significant impact in determining a parcel land value. The hazards variables of *Coastal Flooding* (-83%) and the *L31 Canal* (-67%) were negative and reduced price by a combined 150%. The location amenity of an oceanside property increased price by 95%.

Although the measured distance variables dramatically improved the explanatory power of Model 2, the seven significant measured distance variables (*Dist_CBD*, *Dist_Canal*, *Dist_Water*, *Dist_HWY*, *Dist_Everglades*, *Dist_Scndry_CBD*, and *Dist_EEL*) had a combined effect that increased price by a little more than 2%. Even though the magnitude of the coefficients were relatively small, inspection of the effect of a variable such as the *Distance to the Miami CBD* (-8%) can be substantial. Because the bid-price gradient declined rapidly as the parcel distance increased from the Miami CBD (Exhibit 14.8), this steep decline had a profound effect on land price.

As shown in Exhibit 14.9, the temporal effect of real estate market conditions on land price is manifested as a trend in the county land price index. The effect of the recent real estate boom on South Florida's land values is visually discernible. From 1970 to 2000, the index rose by 300%, or an average of 10% per annum. From 2000 to 2006, the index rose 275%, or an average of 40% per annum. While the effect of the recent speculation in south Florida real estate and the consequent rise in land prices has abated, there has been a widespread self-correction of land prices from the previous overinflated levels. This trend is currently manifesting itself in the region's declining house values and increasing foreclosures at the time of this writing.

55. If scaled up to represent an addition to the size of a lot, for example, to add another 1,000 square feet to a parcel, multiply by 1,000, and the price per square foot declines by 0.7%.

Exhibit 14.9 **Miami-Dade County Land Price Indices (1970-2006)**

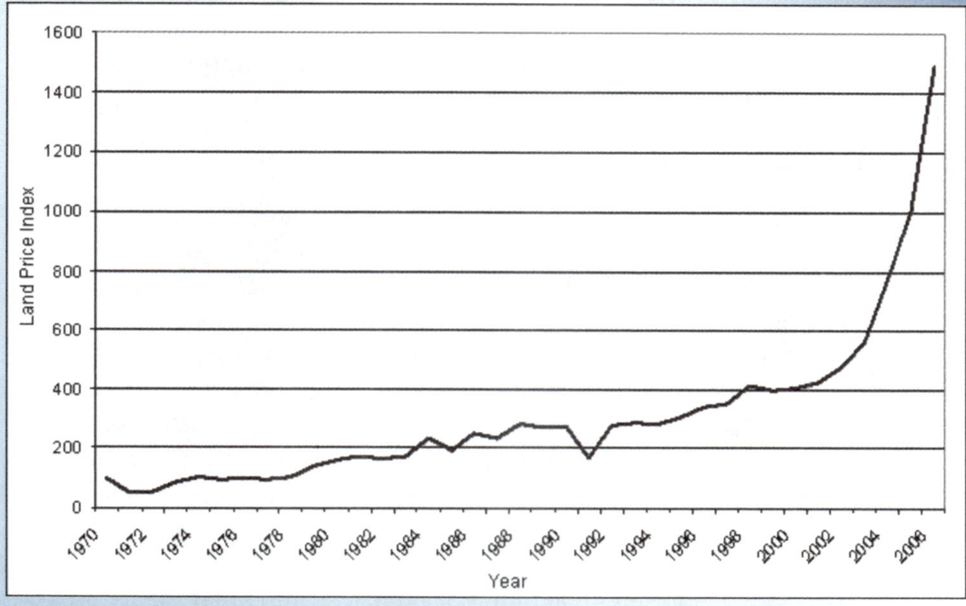

The results showed that, as predicted by theory, land values inside the growth boundary were significantly higher than in the rural area outside the boundary. People were willing to pay more for parcels within the UDB with a view of the countryside and proximity to the pleasures of rural amenities as demonstrated by the variables measuring distances to the UDB, EEL, and Everglades National Park. At various times, developers petition for UDB boundary expansion or request EEL rezoning to expand urban development. Changes in boundaries or zoning status must be reviewed by government agencies. A decision to expand the UDB comes with considerable costs to provide new infrastructure. Farmers outside and close to the boundary have been observed to neglect their lands. Even though lands could be neglected, bringing lower prices, land owners seemed unconcerned because they anticipate urban expansion and selling their land to developers was imminent.[56] Transactions involving EELs identified by the county had lower values than surrounding properties, reflecting the unsuitability for development or the effect of being listed as an EEL. These environmental regulatory variables were significant and behaved as expected in the model.

56. Arthur C. Nelson, "Preserving Prime Farmland in the Face of Urbanization: Lessons from Oregon," *Journal of the American Planning Association*, vol. 58, no. 4 (1992): 467-488.

The spatial variables in Model 2 showed how the measurement of distances to the CBD, and to highways, canals, inland water, and Everglades National Park improved land value estimates. The explanatory power of the hedonic price function was improved signficantly when these variables were included.

Summary and Conclusion

The hedonic pricing function has been used to estimate the value of urban and environmental amenities that affect prices of marketed goods. The method is based on the assumption that people value the characteristics of a good, or the services people consider important when purchasing the good. The hedonic pricing method has been used extensively to estimate economic benefits or costs, as expressed in the marginal price of a specific parcel characteristic, associated with environmental quality, including air pollution, water pollution, soil characteristics, proximity to recreation sites and downtown, and environmental amenities, such as aesthetic views.

The hedonic pricing method is relatively straightforward to apply because it is based on actual market prices and fairly easily measured data. If data is readily available, the model can be relatively inexpensive to apply, however if data must be compiled, the cost of an application can increase substantially.

The hedonic pricing method has several strengths.[57] The method's first and most important strength is that it can be used to estimate values based on actual choice criteria. The property market is efficient in responding to information and provides an accurate indicator of land value. Second, property records are reliable data sets. The method is versatile, and can be adapted to consider several possible interactions between market goods and environmental quality. Using the hedonic pricing method also has several weaknesses related to nonmarket characteristics or externalities. The method will only capture the willingness to pay for perceived differences in environmental characteristics and their direct consequences. The method assumes that, given their income, market participants have the opportunity to select the combination of features they prefer. The method is relatively complex to implement and interpret, requiring a high degree of statistical expertise. The results depend heavily on model specification.

In the application discussed here, the hedonic price function has been used to value vacant land that will come under pressure for development as the population of Miami-Dade County increases

57. Ecosystem valuation Web site at http://www.ecosystemvaluation.org/hedonic_pricing.htm#advantages/.

over the next several decades. Urban expansion most likely will move beyond the UDB if preservation of land is not considered in development planning as specified in the Miami-Dade County Comprehensive Plan. We have demonstrated that the real estate market is influenced significantly by the environmental regulations that have been enacted. We have demonstrated that including spatial explanatory variables in the hedonic equation is a significant improvement over retaining only nonspatial characteristics in the model. It is clear that the application of GIS in property valuation is important and that rigorous statistical methods such as the hedonic pricing function can make good use of spatial data. An improved, more statistically accurate hedonic pricing model that includes environmental and spatial explanatory variables helps to explain consumer preferences for location-specific amenities. Finally, this type of hedonic pricing function would be an important tool to better inform market analysts in Miami-Dade County and perhaps to people in other markets who could apply the lessons learned here to their situations.

ACKNOWLEDGEMENTS

The authors gratefully acknowledge the contributions of Paul Amos, Jared Lang, and Pravin Mathur at the University of Pennsylvania, and Bill Labiosa, Dave Strong of the USGS. We thank Will Forney and Paul Hearn of the USGS for helpful reviews of the manuscript. We also appreciate the helpful input from the Everglades and Biscayne National Park managers. In particular, we thank Sarah Bellmund and David Hallec for valued counsel. We thank Subrata Basu and other members of the Miami-Dade County Department of Planning and Zoning staff for their suggestions and insights related to county rules and regulations.

REFERENCES

Bartik, Timothy J., and V. Kerry Smith. "Urban Amenities and Public Policy," in Edwin S. Mills, ed. *Handbook of Regional and Urban Economics, Volume II: Urban Economics*. Lausanne; Amsterdan; New York: North-Holland, 1987.

Beaton, W. Patrick, and Marcus Pollock. "Economic Impact of Growth Management Policies Surrounding the Chesapeake Bay." *Land Economics*, vol. 68, no. 4 (1992): 434-453.

Bernknopf, R., D. Brookshire, and M. Thayer. "Earthquake and Volcano Alerts: An Economic Evaluation of Risk Perception Changes." *Journal of Environmental Economics and Management*, vol. 18, no. 1 (1990): 35-49.

Bernknopf, R., R. Champion, E. Schwerer-Duffie, E. Graffy, P. Ng, and R. Taketa. "Final Report to TRPA: Tahoe Constrained Optimization Model, Proof of Concept, Initial Results, and Application to Lake Tahoe Management." *USGS Administrative Report*. (2003).

Boyle, Kevin J., S. Lawson, H. Michael, and R. Bouchard. *Lakefront Property Owner's Economic Demand for Water Clarity in Maine Lakes*, Miscellaneous Report 410. Orono, ME: Maine Agricultural and Forest Experiment Station, 1998.

Brookshire, D., M. Thayer, J. Tschirhart, and W. Schulze. "A Test of the Expected Utility Market: Evidence from Earthquake Risks." *Journal of Political Economy*, vol. 93, no. 2 (1985): 369-389.

Brookshire, David S., William D. Schulze, Mark A. Thayer, and Ralph C. d'Arge. "Valuing Public Goods: A Comparison of Survey and Hedonic Approaches." *The American Economic Review*, vol. 72, no. 1 (1982): 165-177.

Court, A. "Hedonic Price Indexes with Automotive Examples," in *The Dynamics of Automobile Demand*. New York: General Motors Corporation, 1939.

Department of Conservation, State of California. http://www.conservation.ca.gov/dlrp/lca/Pages/Index.aspx/.

Dwyer, C. Miami-Dade County Department of Planning and Zoning, personal communication (2007).

Freeman, A. Myrick III. *The Measurement of Environmental and Resource Values: Theory and Methods*. Washington, D.C.: Resources for the Future, 1993.

REFERENCES *(continued)*

Grilliches, Z. "Hedonic Price Indexes for Automobiles: An Econometric Analysis of Quality Change." *The Price Statistics of the Federal Government, General Series No. 73*. New York: National Bureau of Economic Research, 1961.

Hallec, D. U.S. Department of Interior, National Park Service, personal communication (2006).

Hanley, Nick, Jason F. Shogren, and Ben White. *Environmental Economics in Theory and Practice*. New York: Oxford University Press, 1997.

Hellerstein, D., C. Nickerson, J. Cooper, P. Feather, D. Gadsby, D. Mullarkey, A. Tegene, and C. Barnard. *Farmland Protection: The Role of Public Preferences for Rural Amenities, Agricultural Economic Report No. 815*. Washington, D.C.: U.S. Department of Agriculture, Economic Research Service, 2002.

Labiosa, W., R. Bernknopf, P. Hearn, D. Hogan, D. Strong, L. Perlstine, A. Wein, K. Gillen, and S. Wachter. "The South Florida Ecosystem Portfolio Model: An Integrated Ecological, Economic, and Geographic Land Use Planning Tool," USGS Open-file Report, unpublished manuscript (2008).

Lancaster, Kevin J. "A New Approach to Consumer Theory." *Journal of Political Economy*, vol. 74, no. 2 (1966): 132-157.

Loomis, John B., and Gloria E. Helfand. *Environmental Policy Analysis for Decision Making*. Dordrecht, Netherlands; Boston: Kluwer Academic Press, 2001.

McFadden, Robert. Presentation at the 2nd World Congress of Environmental and Resource Economists, Monterey, CA (2002).

Millennium Ecosystem Assessment. *Ecosystems and Human Well-being: A Framework for Assessment*. Washington, D.C.: Island Press, 2003.

Miranowski, J., and M. Cochran. "Economics of Land in Agriculture," in G. Carlson, D. Zilberman, and J. Miranowski, eds. *Agricultural and Environmental Resource Economics*. New York: Oxford University Press, 1993.

Miranowski, John A., and Brian D. Hammes. "Implicit Prices of Soil Characteristics for Farmland in Iowa." *American Journal of Agricultural Economics*, vol. 66, no. 5 (1984): 745-789.

Nelson, Arthur C. "Demand, Segmentation, and Timing Effects of an Urban Containment Program on Urban Fringe Land Values." *Urban Studies*, vol. 22, no. 5 (1985): 439-443.

____. "Preserving Prime Farmland in the Face of Urbanization: Lessons from Oregon." *Journal of the American Planning Association*, vol. 58, no. 4 (1992): 467-488.

Palmquist, Raymond B. "Land as a Differentiated Factor of Production: A Hedonic Model and Its Implications for Welfare Measurement." *Land Economics*, vol. 65, no. 1 (1989): 23-28.

Palmquist, R., and L. Danielson. "A Hedonic Study of the Effects of Erosion Control and Damage on Farmland Values." *American Journal of Agricultural Economics*, vol. 71, no. 1 (1989): 55-62.

Redfearn, C. "Land Markets and Terrorism: Uncovering Perceptions of Risk by Examining Land Price Changes Following 9/11," in Harry Ward Richardson, Peter Gordon, and James Elliott Moore II, eds. *The Economic Impacts of Terrorist Attacks*. Cheltenham, U.K.; Northampton, MA: Edward Elgar, 2005.

Rosen, Sherwin. "Hedonic Prices and Implicit Markets: Product Differentiation in Pure Competition." *Journal of Political Economy*, vol. 82, no. 1 (1974): 34-55.

Sandberg, Krister. "Hedonic Prices, Economic Growth, and Spatial Dependence," PhD thesis (2004).

Smith, V. K., and D. Hallstrom. "Designing Benefit-Cost Analyses for Homeland Security Policies," in Harry Ward Richardson, Peter Gordon, and James Elliott Moore II, eds. *The Economic Impacts of Terrorist Attacks*. Cheltenham, U.K.; Northampton, MA: Edward Elgar, 2005.

Tietenberg, Thomas H. *Environmental Economics and Policy*, 2nd ed. Reading, MA: Addison-Wesley, 1998.

U.S. Census Bureau. U.S. Census, Miami-Dade County QuickFacts (2007). http://quickfacts.census.gov/qfd/states/12/12086.html/.

Xu, Baifu. "An Hedonic Analysis of Southwestern Louisiana Wetland Prices Using GIS," M.S. thesis (2007).

Zilberman, D., and M. Marra. "Agricultural Externalities," in G. Carlson, D. Zilberman, and J. Miranowski, eds. *Agricultural and Environmental Resource Economics*. New York: Oxford University Press, 1993.

15 Coastal Property Analysis Through a Hazard Exposure Lens

Ann-Margaret Esnard, PhD, and Ana Puszkin-Chevlin, PhD

OVERVIEW

A broad range of professionals (e.g., planners, environmental advocates, regulators, insurers, attorneys, appraisers, lenders, developers, and real estate professionals) play key roles in coastal land development and redevelopment, but from different vantage points. The goals of this chapter are to

1. Provide valuation experts and appraisers with an overview of coastal redevelopment issues and challenges from these multiple vantage points
2. Demonstrate ways in which GIS technology is being used to analyze coastal property through a hazard and disaster vulnerability lens

To fulfill these goals, the concepts, analyses, and applications discussed in the chapter are presented at four levels, or scales:

- Parcel
- Portfolio
- Municipal
- Regulatory

Examples in the chapter are drawn from professional and academic sources, documented case studies, and our academic research.

Although the chapter presents an academic view, it will be useful to professional valuation experts and appraisers who may not understand how their role in valuing properties influences (albeit indirectly) the vulnerability of property and people to hurricanes and other coastal hazards. It is also important that valuation experts and appraisers understand the variety of vantage points and types of spatial data and GIS analyses used by the many professionals who shape and reshape coastal communities.

Coastal Development Trends and Property Accumulation

The physical and social aspects of vulnerability have a well-documented relationship, although these vulnerability factors may not always be spatially colocated. Physical vulnerability results from location and exposure to potential hazards, poor or inadequate construction of buildings, and age of structures, while social vulnerability involves the relative ability of an individual, household, or community to respond appropriately to threatening conditions. Coastal communities are affected by a number of natural hazards, including floods, high winds, storm surges, and beach erosion. Disasters occur when people and property are socially and physically vulnerable to these hazards and cannot recover from a storm's impact.

Rising post-disaster reconstruction costs provide evidence of what Mileti refers to as the increased exposure of the nation's "expanding capital stock."[1] In 2003, approximately 153 million people (53% of the nation's population) lived in the 673 U.S. coastal counties, an increase of 33 million people since 1980.[2] Exhibit 15.1 shows a snapshot of population change for the years 1970-2000 for a sample of coastal counties in the United States.

Population growth and property development and redevelopment along beachfronts and on barrier islands continue despite concerns about the possible recurrence of a catastrophic hurricane like Katrina, which destroyed fragile ecosystems, required huge investments to reconstruct private property and public infrastructure, and

Exhibit 15.1 Population Change for Gulf and Atlantic Coastal Counties (1970-2000)

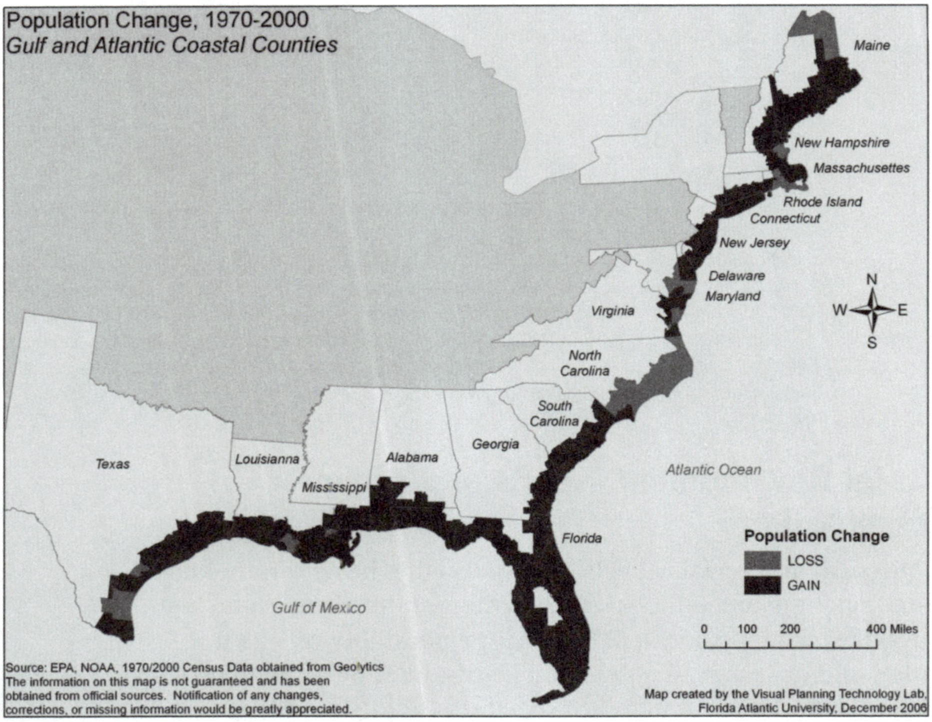

1. D. S. Mileti, *Disasters by Design: A Reassessment of Natural Hazards in the United States* (Washington, D.C.: Joseph Henry Press, 1999).
2. Kristen M. Crossett, T. J. Culliton, P. C. Wiley, and T. R. Goodspeed, *Population Trends along the Coastal United States: 1980-2008* (National Oceanic and Atmospheric Administration, NOAA's National Ocean Service Management and Budget Office, 2004), retrieved December 5, 2007, from http://marineeconomics.noaa.gov/socioeconomics/assessment/population/pdf/1_front_matter_intro.pdf/.

increased demands for county services and beach nourishment. The development of coastal property persists even against the backdrop of emergent community planning, equity problems like a lack of affordable housing, and long-term sea level rise predictions. From 1995 to 2005, the value of coastal assets soared in conjunction with rapid growth of real estate in coastal areas. Based on a Gannett News Service analysis of federal records, the value of property insured (as of 2006) was estimated at $555 billion, more than five times what it was in the 1970s.[3] In a separate study, AIR Worldwide corporation estimated the insured value of coastal (county)[4] residential and commercial properties at $8,849 billion and $10,192 billion respectively for the year 2004.[5] South Florida, one of the nation's fast-growing urban areas, had an estimated $434.4 billion in taxable property value in 2005, a number that underrepresents the actual market value.[6]

At times, sleepy, undeveloped towns can experience real estate growth trends. Wheeler et al. report that "in 2000, the taxable value of all land in Gulf County, Florida, was $630,000 . . . in 2005, that value had climbed to $2.6 billion, almost all of the growth attributable to the incredible churn of real estate on the beach."[7] Coastal growth and redevelopment is affected by a range of professionals (e.g., planners, environmental advocates, regulators, insurers, appraisers, lenders, developers, real estate professionals) who pursue complex and often contradictory agendas related to economic development or disaster resiliency. For example, when a mobile home park that once served seasonal residents of modest means is redeveloped with high-rise buildings, the benefits of eliminating a fragile housing inventory and replacing it with buildings constructed to a higher building code standard should be noted. However, the redevelopment increases asset exposure, as high rises contain more units and the value of the improvements and household contents is greater. If the new construction is then occupied by year-round residents, the number of households living in areas of greater risk during the summer months with

3. L. Wheeler, "America's Coasts at Risk," Gannett News Service (2006), retrieved December 5, 2007, from the following archives: http://www.great-lakes.net/lists/enviro-mich/2006-02/msg00263.html/.
4. The coastal properties were defined as properties located in counties in the following states: Alabama, Connecticut, Delaware, Florida, Georgia, Louisiana, Maine, Maryland, Massachusetts, Mississippi, New Hampshire, New Jersey, New York, North Carolina, Rhode Island, South Carolina, Texas, and Virginia.
5. AIR Worldwide Corporation, *The Coastline at Risk: Estimated Insured Value of Coastal Properties* (Boston: The Corporation, 2005), 1.
6. D. Lunsford and S. Stabley, "A Major Hurricane Here Could Dwarf Cost in New Orleans." *The South Florida Business Journal* (September 26, 2005), retrieved September 27, 2005, from http:southflorida.bizjournals.com/southflorida/stories/2005/09/26/story1.html/.
7. Wheeler et al. (2006).

higher hurricane activity also increases. This in turn affects evacuation times and shelter demands, and could represent increased social vulnerability depending on the nature of the resident population.

A second example pertains to the trend for infill redevelopment of coastal urban centers. Planners and policymakers often desire to curtail the spatial expansion of low-density suburbs, and pursue job growth and growth of the municipal tax base by rebuilding older waterfront areas with existing transportation and infrastructure. However, Levin et al. observe that "planners are now promoting the benefits of higher-density living without a full awareness of the implications of such policies on the potential for mass displacement of people, especially renters, from coastal areas."[8] Exhibits 15.2 and 15.3 show images of post-disaster redevelopment in Biloxi, Mississippi.

Exhibit 15.2 **New Beachfront Luxury Apartment Building Along Gulf of Mexico**

(Photo taken by Ana Puszkin-Chevlin, Biloxi, Mississippi, on September 19, 2007)

Exhibit 15.3 **Future Redevelopment of Commerical Site on Coastal Highway 90**

(Photo taken by Ana Puszkin-Chevlin, Biloxi, Mississippi, on September 19, 2007)

Despite the differences and discrepancies discussed above, "coastal location" is a common thread in the work that is being done

8. Joyce Levine, Ann-Margaret Esnard, and Alka Sapat, "Population Displacement and Housing Dilemmas due to Catastrophic Hurricanes," *Journal of Planning Literature*, vol. 22, no. 1 (2007): 9.

by coastal redevelopment professionals. Ultimately spatial thinking matters in property planning, development, appraisal, valuation, and related disciplines. In that context, GIS technology can offer several advantages and can reveal trends or patterns that might not be observed if paper maps are manually reviewed or if geographic information is presented in a spreadsheet.

Multiscalar GIS Applications

A review of the literature shows that environmental planning professionals have embraced GIS technology, especially for coastal planning and policy making and for addressing hazard mitigation and disaster response. This does not seem to be the case for other coastal property professionals, despite the fact that GIS can be used for a variety of real estate and valuation analyses and that parcel data, tax assessment data, and a variety of data sets already formatted for use in GIS are readily available. (See Exhibit 15.4 for a list of basic data needs.)

What follows are descriptions of a range of GIS applications at four scales:

1. Parcel
2. Portfolio
3. Municipality
4. Regulatory

The parcel scale refers to data and information used in assessing the characteristics of a specific land parcel. The portfolio scale refers to data gathered on a group of parcels that may not be colocated but are analyzed as a discrete subset. The municipal scale includes large numbers of parcels in a community or region that must be analyzed as a whole, or where parts are compared to others in the group. The regulatory scale is not geographically defined, but rather refers to analysis of policy that influences development at various geospatial levels.

Parcel Scale: Analyzing Property Characteristics

Comparable Sales Analysis and Empirical Assessment of Locational Attributes

Comparable sales analysis involves identifying properties similar to the subject and estimating market value based on sale price. GIS can be used to map current and recently retired listings for a more accurate representation of sales information for comparison.[9] Estimation of the value associated with attributes of location remains a challenge. For ex-

9. ESRI, "GIS Best Practices: GIS for Real Estate," retrieved December 12, 2007, from http://www.esri.com/library/bestpractices/real-estate.pdf/.

Exhibit 15.4 Local Data Needs for Hazard Analysis

What is needed for local level analysis?

Databases should capture *natural and environmental hazards* at the local level, including

- Critical/sensitive environmental characteristics: steep slopes, stream corridors, eroded streambanks, aquifer recharge and discharge areas, wetlands and wetland transition areas, watershed basins and subbasins, and habitats for endangered species
- Known natural hazard areas: floodplains, earthquake fault lines, soils classified as highly erodible and subject to landslides, mature stands of natural vegetation prone to wildfires, tornadoes, and coastal storms

Databases should also capture *physical, social, economic, environmental, political, fiscal, and regulatory attributes* and locations of vulnerable resources, zones, facilities, and institutions including

- Property (with attributes on parcel size, ownership, property use, land value and property value, and year structure built)
- Demographic and socioeconomic variables, and population projections
- Special policy zones (e.g., Housing and Urban Development-funded affordable housing)
- Land use regulations (e.g., zoning maps and ordinances)
- Historic landmarks and cultural resources
- Business and other economic hubs (banks, research parks, retail and commercial centers, small businesses, hotels, major employers and financial centers, grocery stores, hardware stores, gas stations)
- Location of expected growth (from a comprehensive plan or based on discussion with community officials)
- Critical facilities and lifelines (as advised by FEMA) such as
  - Essential facilities for health and welfare of the whole population (e.g., hospitals, police and fire stations, emergency operation centers, evacuation shelters, schools)
  - Transportation systems, including airways, highways, railways, and waterways
  - Lifeline utility systems, such as potable water, wastewater, oil, natural gas, electric power, and communication systems
  - High potential loss facilities, such as nuclear power plants, dams, and military installations
  - Hazardous material facilities, including facilities that house industrial/hazardous materials (e.g., corrosives, explosives, flammable materials, radioactive materials, toxins)

Source: Ann-Margaret Esnard, "The Nexus of Hazard Assessment: GeoSpatial Technologies, and Holistic Community Planning Strategies," (Chapter 5) in *Losing Ground: A Nation on Edge* (Washington, D.C.: Environmental Law Institute, 2007), 183-184.

ample, how does an appraiser account for the premium of the lakeside parcel or the depreciation of the parcel that backs up on the busy road? While much of this process is still left up to the interpretation of the appraiser, GIS facilitates empirical assessment of these issues by creating an easy way to identify parcels whose locations have similar characteristics. Using tools that filter tax parcel data by distance, adjacency, or intersection with geographic features or other locations, an appraiser can collect information on property classes, such as waterfront lots or lots that share a boundary with a highway. Statistical analysis can model this data to derive a coefficient, which can be used by appraisers to better adjust their comparable sales analysis of property value.

Determination of Flood Zone Status—In or Out?

Disclosure of flood zone status is mandatory for nearly all sales transactions that require financing. Specifically, owners of buildings located within an "A" or "V" zone are required to carry flood insurance. Insurance agents calculate the appropriate flood insurance premium based on the type of "A" flood zone or "V" flood zone, date of construction, and elevation of the building's lowest floor. Gathering this information manually on a parcel-by-parcel basis can be time-consuming, especially if the scope of work of the assignment involves reviewing a large number of properties. One of the most common uses of GIS is to determine where a parcel or, more specifically, a structure on a parcel is located within a flood zone. Real estate investors, developers, or occupants seeking land within a particular geography can scan multiple properties simultaneously by overlaying FEMA flood maps over parcel and digital aerial imagery.

Taking this type of analysis one step further, appraisers may wish to estimate the effects of flood hazard on residential property values. In a study by Bin and Kruse, the effects of flood hazard on residential property values were estimated using a hedonic property price method.[10] The researchers used GIS to measure the distance to the nearest coastline for each residential property and to analyze digital flood maps coupled with extensive residential property sales records between September 2000 and September 2004 from Carteret County, North Carolina.

Mapping Multiple Land Characteristics and Regulatory Delineations

GIS can be used to map multiple land characteristics and regulatory delineations simultaneously, and to identify the portions of parcels most suitable for construction or conservation. By integrating data about hydrology, vegetative coverage, or wildlife habitats (including turtle nesting sites, for example), an appraiser can identify environmentally sensitive areas that should not be disturbed, such as mangrove forests or tidal wetlands. Unlike paper maps, from which calculating the square footage of irregular shapes is complex, a geographic information system can help an appraiser easily quantify the amount of land in each category and the amount of land that lies in overlapping categories, regardless of shape.

Portfolio Scale: Evaluation of Asset Exposure

Real estate investment trusts (REITs) and other real estate entities with multiple property holdings have gained a competitive edge over the less savvy developer-owner companies by using GIS to quantify the investment potential of portfolios. GIS technology makes it pos-

10. O. Bin and J. B. Kruse, "Real Estate Market Response to Coastal Flood Hazards," *Natural Hazards Review*, vol. 7, no. 4 (2005): 137-144.

sible to show properties in the context of potential customers/tenants, proximity to competitors, inventories of similar property, and labor pools. GIS also allows portfolio managers to assess exposure of properties to different types of hazard risks, such as earthquake, wildfire, hurricanes, or flooding, which are prevalent in different geographies. By gathering data on hazard exposure and comparing it to the geographic distribution of the portfolio assets, appraisers can assess the exposure to property loss and even analyze seasonal vulnerability, thus adjusting revenue and expenditure estimates accordingly. The owner of a hotel portfolio, for example, would like to minimize, or otherwise balance, the properties exposed to different natural hazard threats such as hurricanes, blizzards, floods, or even heat waves as this type of threat would affect occupancy rates and revenue.

Insurance

Location is fundamental to underwriting insurance policies because a wide array of geographical factors (such as zip code) are used to determine risk. Insurance companies use GIS to balance their liability exposure throughout the market. By mapping vulnerabilities to multiple hazards present in multiple unrelated locations, real estate portfolio managers can better manage insurance coverage costs and reserve funds across the portfolio. A manual process limits the number of these factors that can be adduced. O'Donnell reported that since Hurricane Andrew in 1992, insurers have worked with GIS vendor partners to better understand the regional distribution of risk.[11]

Municipality Scale: Assessing Community Vulnerability

In the aftermath of a hurricane disaster, the use of GIS to help allocate assistance to affected residents of a county would be very useful to an emergency operation center. Indeed, from an emergency management perspective, GIS facilitates visualization and assessment of a locality's proximity to evacuation routes and transportation infrastructure like bridges or tunnels. It helps identify the location of nearby critical facilities, such as shelters and hospitals, and proximity to potential sources of environmental contamination, including fuel storage tanks and sewage pumping stations. Documentation and mapping of these facilities is an important first step in vulnerability analysis. In fact, GIS analysis is a critically important tool in the preventive and mitigation planning that should routinely take place well before disaster scenarios arise, although this is sometimes overlooked.

11. Anthony O'Donnell, "Location, Location, Location," *Insurance and Technology* (2006), retrieved December 5, 2007, from http://www.insurancetech.com/showArticle.jhtml;jsessionid=YYLVZF2B1T4QGQSNDLPCKHOCJUNN2JVN?articleID=196604000/.

Prediction of the Likelihood of Development as a Preventative Measure

GIS can be used as an integrative platform to evaluate changes in land use over time. This allows municipal officials to project forward the spatial distribution of assets in their community. An example of such an analysis was reported by Allen et al., who described the development of a GIS database (consisting of building permits, parcel maps, and other miscellaneous data) as part of a Wetland Research collaborative project between Clemson University, the University of South Carolina, and the College of Charleston.[12] A logistic regression model was developed to assess parcel use change (both residential and commercial) in the past and predict land use change in the future.

Evaluation of Municipal Tax Base at Risk

The municipal official needs to understand the value of the real estate and municipal tax base at risk from destructive storms. In the aftermath of an extreme storm event, damage or destroyed properties may receive abatements, reflecting the depreciated value of the asset. The decline in value and hence tax revenue comes at a time when municipalities have increased expenses related to recovery. Without sufficient reserves, municipalities may experience shortfalls or become bankrupt.

In the late 1990s, Nags Head planners created a GIS database to derive their own loss estimates for land-use planning purposes. GIS technology was used to assess

1. The built environment and the extent of development and tax base in the danger zones (floodplains, oceanfront hazard zone, incipient inlets)
2. The status of pre-FIRM structures[13] on developed parcels
3. Vacant land in relation to zoning
4. Properties and land uses within the Ocean Erodible Area of Environmental Concerns[14]

---

12. Jeffrey S. Allen, Kang Shou Lu, and Thomas Delette Potts, "A GIS-Based Analysis and Prediction of Parcel Land Use Change in a Coastal Tourism Destination Area," presented at the 1999 World Congress on Coastal and Marine Tourism. British Columbia: Vancouver (1999).
13. Structures built prior to Nags Head's regular participation in the Flood Insurance Rate Mapping program (i.e., 1978) are generally termed *pre-FIRM*. These structures were not subject to the more stringent building standards, such as elevating or flood-proofing that newer structures must comply with.
14. A.-M. Esnard, D. Brower, and B. Bortz, "Coastal Hazards and the Built Environment on Barrier Islands: A Retrospective View of Nags Head in the Late 1990s," *Coastal Management*, vol. 29, no. 1 (2001): 53-72.

Standardized Damage Estimating Models

Vulnerability analysis can be further enhanced if a municipality is able to predict the extent of damage that can occur to structures and facilities within its borders. In that regard, TAOS (The Arbiter of Storms) and HAZUS-MH (Hazards US-Multi Hazards) are useful software products for predicting damage based on user-defined hazard and storm scenarios and characteristics of the built environment. For example, in 2004 Florida's Department of Emergency Management provided counties with TAOS hazards assessment information, which included estimates of aggregate annualized dollar losses from wind, flooding, sinkholes, earthquakes, and wildfire, as well as estimates for each land use code included in the state department of revenue's property appraiser database.[15]

Regulatory Scale: Policy Analysis

Florida Coastal High Hazard Area Research Study

To address development trends in vulnerable coastal areas, some states and counties designate Coastal High Hazard Areas (CHHA) and regulate or manage development within these zones. The definition of a coastal high hazard varies substantially among states. Florida currently defines its CHHA as the area below the Category 1 storm surge line established by SLOSH (Sea Lake and Overland Surge from Hurricanes). Many other states have chosen to adopt the definition used by FEMA's National Flood Insurance Program (NFIP)

> (A)n area of special flood hazard extending from offshore to the inland limit of a primary frontal dune along an open coast and any other area subject to high-velocity wave action from storms or seismic sources. On a Flood Insurance Rate Map, the Coastal High Hazard Area is designated Zone V, VE, or V1-V30. These zones designate areas subject to inundation by the base flood where wave heights or wave runup depths are greater than or equal to 3.0 feet.[16]

Within these zones, municipalities may adopt specific development restrictions that reduce building density, limit building heights, mandate specific building designs or standards, or require on-site shelter or other mitigation interventions. The additional requirements associated with building or owning property within these zones should be taken into account in property valuation. Hence, coastal development professionals may use GIS analysis to better identify the regulatory context influencing their property, noting that these

15. Florida Department of Community Affairs (FDCA), "Protecting Florida's Communities: Land Use Planning Strategies and Best Development Practices for Minimizing Vulnerability to Flooding and Coastal Storms" (2005), 42, retrieved December 12, 2007, from http://www.dca.state.fl.us/fdcp/DCP/publications/Files/hazmitbp.pdf/.
16. NOAA at http://www.csc.noaa.gov/rvat/glossary.html/.

regulations and the designations are dynamic and subject to periodic revision. An evaluation of the spatial effects of the change to CHHA language has been completed by Puszkin-Chevlin and Esnard.[17]

In that study, GIS was used to document changes in the geographic extent of the old and new CHHA geographic boundaries and to quantify changes in the acreage, value, and land uses of the affected coastal property. GIS was also effectively used to illustrate the total number of parcels affected, distinguishing the parcels added and removed from the CHHA in different colors and showing spatial location and distribution. By contrast, numerical data presented in bar charts typically illustrates only net impact (i.e., the difference in land added and removed to the CHHA). Presenting the removed and gained parcels would require that the bar chart include positive and negative values, thus making it hard to see the net difference. GIS was also effective in evaluating the age of the building inventory in conjunction with land use to assess susceptibility to redevelopment and upzoning pressure (Exhibit 15.5). Findings from the research showed that the new CHHA dropped a large inventory of the structures built in the 1970s and 1980s, as well as several mobile home parks. This alerts

Exhibit 15.5 Building Age Inventory for Parcels without CHHA designation (Martin County, Florida)

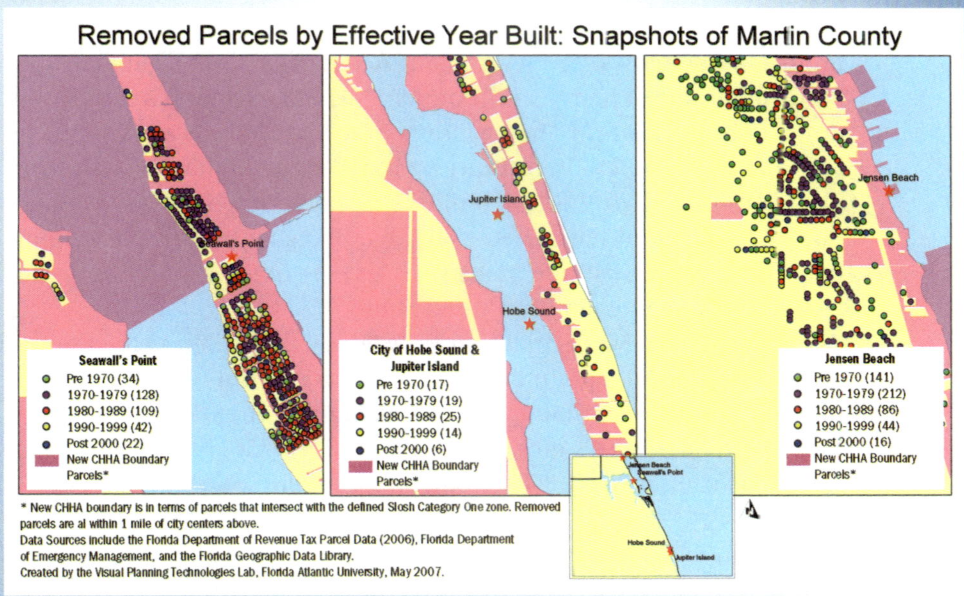

---

17. Ana Puszkin-Chevlin and Ann-Margaret Esnard, "Evaluating Spatial Impacts of Change to Coastal Hazard Policy Language," *URISA Journal*, vol. 21, no. 1 (2009): 41-51.

planners to which properties or neighborhoods might be adversely affected by the change in the way the policy boundaries are depicted.

Focus on Hazard Zones

Hazard and disaster risks must be considered by private as well as public-sector interests involved in coastal land development, financing, and related services. Long-term challenges presented especially by flooding, erosion, and climate change must also be considered if real estate agents and valuation experts want to make sound decisions based on accurate and empirical data and findings. The following discussion highlights the importance of accounting for flood and erosion hazards and risks and provides useful resources.

Flood Hazards

Flood damage is the most common form of natural hazard loss, with approximately 45 flood-related FEMA disaster declarations made in the United States in 2007 alone.[18] However, because flood risk affects vast amounts of land, restricting development of flood-prone land is impractical, if not politically infeasible. Instead, U.S. federal policy has delegated land development regulation to local governments and mitigated flood hazard loss by making subsidized flood insurance available to all communities. The National Flood Insurance Program (NFIP), currently managed by FEMA, was created to prevent market failure of the private property insurance market, and as of February 2007 the program had 5.4 million policies in force with a total coverage of $1.0 trillion.[19] The program was also charged with helping to reduce flood vulnerability by mapping flood hazard risk for all communities, promoting building design guidelines that reduce flood damage, and when necessary acquiring repetitive loss properties. Flood insurance rate maps (FIRMs) are available for more than 20,000 communities in the United States.[20]

FEMA's Map Modernization Program is an ongoing effort to make GIS-formatted floodplain data available on a countywide basis. These maps are usually available from a local government's GIS department or tax assessor's office, or online at FEMA's Map Service Center (http://msc.fema.gov/).[21] The Map Service Center also contains infor-

18. FEMA (2007) at http://www.fema.gov/news/disasters.fema/.
19. Congressional Budget Office, *The Value of Properties in the National Flood Insurance Program* (2007), available online at http://www.cbo.gov/ftpdocs/82xx/doc8256/06-25-FloodInsurance.pdf/.
20. FEMA flood maps are available at http://www.fema.gov/hazard/map/flood.shtm/. This site offers online training and information about how to read and use flood map data.
21. Nick Delmedico, "Take Advantage of New FloodPlain Data: Understanding dFIRMs Necessary for Intelligently Mapping Hazards," *ArcUser Magazine*, vol. 10, no. 3 (2007): 22-25.

mation useful for homeowners and renters, real estate and flood determination agents (e.g., individuals who need to determine whether a house is in a flood hazard area), and insurers (e.g., an underwriter checking on a community's participation status in the NFIP's Community Rating Program and searching for previously effective flood maps for grandfathering an insurance rating).

Determining the flood zone designation of a parcel and property is of critical importance to developers, owners, and lenders. Properties in the highest risk zones are eligible to purchase subsidized flood insurance through NFIP, but the additional cost of this insurance must be factored into operating costs. Additionally, properties in these zones must be elevated above the base flood elevation (BFE). These and other design requirements may add to the construction costs and must be considered in a development pro forma.

Properties constructed prior to the adoption of a community's FIRM, often termed pre-FIRM, are also eligible for NFIP insurance. However, if such a property experiences damage greater than 50% of its market value, and if the owner chooses to repair or rebuild, the owner must upgrade the entire structure to meet current NFIP guidelines. Determining the 50% value is a judgment call, depending on the market value assigned to the property and the estimated cost of repairs or improvements to correct the damage. As this determination can affect the cost of recovery significantly, appraisers must be judicious in their assessment and use the best empirical data. GIS mapping of property values and damage levels in the surrounding area helps clarify and evaluate appraiser and contractor estimates. In this way appraised values and appraisers play a key part in advancing the future resiliency of a community.

One pervasive question regarding flood risk is whether it impacts property values. The current literature in the United States, the United Kingdom, and Australia yields different results depending on the area studied and the way the variables are defined. An extensive Australian study, which included review of literature in the United States, concluded that a flood is likely to have a greater effect on property values than floodplain designation.[22] In a study of the effects of flood hazard on residential property values for Carteret, North Carolina, Bin and Kruse found that average property values are 5%-10% lower if a house is located within a flood zone that is not subject to wave action.[23] The researchers also found that

22. Stephen Yeo, "Effects of Disclosure of Flood liability on Residential Property Values," Risk Frontiers-NHRC Macquarie University, Sydney, Australia (undated), retrieved December 11, 2007, from http://www.riskfrontiers.com/yeopropertyvalues/propertyvaluereport.htm/.

23. Bin and Kruse (2005).

1. Location within a flood zone that is vulnerable to wave action is associated with higher property values, as there is a premium for proximity to beachfront, as opposed to low-lying areas prone to damage that are not near the waterfront.
2. In most inland areas, price differentials for flood risk were roughly equivalent to the capitalized value of flood insurance premiums.

Overall, the risk of coastal flooding was inconsequential in determining property values due to the substantial value premiums that appear to be associated with proximity to coastal water.

Erosion Hazard

Erosion and accretion are dynamic natural processes affecting shorelines. A Heinz Center report noted that development in several high-risk coastal areas had increased by more than 60% between 1980 to 2000.[24] The report also pointed out that about 25% of the homes located within 500 feet of the coast (excluding urban centers) are expected to suffer the effects of storm-related erosion within the next 60 years. Although many communities with significant beach erosion have implemented beach nourishment programs and other coastal engineering projects to stave off erosion, these costly interventions are not comprehensive solutions and are not applicable to all coastal areas or communities.

Projections of erosion or accretion must therefore be considered by appraisers and other real estate professionals. While these estimates are typically prepared by coastal engineers and marine scientists, familiarity with the terminology and concepts allows market participants to understand the validity of the data and permits them to incorporate important information into development and financial decisions.

Combining information on the proximity of the building to the waterline and the erosion rate can provide information on the building's useful lifespan and risk to damage, both of which affect real estate value. Property subject to rapid erosion will lose value for the following reasons:

- Although proximity to the water usually increases value, buildings positioned too close to the water in relation to erosion may be worth less when 1) the public easement along the beach moves landward, 2) access to the property is compromised, or 3) foundations suffer damage.
- The size of the parcel actually gets smaller over time, so less land is available to be sold in the future.

24. The H. John Heinz III Center for Science, Economics and the Environment, "Evaluation of Erosion Hazards" (April 2000) retrieved May 2010 from http://www.heinzctr.org/publications/PDF/erosnrpt.pdf/.

- The developable square footage is often based on lot size. Erosion can impact the amount of square footage that can be developed.

In these cases, a property owner or developer must reconsider the time period for which rent (or utility and enjoyment) can be collected. The limited or shortened period of the income stream would affect the value of the property using the income capitalization approach. It may also suggest that the depreciation taken for tax or other purposes be accelerated or the period of the financing shortened. Understanding the erosion rate and its spatial components may also help owners plan the timing of property maintenance and capital expenditures.

GIS is a key tool for documenting and assessing erosion risk. Geoscientists typically document erosion rates by observing the changing beach profile and the location of the seasonal mean high water line and calculating differences in the sand volume. The mean high water line can be mapped in a GIS, and distances between the mean high water line, dune fields, and building structures can be measured. Over time, erosion will cause the mean high water line to advance landward. Once the rate of change is estimated, it can be used to project shoreline advance in the future. In Florida, the 30-year projection is one of the setback benchmarks used to site buildings on beachfront lots. A 30-year term is used because it coincides with the length of the typical long-term mortgage instrument and covers much of building's primary lifespan. Nonetheless, using a 50- to 60-year erosion projection may be prudent, especially for facilities with high capital assets, such as ports with their durable quays and container-handling gantry cranes or sewage treatment facilities.

Similarly, sea level rise caused by global warming will shift the mean high water line landward and result in flooding of low-lying areas. Different sea level rise scenarios have been developed in GIS formats to estimate what areas will become submerged or intermittently flooded, and that information is now becoming available to the public. Increasingly, this risk parameter will be incorporated into building siting decisions as well decisions regarding the term of real investment and asset disposition. For example, Climate Appraisal Services was established in 2006 to provide address-based climate reporting services, which help homebuyers and property owners assess how climate change and other environmental risks (including shoreline reduction, earthquakes, hurricanes, tornadoes, wildfire, drought, flood, health, and disease) could affect their property.[25] The organization's services suggest that real estate developers are interested in climate change information.

25. Climate Appraisal Services, retrieved December 12, 2007, from: http://www.climateappraisal.com/index.aspx#Scene_1/.

Conclusions

A broad range of professionals influence coastal land development and redevelopment trends and, in turn, the vulnerability of property and people to hurricanes and other coastal hazards. These professionals view coastal land from different vantage points. Objectives range from conservation of natural resources, public access, and recreation to transportation, tourism, and development resulting in the highest financial return. An additional issue underlies those competing objectives: natural hazard risk, which until recently was ignored because the effects of infrequent and periodic catastrophic events were mitigated by engineered hardening solutions such as armored shorelines, levees, and jetties. In the past, the overall level of damage was more moderate as fewer and less valuable assets were at stake. However, it has become increasingly clear that hazards (including floods and erosion) have both short-term and long-term negative implications that must be accounted for by all coastal development professionals.

GLOSSARY OF TERMS

A Flood Zone. Areas with a 1% annual chance of flooding and a 26% chance of flooding over the life of a 30-year mortgage. Because detailed analyses are not performed for such areas, no depths or base flood elevations are shown within these zones.

Base Flood Elevations (BFE). Expected elevation of flood waters and wave effects during the 100-year flood (also known as the "Base Flood").

Coastal High Hazard Area (CHHA). An area of special flood hazard extending from off shore to the inland limit of a primary frontal dune along an open coast and any other area subject to high-velocity wave action from storms or seismic sources (Source: FEMA).

FEMA. Federal Emergency Management Agency.

Flood Insurance Rate Map (FIRM). Official map of a community on which FEMA has delineated both the special hazard areas and the risk premium zones applicable to the community.

Hazards US–MultiHazards (HAZUS-MH). A risk assessment software program for analyzing potential losses from floods, hurricane winds, and earthquakes

NFIP. National Flood Insurance Program.

Special Flood Hazard Areas (SFHA). Land areas subject to inundation by a flood that has a 1% probability of being equaled or exceeded in any given year (hence the terms *1% annual chance flood* and *100-year flood.*)

Sea, Lake, and Overland Surge from Hurricanes (SLOSH). A computerized model used by the National Weather Service to calculate potential surge heights from hurricanes.

The Arbiter of Storms (TAOS). A proprietary modular numerical storm hazard modeling program developed by Kinetic Analysis Corporation.

VE Flood Zone. Coastal areas with a 1% or greater chance of flooding and an additional hazard associated with storm waves. These areas have a 26% chance of flooding over the life of a 30-year mortgage. No base flood elevations are shown within these zones.

REFERENCES

AIR Worldwide Corporation. *The Coastline at Risk: Estimated Insured Value of Coastal Properties*. Boston: The Corporation, 2005.

Allen, Jeffrey S., Kang Shou Lu, and Thomas Delette Potts. "A GIS-Based Analysis and Prediction of Parcel Land Use Change in a Coastal Tourism Destination Area," presented at the 1999 World Congress on Coastal and Marine Tourism. British Columbia: Vancouver (1999).

Bin, O., and J. B. Kruse. "Real Estate Market Response to Coastal Flood Hazards." *Natural Hazards Review*, vol. 7, no. 4 (2005): 137-144.

Climate Appraisal Services. "Climate Appraisal Mission," retrieved December 12, 2007, from: http://www.climateappraisal.com/index.aspx#Scene_1/.

Congressional Budget Office. *The Value of Properties in the National Flood Insurance Program*. (2007), available online at http://www.cbo.gov/ftpdocs/82xx/doc8256/06-25-FloodInsurance.pdf/.

Crossett, Kristen M., T. J. Culliton, P. C. Wiley, and T. R. Goodspeed. *Population Trends along the Coastal United States: 1980-2008*. National Oceanic and Atmospheric Administration, NOAA's National Ocean Service Management and Budget Office, 2004, retrieved December 5, 2007, from http://marineeconomics.noaa.gov/socioeconomics/assessment/population/pdf/1_front_matter_intro.pdf/.

Delmedico, Nick. "Take Advantage of New FloodPlain Data: Understanding dFIRMs Necessary for Intelligently Mapping Hazards." *ArcUser Magazine*, vol. 10, no. 3 (2007): 22-25.

Esnard, A.-M. "The Nexus of Hazard Assessment: GeoSpatial Technologies, and Holistic Community Planning Strategies." (Chapter 5) in *Losing Ground: A Nation on Edge*. Washington, D.C.: Environmental Law Institute, 2007.

Esnard, A.-M., D. Brower, and B. Bortz. "Coastal Hazards and the Built Environment on Barrier Islands: A Retrospective View of Nags Head in the Late 1990s." *Coastal Management*, vol. 29, no. 1 (2001): 53-72.

ESRI. "GIS Best Practices: GIS for Real Estate," retrieved December 12, 2007, from http://www.esri.com/library/bestpractices/real-estate.pdf/.

Florida Department of Community Affairs. "Protecting Florida's Communities: Land Use Planning Strategies and Best Development Practices for Minimizing Vulnerability to Flooding and Coastal Storms." (2005), retrieved December 12, 2007, from http://www.dca.state.fl.us/fdcp/DCP/publications/Files/hazmitbp.pdf/.

____. *Coastal High Hazard Areas: Division of Community Planning Technical Report 2007-3*. (2007).

Levine, Joyce, Ann-Margaret Esnard, and Alka Sapat. "Population Displacement and Housing Dilemmas due to Catastrophic Hurricanes." *Journal of Planning Literature*, vol. 22, no. 1 (2007): 3-15.

Lunsford, D., and S. Stabley. "A Major Hurricane Here Could Dwarf Cost in New Orleans." *The South Florida Business Journal* (September 26, 2005), retrieved September 27, 2005, from http://southflorida.bizjournals.com/southflorida/stories/2005/09/26/story1.html/.

Mileti, D. S. *Disasters by Design: A Reassessment of Natural Hazards in the United States*. Washington, D.C.: Joseph Henry Press, 1999.

O'Donnell, Anthony. "Location, Location, Location." *Insurance and Technology* (2006), retrieved December 5, 2007, from http://www.insurancetech.com/showArticle.jhtml;jsessionid=YYLVZF2B1T4QGQSNDLPCKHOCJUNN2JVN?articleID=196604000/.

Puszkin-Chevlin, Ana. "Determinants of Local Hazard Mitigation Policy and Built Environment Vulnerability: Three Case Studies from Florida's Treasure Coast," Dissertation, Columbia University, Graduate School of Arts and Sciences, New York (2007).

____. *Living on the Edge: Coastal Storm Vulnerability of the Treasure Coast Barrier Islands*. Center for Urban and Environmental Solutions, Florida Atlantic University, 2007, available at http://docs.caupa.fau.edu/cues/LivingontheEdgeFinalacp-printversion.pdf/.

Puszkin-Chevlin, Ana, Debra Hernandez, and James Murley. "Overview of Inundation Events and Responses—Land Use Planning and its Potential to Reduce Hazard Vulnerability: Current Practice and Future Possibilities." *Marine Technology Society Journal*, vol. 40, no. 4 (2007): 7-15.

Puszkin-Chevlin, Ana, and Ann-Margaret Esnard. "Evaluating Spatial Impacts of Change to Coastal Hazard Policy Language." *URISA Journal*, vol. 21, no. 1 (2009): 41-51.

REFERENCES *(continued)*

The H. John Heinz III Center for Science, Economics and the Environment. "Evaluation of Erosion Hazards." (April 2000). Retrieved May 2010 from http://www.heinzctr.org/publications/PDF/erosnrpt.pdf/.

Thompson, M. M., and Ann-Margaret Esnard. "Property Maps in Modern Cadastre." In American Planning Association. *Planning and Urban Design Standards*. New York: John Wiley and Sons, 2006.

Wheeler, L. "America's Coasts at Risk." Gannett News Service (2006), retrieved December 5, 2007, from the following Archives: http://www.great-lakes.net/lists/enviro-mich/2006-02/msg00263.html/.

Yeo, Stephen. "Effects of Disclosure of Flood liability on Residential Property Values." Risk Frontiers-NHRC Macquarie University, Sydney, Australia (undated), retrieved December 11, 2007, from http://www.riskfrontiers.com/yeopropertyvalues/propertyvaluereport.htm/.